EVOLVED PACKET SYSTEM (EPS)

EVOLVED PACKET SYSTEM (EPS)

THE LTE AND SAE EVOLUTION OF 3G UMTS

Pierre Lescuyer and Thierry Lucidarme

*Both of
Alcatel-Lucent, France*

John Wiley & Sons, Ltd

Copyright © 2008 John Wiley & Sons Ltd, The Atrium, Southern Gate, Chichester,
West Sussex PO19 8SQ, England
Telephone (+44) 1243 779777

Email (for orders and customer service enquiries): cs-books@wiley.co.uk
Visit our Home Page on www.wiley.com

All Rights Reserved. No part of this publication may be reproduced, stored in a retrieval system or transmitted in any form or by any means, electronic, mechanical, photocopying, recording, scanning or otherwise, except under the terms of the Copyright, Designs and Patents Act 1988 or under the terms of a licence issued by the Copyright Licensing Agency Ltd, 90 Tottenham Court Road, London W1T 4LP, UK, without the permission in writing of the Publisher. Requests to the Publisher should be addressed to the Permissions Department, John Wiley & Sons Ltd, The Atrium, Southern Gate, Chichester, West Sussex PO19 8SQ, England, or emailed to permreq@wiley.co.uk, or faxed to (+44) 1243 770620.

Designations used by companies to distinguish their products are often claimed as trademarks. All brand names and product names used in this book are trade names, service marks, trademarks or registered trademarks of their respective owners. The Publisher is not associated with any product or vendor mentioned in this book.
All trademarks referred to in the text of this publication are the property of their respective owners.

This publication is designed to provide accurate and authoritative information in regard to the subject matter covered. It is sold on the understanding that the Publisher is not engaged in rendering professional services. If professional advice or other expert assistance is required, the services of a competent professional should be sought.

Other Wiley Editorial Offices

John Wiley & Sons Inc., 111 River Street, Hoboken, NJ 07030, USA

Jossey-Bass, 989 Market Street, San Francisco, CA 94103-1741, USA

Wiley-VCH Verlag GmbH, Boschstr. 12, D-69469 Weinheim, Germany

John Wiley & Sons Australia Ltd, 42 McDougall Street, Milton, Queensland 4064, Australia

John Wiley & Sons (Asia) Pte Ltd, 2 Clementi Loop #02-01, Jin Xing Distripark, Singapore 129809

John Wiley & Sons Canada Ltd, 6045 Freemont Blvd, Mississauga, ONT, L5R 4J3, Canada

Wiley also publishes its books in a variety of electronic formats. Some content that appears in print may not be available in electronic books.

Library of Congress Cataloging-in-Publication Data

Lescuyer, Pierre, 1967-
 Evolved packet system (EPS): the LTE and SAE evolution of 3G UMTS / Pierre Lescuyer and Thierry Lucidarme.
    p. cm.
 Includes index.
 ISBN 978-0-470-05976-0 (cloth)
 1. Universal Mobile Telecommunications System. 2. Wireless communication systems. I. Lucidarme, Thierry. II. Title. III. Title: Evolved packet system, the long term evolution and system architecture evolution of 3G Universal Mobile Telecommunications System.
  TK5103.483.L47 2008
  621.384–dc22   2007033388

British Library Cataloguing in Publication Data

A catalogue record for this book is available from the British Library

ISBN 978-0-470-05976-0 (HB)

Typeset by 10/12pt Times by Thomson Digital Noida, India
Printed and bound in Great Britain by Antony Rowe Ltd, Chippenham, England.
This book is printed on acid-free paper responsibly manufactured from sustainable forestry
in which at least two trees are planted for each one used for paper production.

Contents

Preface   xi

1 Introduction   1

   1.1 Wireless World Picture   1
   1.2 About Technologies   3
       1.2.1 Heterogeneous 2G Systems   4
       1.2.2 'MAP' and 'IS-41' Systems   4
       1.2.3 The MAP Technologies   6
       1.2.4 The IS-41 Technologies   9
   1.3 Standards and Organizations   12
       1.3.1 The Role of ITU   12
       1.3.2 3G Cross-Country Standardization Bodies   13
       1.3.3 The Structure of 3GPP   14
       1.3.4 The NGN Evolution   17
       1.3.5 The NGMN Initiative   18
   1.4 Spectrum   20
   1.5 The Evolution of UMTS   21
       1.5.1 1st Evolution Driver: The Move towards Data Applications   21
       1.5.2 2nd Evolution Driver: Enhanced Radio Interface Capabilities   23
       1.5.3 What Will Change Within the Network?   23
       1.5.4 What is Described in this Book?   24
   1.6 Links and Documents   24
       1.6.1 Useful Web Sites   24
       1.6.2 Evolved UMTS Specifications   24

2 Evolved UMTS Overview   27

   2.1 The Access Network Requirements   27
       2.1.1 Radio Interface Throughput   28
       2.1.2 Data Transmission Latency   28
       2.1.3 Terminal State Transition   29
       2.1.4 Mobility   30
       2.1.5 Spectrum Flexibility   30
       2.1.6 Co-existence and Inter-Working with Existing UMTS   31

	2.2	Evolved UMTS Concepts	31
		2.2.1 A Packet-Only Architecture	32
		2.2.2 A Shared Radio Interface	35
		2.2.3 Other Access Technologies	35
	2.3	Overall Evolved UMTS Architecture	36
		2.3.1 E-UTRAN: The Evolved Access Network	37
		2.3.2 EPC: The Evolved Packet Core Network	39
		2.3.3 The HSS	47
	2.4	The IMS Subsystem	50
		2.4.1 The Session Control Function	50
		2.4.2 The Media Gateway Nodes	52
	2.5	Policy Control and Charging	53
		2.5.1 Policy Control in UMTS	53
		2.5.2 Evolved UMTS Policy Control	57
		2.5.3 The Charging Architecture	57
	2.6	The Terminal	61
		2.6.1 The User Device Architecture	61
		2.6.2 Terminal Capabilities	63
		2.6.3 The Subscriber Module	63
	2.7	The Evolved UMTS Interfaces	68
	2.8	Major Disruptions with 3G UTRAN-FDD Networks	68
		2.8.1 About Soft Handover	68
		2.8.2 About Compressed Mode	71
		2.8.3 About Dedicated Channels	72
3	**Physical Layer of E-UTRAN**		**75**
	3.1	Basic Concepts of Evolved 3G Radio Interface	75
	3.2	OFDM (Orthogonal Frequency Division Multiplex)	76
		3.2.1 OFDMA Multiple Access	80
		3.2.2 MC-CDMA Multiple Access	82
		3.2.3 Common Points between OFDM, CDMA, MC-CDMA, etc.	82
		3.2.4 Frequency Stability Considerations for OFDM Systems	84
		3.2.5 System Load in OFDMA Systems	84
		3.2.6 SC-FDMA: The PAPR (Peak-Average-Power-Ratio) Problem	85
		3.2.7 Dimensioning an OFDM System	89
	3.3	MIMO (Multiple Input Multiple Output)	91
		3.3.1 Traditional Beamforming	91
		3.3.2 MIMO Channel and Capacity	92
		3.3.3 A Simplified View of MIMO 2.2	96
		3.3.4 The Harmonious Coupling between OFDM and MIMO	97
		3.3.5 MIMO: A Classification Attempt	98
		3.3.6 Some Classical Open Loop MIMO Schemes	99
		3.3.7 Notions of Cyclic Delay Diversity (CDD)	102
		3.3.8 MIMO Schemes and Link Adaptation	103
		3.3.9 Improving MIMO with Some Feedback	104

	3.3.10 MU-MIMO, Virtual MIMO and Transmit Diversity	107
	3.3.11 Towards a Generalized Downlink Scheme	108
3.4	Architecture of the Base Station	109
	3.4.1 The Block Scheme of the Base Station	109
	3.4.2 The Analogue-to-Digital Conversion	111
	3.4.3 Power Amplification (PA) Basics	113
	3.4.4 Cellular Antennas Basics	114
3.5	The E-UTRAN Physical Layer Standard	118
3.6	FDD and TDD Arrangement for E-UTRAN	118
	3.6.1 A Word about Interferences in TDD Mode	119
	3.6.2 Some Basic Physical Parameters	120
	3.6.3 TDD and Existing UTRAN Compatibility	121
	3.6.4 Combined FDD-TDD Mode	122
3.7	Downlink Scheme: OFDMA (FDD/TDD)	122
	3.7.1 Downlink Physical Channels and Signals	124
	3.7.2 Physical Signal Transmitter Architecture	125
	3.7.3 Downlink Data Multiplexing	126
	3.7.4 Scrambling	130
	3.7.5 Modulation Scheme	130
	3.7.6 Downlink Scheduling Information and Uplink Grant	132
	3.7.7 Channel Coding	132
	3.7.8 OFDM Signal Generation	132
	3.7.9 Downlink MIMO	133
	3.7.10 Channels Layer Mapping, Precoding and Mapping to Resource Elements	137
	3.7.11 E-MBMS Concepts	140
	3.7.12 Downlink Link Adaptation	143
	3.7.13 HARQ	143
	3.7.14 Downlink Packet Scheduling	146
	3.7.15 Cell Search and Acquisition	148
	3.7.16 Methods of Limiting the Inter-Cell Interference	153
	3.7.17 Downlink Physical Layer Measurements	155
3.8	Uplink Scheme: SC-FDMA (FDD/TDD)	156
	3.8.1 Uplink Physical Channel and Signals	156
	3.8.2 SC-FDMA	156
	3.8.3 Uplink Subframe Structure	157
	3.8.4 Resource Grid	159
	3.8.5 PUSCH Physical Characteristics	160
	3.8.6 PUCCH Physical Characteristics	161
	3.8.7 Uplink Multiplexing Including Reference Signals	162
	3.8.8 Reference Signals	162
	3.8.9 Multiplexing of L1/L2 Control Signalling	163
	3.8.10 Channel Coding and Physical Channel Mapping	164
	3.8.11 SC-FDMA Signal Generation	164
	3.8.12 The Random Access Channel	164
	3.8.13 Uplink-Downlink Frame Timing	168

	3.8.14 Scheduling	168
	3.8.15 Link Adaptation	168
	3.8.16 Uplink HARQ	169

4 Evolved UMTS Architecture — 171

- 4.1 Overall Architecture — 171
  - *4.1.1 Evolved UMTS Node Features* — 172
  - *4.1.2 E-UTRAN Network Interfaces* — 176
  - *4.1.3 S1 Interface* — 177
  - *4.1.4 S1 Flexibility* — 181
  - *4.1.5 X2 Interface* — 183
- 4.2 User and Control Planes — 184
  - *4.2.1 User Plane Architecture* — 184
  - *4.2.2 Control Plane Architecture* — 188
- 4.3 Radio Interface Protocols — 189
  - *4.3.1 The E-UTRAN Radio Layered Architecture* — 189
  - *4.3.2 The Radio Channels* — 190
  - *4.3.3 PHY* — 194
  - *4.3.4 MAC* — 196
  - *4.3.5 RLC* — 197
  - *4.3.6 RRC* — 198
  - *4.3.7 PDCP* — 200
  - *4.3.8 NAS Protocols* — 206
- 4.4 IMS Protocols — 209
  - *4.4.1 The IMS Protocol Stack* — 210
  - *4.4.2 SIP* — 210
  - *4.4.3 SDP* — 220
  - *4.4.4 RTP* — 223
  - *4.4.5 A SIP/SDP IMS Example* — 227

5 Life in EPS Networks — 229

- 5.1 Network Attachment — 229
  - *5.1.1 Broadcast of System Information* — 230
  - *5.1.2 Cell Selection* — 231
  - *5.1.3 The Initial Access* — 232
  - *5.1.4 Registration* — 236
  - *5.1.5 De-registration* — 240
- 5.2 Communication Sessions — 241
  - *5.2.1 Terminal States* — 241
  - *5.2.2 Quality of Service in Evolved UMTS* — 245
  - *5.2.3 Security Overview* — 249
  - *5.2.4 User Security in EPS* — 253
  - *5.2.5 User Security in IMS* — 260
  - *5.2.6 Session Setup* — 261
  - *5.2.7 Data Transmission* — 265

	5.3	Mobility in IDLE Mode	266
	5.3.1	Cell Reselection Principles	266
	5.3.2	Terminal Location Management	266
	5.3.3	Tracking Area Update	269
	5.4	Mobility in ACTIVE Mode	270
	5.4.1	Intra-E-UTRAN Mobility with X2 Support	272
	5.4.2	Intra-E-UTRAN Mobility without X2 Support	274
	5.4.3	Intra-E-UTRAN Mobility with EPC Node Relocation	276
	5.4.4	Mobility between 2G/3G Packet and E-UTRAN	278

6 The Services 281

6.1	The Role of OMA		281
6.2	Push-to-talk Over Cellular		282
	6.2.1	Service Architecture	284
	6.2.2	PoC Protocol Suite	287
	6.2.3	An Example of PoC Session Setup	289
	6.2.4	Charging Aspects	292
6.3	Presence		294
	6.3.1	Service Architecture	294
	6.3.2	An Example of a Presence Session	295
	6.3.3	Charging Aspects	297
6.4	Broadcast and Multicast		298
	6.4.1	Some Definitions	298
	6.4.2	Typical Applications	299
	6.4.3	Service Architecture	299
	6.4.4	MBMS Security	303
	6.4.5	The MBMS Service Steps	305
	6.4.6	The E-UTRAN Aspects of MBMS	307
	6.4.7	Charging Aspects	307
6.5	Voice and Multimedia Telephony		309
	6.5.1	About Circuit and Packet Voice Support	309
	6.5.2	Service Architecture	312
	6.5.3	About Information Coding	313
	6.5.4	About Supplementary Services	317
	6.5.5	Multimedia Services in EPS Systems	320

Glossary 323

Index 335

Preface

With more than two billion customers, there is no doubt that 2G GSM and 3G UMTS cellular technologies are a worldwide success, adopted by most countries and network operators. The 3G UMTS technology has significantly evolved since the first declination. The first release of the standard, published in 1999, was mostly oriented towards dedicated channel allocation, and circuit-switched service support. Later on, the standard evolved to high-speed packet radio interface for downlink transmission (HSDPA for High Speed Downlink Packet Access) and uplink transmission HSUPA as a clear orientation towards IMS (IP Multimedia Subsystem) and IP-based services.

EPS (Evolved Packet System) represents the very latest evolution of the UMTS standard. EPS is also known by other acronyms related to technical study items being worked on at 3GPP standard committees: **LTE** (Long Term Evolution), which is dedicated to the evolution of the radio interface, and **SAE** (System Architecture Evolution), which focuses on Core Network architecture evolution.

Although still a 3G-related standard, EPS proposes a significant improvement step, with a brand new radio interface and an evolved architecture for both the Access and the Core Network parts. The two major disruptions brought by EPS are:

- **Improved performances** – characterized by a spectrum efficiency which is twice as large as HSDPA/HSUPA.
- **A packet-only system** – resulting in a unified and simplified architecture.

EPS is specified as part of the 3GPP family and, from that perspective, EPS will benefit from the same ecosystem that made the success of GSM and UMTS technologies. In addition, it is believed that technical and architectural evolutions brought by EPS prefigure future 4G networks (also known as IMT-Advanced networks).

This book presents the EPS evolution, as introduced in Release 8 of the 3GPP standard. It is not a substitute to the 3GPP standard, and advanced readers willing to dig into any specific domain of EPS are encouraged to consult the 3GPP specification documents which are referenced, when appropriate, through the different chapters.

The objective here is rather to provide a comprehensive system end-to-end vision of EPS, from the radio interface to the service level, including network architecture, radio protocols, as well as subscriber and session management. As EPS was not thought of as a completely new and standalone technology, the authors have also tried to show the inheritance and relations with 2G GSM and early 3G UMTS in terms of ground principles and technical aspects.

The technical content of this book is based on early documents and standards available at the time of writing. For that reason, the view presented here might be slightly different from the actual reference standard. This should, however, be constrained to very limited parts or specific details of this book.

1

Introduction

This chapter is an introduction to the evolution of UMTS systems, also known as EPS (Evolved Packet System). It provides a picture of current wireless and cellular communications, as an introduction to the requirements and motivations for Evolved 3G systems, which are the subject of the next chapter.

This chapter presents the following elements:

- A brief history of digital cellular systems, from 2G to the latest 3G evolutions.
- The evolution of the subscriber base.
- The various organizations which are supporting 3G and Evolved 3G system specifications.
- An overview of the spectrum usage.
- A list of Web links and documents directly connected to Evolved UMTS.

1.1 Wireless World Picture

Wireless cellular communication is certainly one of the major evolutions provided to the telecommunication world, experiencing an exponential growth from the early 1990s.

Wireless communication systems started to emerge in the mid-1980s, first based on so-called 1G (first-generation) analogue technologies like AMPS (Advanced Mobile Phone System) in the United States or NMT (Nordic Mobile Telephone) in Northern Europe. Those systems have evolved to 2G (second-generation) digital radio – providing robustness and better spectral efficiency – and, ultimately, to 3G, so as to offer global mobility and improved end-user experience over a wide range of services.

The unprecedented success of wireless communication has multiple business repercussions, by developing the potential for voice traffic and added-value services like Instant text and Voice Messaging, Multimedia Messaging (MMS), high-value content delivery or streaming, location-based services, etc.

As of mid-2006, there were:

- 2.3 billion mobile subscribers worldwide.
- 1.8 billion GSM mobile subscribers – GSM represented a 78% market share of cellular subscribers.

Evolved Packet System (EPS) P. Lescuyer and T. Lucidarme
Copyright © 2008 John Wiley & Sons, Ltd.

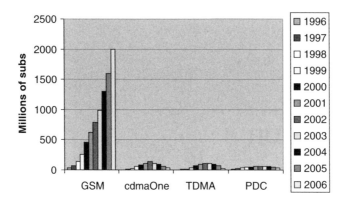

Figure 1.1 Evolution of 2G technologies.

Figure 1.1 describes the evolution of the main 2G technologies during the past few years:

- **GSM** (Global System for Mobile communications), originated from Europe and worldwide deployed.
- **cdmaOne**, corresponding the IS-95 North American standard. This technology is mainly used in Asia-Pacific, North and Latin America.
- **TDMA** (Time Division Multiple Access), corresponding to the IS-136 North American standard, mostly used in North America. This system is also called D-AMPS (Digital Advanced Mobile Phone System), as it is an evolution to AMPS, an analogue 1G cellular system.
- **PDC** (Personal Digital Cellular), the 2G standard developed and used exclusively in Japan.

It can be observed that TDMA and PDC started to decline rapidly as 3G cdma2000 and UMTS technologies became commercially available. This is true especially in Japan, where UMTS was commercially offered at the end of 2001 under the commercial name of FOMA (Freedom of Mobile Multimedia Access), and in North America, where 2G EDGE (Enhanced Data rates for GSM Evolution) and 3G cdma2000 services were been released at the end of 2000. The same also applies to cdmaOne networks, which progressively migrated towards early cdma2000 3G technology in 2002.

However, although new services like video-telephony and content streaming have been proposed as 3G started to be commercial, GSM and cdmaOne have continued to grow substantially along with first 3G network deployments, thanks to the remaining potential of voice services.

Figure 1.2 presents the 3G subscriber evolution from 2002 to 2006 for the two main 3G systems being deployed worldwide: the UMTS (Universal Mobile Telecommunications System) and the cdma2000. This figure shows that the cdma2000 is ahead of UMTS over this period of time in terms of subscribers. This can be explained by the fact that cdma2000 was released at the end of 2000 – earlier than UMTS, which only reached commercial availability in 2003. However, because of the GSM subscriber base prevalence, it can be expected that UMTS will follow the GSM trend within the next few years.

Introduction

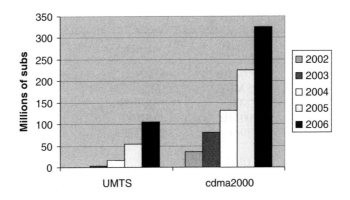

Figure 1.2 3G subscriber evolution.

1.2 About Technologies

This section provides an overview of the main 2G and 3G technologies, their evolutions, and how they relate to each other. At first glance, the picture represented by Figure 1.3 contains lots of different systems, but most of them actually fall into two main families: the 'MAP' and the 'IS-41'. In mid-2006, MAP (Mobile Application Part) systems were adopted by 80% of the subscriber base, while the IS-41 family captured the remaining 20%.

Most of the networks which were using PDC and TDMA moved towards MAP systems (Japan PDC was replaced by UMTS in 2001 and most of the TDMA networks have been

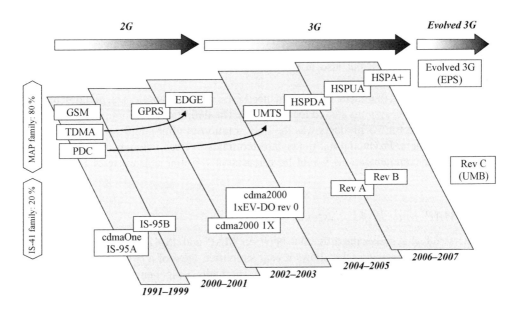

Figure 1.3 Evolutionary path of 2G and 3G technologies.

Table 1.1 Core parameters of the main 2G systems.

	GSM	cdmaOne	TDMA	PDC
Initial frequency band	900 MHz	800 MHz	800 MHz	900 MHz
Duplex separation	45 MHz	45 MHz	45 MHz	130 MHz
Modulation	GMSK	QPSK/BPSK	QPSK	QPSK
RF carrier spacing	200 kHz	1.25 MHz	30 kHz	25 kHz
Carrier modulation rate	270 kbits/s	1.2288 Mchip/s	48.6 kbits/s	42 kbits/s
Traffic channel per carrier	8	61	3	3
Access method	TDMA	CDMA	TDMA	TDMA
Initial data rate	9.6 kbits/s	14.4 kbits/s	28.8 kbits/s	4.8 kbits/s
Speech codec algorithm	RPE-LTP	CELP	VSELP	VSELP
Speech rate	13 kbits/s	13.3 kbits/s	7.95 kbits/s	6.7 kbit/s

upgraded to 2G EDGE). This is the reason why PDC and TDMA systems are considered in this picture as being part of the MAP family, although not being MAP technologies as such.

Both 3G standard family systems are moving towards Evolved 3G technologies. On the MAP side, Evolved 3G systems are known as EPS (Evolved Packet System), described in this book. The IS-41 standard family will move towards cdma2000 Revision C, also known as UMB (Ultra Mobile Broadband).

1.2.1 Heterogeneous 2G Systems

This section provides a very brief description of the main 2G systems' initial characteristics. Table 1.1 highlights the main differences between the four leading 2G technologies, not only in terms of radio basic parameters (such as radio modulation, carrier spacing and radio channel structure), but also at the service level (initial data rate and voice-coding scheme).

This table helps us to understand why the need for a common wireless technology became obvious once the 2G systems started to be popular. The definition of a common specification and product basis for 3G products was the only solution to offer simple global mobility to wireless customers. Producing multi-standard terminals in large volumes only for covering public cellular communications would have represented a dramatic waste of resource and energy.

1.2.2 'MAP' and 'IS-41' Systems

To understand what makes the difference between 'MAP' and 'IS-41' – the two main families of cellular systems – Figure 1.4 shows a very simplified view of a cellular communication system that all 2G and 3G systems comply to. The main components are:

- **The end-user terminal**, generally associated to an integrated circuit card containing subscriber-related information such as identifiers, security keys, etc.

Introduction 5

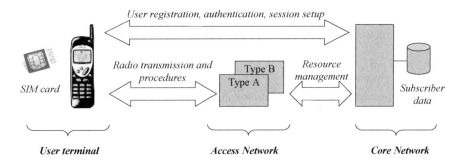

Figure 1.4 A simplified view of cellular communication systems.

- **The Access Network part**, which is responsible for radio-specific related tasks like secure and reliable transmission over the radio interface, radio resource management, handling of radio mobility procedures (this includes radio measurement processing as well as the handover decision process), etc.
- **The Core Network part**, which is responsible for end-to-end session setup, subscriber data management (this later includes authentication, authorization and billing), inter-working with external packet and circuit-switched networks, etc.

As pointed out before, the cellular systems can be distributed into two categories: the 'MAP' one and the 'IS-41' one. The difference between the two is not really about radio interfaces. Of course, radio interfaces are different between the two families, but this happens anyway within a given system, because of technological evolutions. These evolutions usually provide added value, such as better protection over radio transmission errors, increased bit rate, better radio resource usage efficiency, etc. These improvements often require the terminal manufacturers to design multi-mode terminals able to cope with new modulations or new data-coding schemes.

The major differences between those two families actually reside in the two following points:

- **The handling and management of user identities and subscription data** – this refers to the way customers are identified, and how these identities are stored in both network and user terminals.
- **The network procedures** – GSM and other technologies derived from GSM rely on the MAP (Mobile Application Part) protocol, whereas cdmaOne and cdma2000 rely on a completely different IS-41 North American standard. MAP and IS-41 are end-to-end protocols used between the terminal and the Core Network, and also between Core Network entities, for the purpose of user registration and authentication, call or data session setup, mobility management, and management of user subscription data.

In the past, systems from the two families used to be quite different and incompatible. However, as in recent evolutions of the standards, lots of effort has been made to reduce the gap and define synergies between MAP and IS-41 systems, for the benefit of R&D effort and product simplification. This can be observed at many levels when looking into latest detailed

GSM/UMTS and cdma2000 specifications. For illustration purposes, the two following major points are highlighted:

- User modules put in the terminal such as USIM (Universal Subscriber Identity Module) for UMTS or R-UIM (Removable User Identification Module) for cdma2000 have lots of similarities in terms of data organization and management process.
- The all-IP architecture evolution introduced in both UMTS and cdma2000 under different names – IMS (IP Multimedia Subsystem) for UMTS and MMD (MultiMedia Domain) for cdma2000 – are very close to each other in terms of architecture and underlining concepts. Besides, both IMS and MMD make use of SIP (Session Initiation Protocol) to set up communication sessions.

In the following sections, more details are provided about the main steps of the MAP and IS-41 evolutionary paths.

1.2.3 The MAP Technologies

This section briefly describes the technologies part of the MAP family. Figure 1.5 compares the typical user data throughput for each of them.

(i) GSM

The well known GSM (Global System for Mobile communications) entered into commercial service in 1991. This technology uses a TDMA/FDMA (Time Division Multiple Access/ Frequency Division Multiple Access) radio multiplexing scheme and GMSK (Gaussian Minimum Shift Keying) radio modulation. In its early days, GSM was only providing a voice service, SMS (Short Message Service) and low-rate circuit-switched data at 9.6 Kb/s.

(ii) GPRS

GPRS (General Packet Radio Service) is an evolution of GSM, designed for packet data communication. Based on the same GMSK modulation, GPRS provides, however, better user

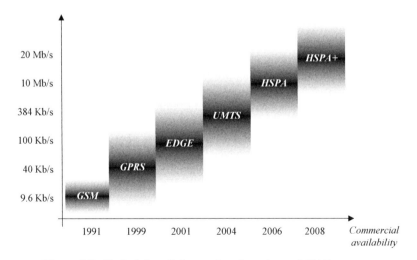

Figure 1.5 Typical downlink user data throughput of MAP systems.

rate by providing the least robust but fastest coding scheme (theoretical maximum bit rate is around 20 Kb/s per radio time slot) and the possibility to aggregate several radio slots for data transmission.

This system was intended to be more suitable and efficient for packet data applications, as it relies on sharing a set of radio channels among several terminals, for both uplink and downlink transmission.

(iii) EDGE

EDGE (Enhanced Data rates for GSM Evolution) introduces the 8-PSK (8 Phase Shift Keying) modulation, providing increased data rate. Using the fastest coding scheme, EDGE provides a maximum theoretical bit rate of around 50 Kb/s per radio time slot.

When combined with GPRS, it is possible to aggregate several EDGE-coded radio slots to increase the peak data rate for packet applications.

(iv) UMTS

UMTS (Universal Mobile Telecommunications System) is based on a standard developed by the 3GPP (3rd Generation Partnership Project), commercially launched in 2001 in Japan – under the name of FOMA (Freedom of Mobile Multimedia Access) – and 2003 in other countries. UMTS supports two variants: a FDD mode, being the most deployed, and a TDD mode, mainly supported for the Chinese market.

UMTS relies on the CDMA (Code Division Multiple Access) multiplexing scheme using a high chip rate direct spread sequence. In its first form, UMTS/FDD advantages were limited to increased data rates (up to 384 Kb/s per user on a single channel), the possibility of simultaneous packet and circuit applications, and improved roaming capabilities.

A theoretical maximum bit rate of 2 Mb/s over dedicated channels has been defined within the specification, but has never been deployed as such in commercial networks.

UMTS is often presented as the 3G evolution of GSM networks. Although the UMTS radio interface is completely different from the GSM/EDGE one, a lot of architectural concepts and procedures have been inherited from GSM. From a practical perspective, this can be expressed as follows:

- UMTS SIM cards are backward-compatible with GSM, meaning that a UMTS SIM can be inserted in a GSM phone, so that the SIM owner can benefit from services available under the GSM coverage.
- The MAP protocol of UMTS networks is an evolution of the GSM MAP as well as the packet and circuit session protocol being used between the terminal and the network. Although new elements and procedures have been defined in 3G, they rely on the same basis of concepts and principles.
- Specific network and user terminal procedures have been defined within the standard to ensure seamless mobility between GSM and UMTS, for both packet and circuit applications.

(v) HSDPA

HSDPA (High Speed Downlink Packet Access) is a UMTS enhancement, commercially available at the end of 2005.

The aim of HSDPA is to increase user throughput for packet downlink transmission (from network to mobile). For this purpose, new modulation has been introduced – 16 QAM (Quadrature Amplitude Modulation) – allowing a theoretical peak rate of 14.4 Mb/s (using the lowest channel protection algorithm). At first, 1.8 or even 3.6 Mb/s was expected as a realistic user experience.

From a radio interface perspective, HSDPA is based on a shared radio scheme and real time (every 2 ms) evaluation and allocation of radio resources, allowing the system to quickly react to data bursts. In addition, HSDPA implements a HARQ (Hybrid Automatic Repeat Request) which is a fast packet retransmission scheme located in the Base Station as close as possible to the radio interface. This allows fast adaptation to a change in radio transmission characteristics.

(vi) HSUPA

HSUPA (High Speed Uplink Packet Access) is the equivalent of HSDPA for uplink (from terminal to network) packet transmission.

HSUPA actually implements the same sort of techniques already used by HSDPA, such as a HARQ packet retransmission scheme providing low latency packet repetition between the terminal and the base station, and a reduced transmission time interval of 2 ms. However, unlike HSDPA, HSUPA is not based on a complete shared channel transmission scheme. Each of the HSUPA channels is actually a dedicated channel with its own physical resources. The actual resource sharing is provided by the Base Station, which allocates transmission power for uplink HSUPA transmission based on resource requests sent by terminals.

In theory, HSUPA can provide up to 5.7 Mb/s, using the top-level mobile category and larger transmission resources than can be allocated to a single terminal.

HSUPA may be combined with HSDPA – the association of the two is often referred to as HSPA (High Speed Packet Access) – so that data sessions can benefit from an increased data rate for both uplink and downlink.

As described in the following chapters, it is interesting to note that Evolved 3G makes use of techniques similar to HSDPA and HSUPA for transmission over the radio interface. The main difference is that those techniques are integrated from the early beginning of the definition of the Evolved 3G new interface and are applicable to all data-transmission flows. From a historical perspective, HSDPA and HSUPA have been added on top of a UMTS radio access initially designed for high bit rate, dedicated and circuit-oriented transmission.

(vii) HSPA+

High Speed Packet Access Plus, also known as 'HSPA Evolution', is an enhancement of HSDPA and HSUPA technologies, and intends to provide a fully 3G backward-compatible evolution step, until Evolved UMTS systems are available. HSPA+ is intended to be commercially available in 2008.

It is expected that HSPA+ will be as efficient as Evolved UMTS within the typical 5-MHz WCDMA channel bandwidth, thanks to a number of technical evolutions, including the use of MIMO (Multiple Input Multiple Output) and higher-order modulations (like 64 QAM in downlink and 16 QAM in uplink) for radio transmission and reception. Architectural evolutions are also envisaged to reduce data-transmission latency.

HSPA+ has to be considered as an optional and intermediate step between current HSPA and Evolved UMTS networks, intended to provide increased performance in a 3G backward-compatible way. This offers operators an opportunity to smoothly upgrade their networks and benefit from HSDPA enhancement until Evolved UMTS networks actually become commercially available.

1.2.4 The IS-41 Technologies

The IS-41 technology family contains lots of evolutions. For clarity, Figure 1.6 describes the various versions of cdma2000 standard, in relation to the names used in public and press communications.

As opposed to the MAP standard family, North American CDMA technology has been thought of in terms of a backward-compatible evolutionary path, starting from 2G IS-95 to 3G cdma2000.

The 3G cdma2000 main branch leads to 1xEV-DV (Evolution Data and Voice), which is probably never going to be available commercially on a large scale. The other 1xEV-DO branch (Evolution Data Optimized) is already in service and provides an enhanced user rate. This will eventually lead to the Revision C, which is the cdma2000 equivalent of Evolved UMTS.

The rest of this section briefly describes the technologies part of the IS-41 family. Figure 1.7 compares the typical user data throughput for each of them.

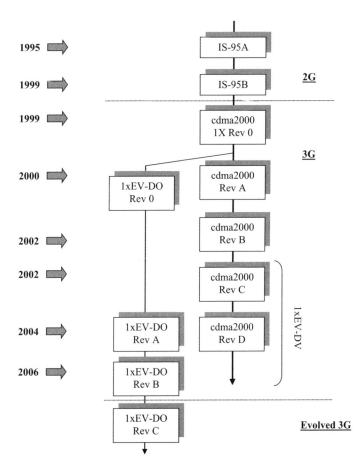

Figure 1.6 IS-41 standard family publication dates.

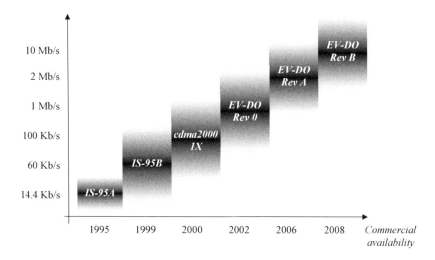

Figure 1.7 Typical downlink user data throughput of IS-41 systems.

(i) cdmaOne (IS-95A)
cdmaOne is the brand name of the first CDMA cellular system, deployed in September 1995, initially based on IS-95A North American standards. This system provides voice services as well as circuit-switched data up to 14.4 Kb/s.

This system was mostly deployed in North and Latin America, as well as other countries, such as South Korea and Australia. This system provided a similar grade of service as the GSM technology which was delivered at that time in other parts of the world.

(ii) cdmaOne (IS-95B)
IS-95B is a standard evolution of IS-95A systems, first deployed in September 1999, offering simultaneous voice and packet data services up to 115 Kb/s (maximum theoretical bit rate).

(iii) cdma2000 1X
cdma2000 1X – commercially deployed in October 2000 – is the first 3G system derived from the IS-95 technology. The cdma2000 standard has been developed by 3GPP2 (3rd Generation Partnership Project 2) and is part of approved radio interfaces for IMT-2000. (Please refer to the section on standard organizations for further details.)

As opposed to the GSM/UMTS transition, the cdma2000 radio interface has been specified as an evolution of the already existing CDMA-based 2G IS-95 standards. This can be seen in many parts of the radio interface definition, such as the structure of the common radio channel or the carrier spacing (the cdma2000 1.25-MHz carrier spacing is inherited from the IS-95 standard). This provides the reason why cdma2000 systems have been commercially available around 3 years before 3G UMTS systems.

The '1X' name comes from the fact that this system relies on a single 1.25-MHz carrier, as opposed to multi-carrier transmission schemes making use of three 1.25-MHz carriers. The 3X MC (or Multi-Carrier) scheme was only referred to as an IMT-2000 candidate technology, but has never been deployed in this form.

Initially, cdma2000 1X was able to provide voice services as well as up to 307 Kb/s downlink packet data and 153 Kb/s on the uplink on a single 1.25-MHz carrier.

More details about the physical layer of this system can be found in document C.S0002, 'Physical Layer Standard for cdma2000 Spread Spectrum Systems', published by the 3GPP2 consortium.

From that initial cdma2000 1X version, two branches have emerged. The first one is based on the evolution of the 1X specifications, leading to the 1xEV-DV (Evolution Data and Voice) and providing a high bit rate as well as simultaneous voice and data services in a fully backward-compatible way with 1X systems. The second branch, known as 1xEV-DO (for 'Evolution Data Only', which was eventually renamed 'Data Optimized'), provides improved data transmission as an overlay technology.

Eventually, EV-DV development was stopped, due to the lack of interest from operators and manufacturers.

(iv) cdma2000 1xEV-DO Revision 0

The 1xEV-DO (Evolution Data Optimized) has been in commercial service since the end of 2002.

This evolution allows the operator to provide simultaneous voice and high-speed packet data at the cost of an additional 1.25-MHz carrier. For that reason, a 1xEV-DO requires terminals to support multi-mode radio interface, in order to be fully backward-compatible with 1X and IS-95 technologies.

EV-DO Revision 0 introduces new 8-PSK and 16-QAM modulations, and is able to provide theoretical peak data speeds of 2.4 Mb/s on the downlink and 153 Kb/s on the uplink.

More information about Revision 0 additions can be found in document C.S0024-0, 'cdma2000 High Rate Packet Data Air Interface Specification', published by the 3GPP2 consortium.

(v) cdma2000 1xEV-DO Revision A

Commercially available in 2H2006, the Revision A objective was to improve the lack of real Quality of Service for packet data transmission and limited uplink capabilities of Revision 0.

As a result, Revision A systems are able to deliver theoretical peak data speeds of 3.1 Mb/s on the downlink and 1.8 Mb/s on the uplink. Moreover, packet prioritization schemes and more robust algorithms for uplink transmission have also been introduced.

More information about Revision A additions can be found in document C.S0024-A, 'cdma2000 High Rate Packet Data Air Interface Specification', published by the 3GPP2 consortium.

(vi) cdma2000 1xEV-DO Revision B

Revision B standard evolution was published in the first half of 2006 for planned commercial availability in 2008. Its main objective is to improve multimedia experience and packet-based delay-sensitive application performance in general.

This evolution introduces a new 64-QAM modulation scheme as well as a multi-carrier transmission scheme. Thanks to these new radio capabilities, Revision B systems will be able to deliver theoretical peak rates of 73.5 Mb/s in the forward link (from network to terminal) and 27 Mb/s in the reverse link (from terminal to network) through the aggregation of 15 1.25-MHz carriers within 20 MHz of bandwidth.

(vii) cdma2000 1xEV-DO Revision C
Revision C, currently under development, is the equivalent of the 3GPP Evolved UMTS. This new step in the EV-DO standard family is also known as UMB (Ultra Mobile Broadband).

It is planned that this evolution will support flexible and dynamic channel bandwidth scalability from 1.25 MHz up to 20 MHz and will be backward-compatible with Revisions A and B.

As for Evolved UMTS, Revision C is oriented towards 'all over IP' service support over a high-speed packet radio interface.

More information about Revision A additions can be found in document C.S0084-000, 'Overview for Ultra Mobile Broadband (UMB) Air Interface Specification', published by the 3GPP2 consortium.

1.3 Standards and Organizations

The development and introduction of worldwide technology cannot be achieved without global standard organizations.

This section describes the structure of the main standard organizations and support bodies involved in future system definition, in an attempt to shed some light on the jungle of acronyms, terms and initiatives emerging around 3G and its evolution.

1.3.1 The Role of ITU

The ITU (International Telecommunication Union), as a worldwide telecommunication standard body, played a key role in 3G definition.

(i) ITU and the IMT-2000 Framework
The IMT-2000 framework (International Mobile Telecommunications 2000) was initially launched by the ITU (International Telecommunication Union), in order to define 3G system and evolution for heterogeneous 2G technologies (Figure 1.8). This was felt the best way to define a common basis for performance, global access and seamless mobility requirements.

The IMT-2000 framework not only aims at defining international standards for 3G, but also cares about frequency spectrum issues, technical specifications for radio and network components, as well as regulatory aspects.

As a result of the call for the submission of candidate IMT-2000 radio interfaces, up to 16 candidates as Radio Access Technologies (RAT) have been proposed by various Regional Standard Organizations. The proposals were related to the terrestrial radio interface, as well as satellite systems. In 1999, a decision was eventually made by IMT-2000 to only accept five of them, as described in document M.1457, 'Detailed Specifications of the Radio Interfaces of International Mobile Telecommunications – 2000', published by the IUT:

- **CDMA DS** (Direct Spread): this corresponds to the FDD mode of UMTS.
- **CDMA MC** (Multi Carrier): this corresponds to the cdma2000 system family.
- **CDMA TDD**: this corresponds to the TDD mode of UMTS. As in UMTS/TDD specifications, the two variants are represented: the low chip rate mode (1.6-MHz bandwidth) and the high chip rate mode (5-MHz bandwidth).

Introduction

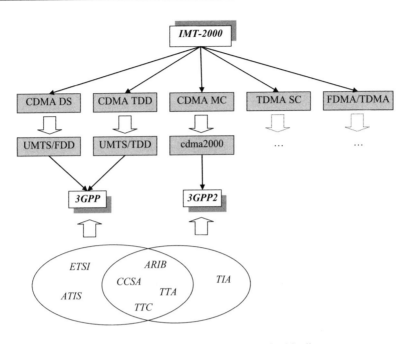

Figure 1.8 3G technologies and standard bodies.

- **TDMA SC** (Single Carrier): this access technology is based on 2G GSM/EDGE interface evolution.
- **FDMA/TDMA**: this corresponds to a high bit rate and packet data evolution of 2G DECT standard (Digital European Cordless Telephone).

Although ITU documents describe five technologies for use in the IMT-2000 terrestrial radio interface, only the CDMA-based first three systems have gained momentum within the industry.

The definition of those three radio systems was eventually handled by two 3G cross-country standardization bodies further described in this chapter: the 3GPP and the 3GPP2.

(ii) The IMT-Advanced Evolution
IMT-Advanced is a concept from the ITU for mobile communication systems with capabilities which go further than that of IMT-2000. IMT-Advanced was previously known as 'systems beyond IMT-2000'. The initial objective of ITU is to enable the commercial availability of IMT-Advanced compliant systems as early as 2011.

1.3.2 3G Cross-Country Standardization Bodies

The concept of 'cross-country organization' emerged in 1998, resulting from IMT-2000 activities. This was felt as the best way to ensure worldwide 3G adoption and success.

Table 1.2 List of regional organizations involved in 3GPP and 3GPP2.

ARIB	Association of Radio Industries and Businesses	Japan
ATIS	Alliance for Telecommunications Industry Solution	North America
CCSA	China Communications Standards Association	China
ETSI	European Telecommunication Standard Institute	Europe
TIA	Telecommunications Industry Association	North America
TTA	Telecommunications Technology Association	Korea
TTC	Telecommunications Technology Committee	Japan

As a result of IMT-2000's call for candidate 3G radio access technologies, two parallel consortiums were created in early 1999: the 3GPP (3rd Generation Partnership Project) and the 3GPP2 (3rd Generation Partnership Project 2), both aiming at producing technical specifications for third-generation mobile systems.

The decision process within these two organizations is driven by the consensus (or by votes in exceptional cases in which consensus is not reached), and a lot of care is given to maintaining as much as possible the balance between regions in the process of attribution of chairman seats of the various technical working groups.

The main difference between the two is that 3GPP is promoting UMTS (Universal Mobile Telecommunications System) – the 3G evolution from the GSM/MAP system family – whereas 3GPP2 is focusing on IS-41 technology evolution.

3GPP and 3GPP2 consortiums involve lots of different organizations like terminal and network manufacturers, the main network operators, all of them representing the various regional standard bodies listed in Table 1.2. The role of the regional organizations listed in this table is to promote locally unified standards and may include product testing and certification. Their scope of activity is not limited to cellular communication and may also include satellite communication, broadcasting systems, private mobile radio systems, etc.

The regional organizations involved in 3GPP are ARIB, ATIS, CCSA, ETSI, TTA and TTC. The regional organizations involved in 3GPP2 are ARIB, CCSA, TIA, TTA and TTC.

1.3.3 The Structure of 3GPP

As the focus of this book is Evolved UMTS, more details are provided about the 3GPP structure and associated working groups (Figure 1.9).

The technical part of the 3GPP consortium is composed of four groups or TSG (Technical Specification Groups):

- **SA** (Services and system Aspects), which is responsible for the overall architecture and service definition.
- **RAN** (Radio Access Networks), which is responsible for the features, interfaces and protocols of the Access Network.
- **CT** (Core Network and Terminals), which is responsible for Core Network specific protocols and end-to-end protocols for call control and session management.
- **GERAN** (GSM EDGE Radio Access Network), which is responsible for the evolution of the GSM/EDGE-based Radio Access Network. As described in previous sections, UMTS

Introduction

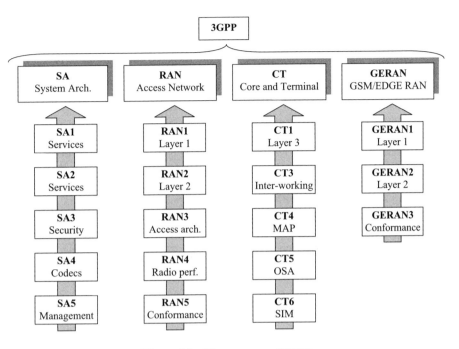

Figure 1.9 The structure of 3GPP.

inheritance from GSM technology is quite significant from a Core Network and Service perspective. This is the reason why this TSG, initially part of the ETSI, was eventually moved to the 3GPP organization in late 2000.

Each of these groups is composed of different working groups (or WG), each of them having responsibility for an area of work requiring a specific level of expertise.

The SA TSG (Services and system Aspects) is composed of the following groups:

- **SA1** is in charge of high-level service and feature requirements. SA1 produces specific documents (called Stage 1 specifications) focused on system and service capabilities and used as a baseline by other groups.
- **SA2** has the responsibility of defining network architecture and the features to be supported by the network entities, based on the stage 1 documents provided by SA1. SA2 produces Stage 2 specifications, which are used as inputs by groups in charge of detailed interface specifications.
- **SA3** is the security group of 3GPP. Its role is to set up security requirements for the overall system, and produce specifications for security algorithms to be applied in the network.
- **SA4** works on the specification of speech, audio, video and multimedia codecs applicable to circuit and packet-based applications.
- **SA5** is focused on network management. SA5 has the responsibility of specifying the architecture, procedures and interface-related issues of network management, including charging, configuration and performance management, inventory, etc.

The RAN TSG (Radio Access Networks) is composed of the following groups:

- **RAN1** is responsible for the definition of the physical (Layer 1) interface of the UTRAN Access Network. This includes the specification of the radio modulation, channel coding, physical layer measurements, etc.
- **RAN2** works on the radio interface protocols used on top of the physical layer. This includes Layer 2 protocols for data transmission as well as the signalling procedures related to radio interface such as radio resource allocation, handover management, etc.
- **RAN3** is in charge of overall UTRAN architecture. This includes the definition of interfaces between entities of the Access Network, as well as the specification of the UTRAN transport network.
- **RAN4** is responsible for the RF conformance aspects of UTRAN and produces test specifications for radio network and terminal equipment regarding RF transmission and reception performance.
- **RAN5** works on radio interface conformance test specifications. RAN5 produces test specifications, based on RAN4 documents, and signalling procedures defined by other groups such as RAN2.

The CT TSG (Core Network and Terminals) is composed of the following groups:

- **CT1** has the responsibility of specifying Layer 3 protocols, being used between the Core Network and terminals to set up communication sessions for circuit, packet and IMS-based applications.
- **CT3** is in charge of the inter-working between 3GPP networks and external packet or circuit networks. This includes signalling or protocol inter-working, as well as possible user plane data adaptation.
- **CT4** is in charge of supplementary service definitions (such as SMS or Call Transfer) as well as the specification of interfaces and protocols between Core Network elements such as the MAP.
- **CT5** works on UMTS OSA (Open Service Access) and produces Application Programming Interfaces to facilitate UMTS service definition and creation.
- **CT6** is in charge of the definition of the format of the Subscriber Identity Module (SIM). This includes the specification of SIM card data content, format and organization, as well as interfaces between the SIM card and external entities.

The GERAN TSG (GSM EDGE Radio Access Network) is composed of the following groups:

- **GERAN1** (the RAN1 equivalent) is responsible for the definition of the physical interface of the GSM/EDGE Access Network.
- **GERAN2** (the RAN2 equivalent) works on the radio interface protocols defined on top of the physical layer.
- **GERAN3** is in charge of conformance test specification for the GERAN access network.

In the scope of 'Evolved UMTS', the structure of 3GPP will not change. All existing TSG and most of their Working Groups will extend their scope in order to cover Evolved UMTS requirements and specifications.

1.3.4 The NGN Evolution

NGN (Next Generation Networks) is a new broad term which refers to emerging network architectures. Briefly, NGN is a concept of converged architecture which natively supports a wide range of services, such as:

- Conversational services (person-to-person communications or voice call).
- Messaging (email, short or multimedia message exchange).
- Content-on-demand (streaming, video on demand, broadcast, multicast).

Beyond this very general and high-level statement, there are actually two main ideas behind NGN. All information is transmitted via packets with a strong support of Quality of Service mechanisms ensuring that the end-to-end service-level agreement is reached. This mandates that the packet transport network is aware of the Quality of Service requirements and supports all the needed features to control and enforce them.

The second principle relates to the decoupling of service and transport layers, allowing them to be offered separately and to evolve independently. This clear separation between services and transport should allow the provision of new services independently from the access technology and transport type.

NGN standardization is actively going on in parallel within two major standardization bodies:

- The Global Standard Initiative (GSI) within the ITU-T.
- The TISPAN technical body (Telecommunication and Internet converged Services and Protocols for Advanced Networking) within the ETSI.

Through a large number of working groups, the two standardization bodies study and define technical aspects of NGN, such as the general architecture, Quality of Service support, network management, signalling protocols, migration from existing network architectures towards NGN, etc.

Initially, NGN has a focus on fixed access such as ADSL. However, wireless access will also be considered in a second step as normal evolution.

It is interesting to note that Evolved UMTS is actually moving in the same direction as NGN. As described in Chapter 2, Core Network simplification introduced by the Evolved UMTS Packet Core is completely in line with NGN objectives. In some respects, IMS already answers parts of the NGN requirements in providing a common framework for the support of a large set of services, from interactive data services to Quality of Service-constrained voice or streaming applications.

For illustration purposes, Figure 1.10 presents the TISPAN NGN overall architecture, as described in specification 282 001, 'NGN Functional Architecture Release 1', from the ETSI. The picture shows a clear separation of the transport and service parts of the network, represented by the dotted horizontal line.

The main purpose of the Transport Layer is to provide IP connectivity to the end-user under the control of Network Attachment and Admission Control subsystems. In principle, these two subsystems are independent from the transport technology used below the IP network layer, so this model is suitable for any type of fixed or wireless access.

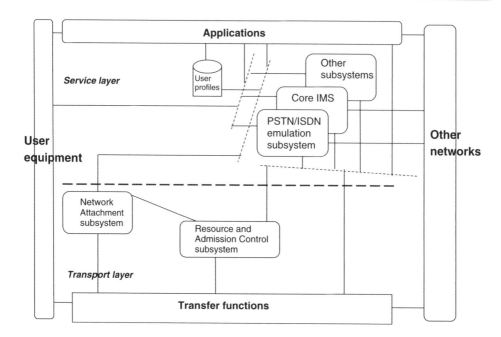

Figure 1.10 TISPAN NGN overall architecture (source: ETSI ES 282 001).

The Service part is represented as an integrated layer supporting applications, user profile management and IMS (IP Multimedia Subsystem) elements. The 'Core IMS' of the NGN is actually a subset of the full 3GPP IMS architecture and re-uses the same principles in terms of SIP-based session signalling and call servers as well as multimedia resource and gateway functions. In addition, the Service Layer also provides access to other networks such as the legacy PSTN (Public Switched Telephony Network).

In addition to basic communication session support, TISPAN NGN has also specified a complete framework for supplementary services, similar to those proposed for circuit-switched speech, and supported by a wireline and wireless legacy circuit infrastructure. This point is further described in Chapter 6.

In order to avoid the development of diverging IMS architectures and services, it was agreed in June 2007 to migrate all TISPAN activity within 3GPP under a new framework known as 'Common IMS'. 'Common IMS' is therefore an extension of 3GPP IMS activity handled by the SA (System Architecture) group, enabling IMS services not only for wireless networks (legacy mobile cellular and WLAN networks) but also for packet cable and fixed networks. Common IMS is part of Release 8 of the 3GPP standard.

1.3.5 The NGMN Initiative

NGMN (Next Generation Mobile Networks) is an initiative led by a group of mobile operators, including NTT DoCoMo, Vodafone, Orange, T-Mobile, China Mobile, Sprint and KPN. The objective of NGMN is to provide a vision for 3G technology evolution for commercial availability planned in 2010. NGMN expects the standardization phase of the next generation to be completed by the end of 2008, so that the trial phase can start in 2009. NGMN is not a competitor to ITU or any other cross-country standardization body. It rather intends to complete and support the work handled by the standardization committees and to make sure that future systems will meet the needs of mobile operators and their customers.

Not surprisingly, the target NGMN performance requirements in terms of data rate, latency and spectrum efficiency are aligned with Evolved UMTS (as described in Chapter 2). The same goes for the overall architecture concepts supported by NGMN, which is based on a Packet Switched Core Network together with an evolved high-speed radio interface.

On the service side, NGMN initiative and Evolved UMTS also share the same interest on three key subjects:

- **Efficient Always-On support:** as most of the new services (such as 'Presence' or 'Push-To-Talk') require the end-user to be 'always connected'.
- **Seamless mobility between systems:** this applies not only to voice calls, but also to all other services. In principle, no interruption or drop in performance should be experienced by end-users when changing systems.
- **Efficient Support of Broadcast and Multicast:** this requirement is a consequence of the growing interest of operators and customers in mobile broadcasting technologies such as DVB-H.

The NGMN initiative is technology-independent, as it actually considers all 3G system migration including UMTS as well as the cdma2000 1xEV-DO family.

Figure 1.11 describes the high-level architecture envisaged by NGMN. It is planned that future systems will support all kinds of services and ultimately integrate all the circuit and packet-switched services through one unique Packet Core network domain. As we will see in Chapter 2, the convergence of all services on a common packet-based network architecture is also a key concept of Evolved UMTS.

For more information about the NGMN initiative, please consult http://www.ngmn-cooperation.com/.

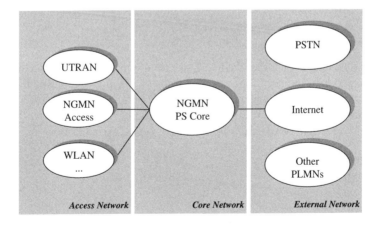

Figure 1.11 High-level NGMN system architecture.

1.4 Spectrum

The radio spectrum is the most critical resource in radio communication systems, making frequency regulation and allocation a complex issue:

- **A scarce resource:** because of physics laws, the amount of spectrum that can actually be used for mobile (possibly high-speed) and high bit rate applications is quite limited. The lower part of the spectrum (starting at 400 MHz) is already overcrowded with digital, analogue, public and military applications. The upper part of the spectrum, above 2 GHz, offers more possibility for extensions and high bit rate-demanding applications, at the cost of a shorter cell range because the propagation loss increases with the frequency.
- **Lack of harmonization:** because of the history of and disparities in regulation rules, spectrum allocation is not consistent between countries. This complicates worldwide system spectrum definition.

Coming back to cellular system history, GSM (the most popular 2G technology) was initially built to work on a single 900-MHz frequency band. The way it was introduced in the core specifications of GSM left quite limited possibilities for extension to other frequency bands. However, as GSM became largely accepted, the need for additional frequencies appeared as a requirement to support the increasing number of customers and operators. GSM was therefore modified to be able to support other frequencies on the request of specific countries, first in the 1800 and 1900-MHz bands, and then in the 450, 480 and 850-MHz bands.

When 3G and IMT-2000 emerged, a lot of attention was paid to the issue of spectrum allocation. One of the key IMT-2000 objectives was to allow global mobility, seamless service access and increased user experience. This was felt as difficult to achieve without a minimum level of spectrum harmonization. For that reason, the decision was made at the WARC-92 (World Administrative Radio Conference of 1992) to identify 230 MHz of spectrum on a worldwide basis for the operation of IMT-2000 networks. This amount of spectrum was shared in two bands:

- 1885–2025 MHz.
- 2110–2200 MHz.

In a second step, and because the WARC-92 spectrum studies were based on the assumption of low bit rate services, the need for additional frequency was raised during the WRC-2000 (World Radiocommunication Conference of 2000). In order for the IMT-2000 to be able to cope with upcoming multimedia services, three additional new bands have been defined, adding close to 500 MHz of spectrum to the initial WARC-92 allocation:

- 806–960 MHz.
- 1710–1885 MHz.
- 2500–2690 MHz.

The IMT-2000 does not specify the way the spectrum is used. This point is left for decision to the local regulator. In Europe, this activity is under the European Radiocommunications

Committee (ERC)'s responsibility, within the CEPT (European Conference of Postal and Telecommunications Administrations).

Figure 1.12 represents an overview of the spectrum allocated to IMT-2000 in comparison with existing communication systems currently using those frequencies in Europe.

Regarding 3G frequencies, the spectrum is shared between the three following technologies:

- UMTS/FDD (Frequency Division Duplex), which mandates paired spectrum allocation: one band for uplink (UL) transmission (from terminal to network) and one band for downlink (DL) transmission (from network to terminal).
- UMTS/TDD (Time Division Duplex), which is designed for unpaired spectrums. Uplink and downlink transmissions are therefore using the same frequencies at different times.
- MSS (Mobile Satellite Services), which is the satellite component of IMT-2000.

In any case, and as opposed to GSM, one improvement point is that UMTS specifications are flexible enough to accommodate a large range of frequencies (theoretically from 0 to 3276.6 MHz).

Figure 1.12 also represents 2G technologies overlapping the IMT-2000 spectrum. As 2G technologies will start to decline, spectrum can therefore easily be re-allocated to IMT-2000 systems. The major 2G systems subject to spectrum re-use are:

- The GSM technology, in the 900 and 1800-MHz bands (Figure 1.12 includes the Primary as well as the Extended GSM 900 band), working in FDD mode and requiring paired spectrum.
- The DECT system (Digital European Cordless Telephone), working in TDD unpaired spectrum.

At this point in time, the spectrum for systems beyond IMT-2000 is not yet identified. This will be a subject for the WRC-07 conference, being held at the end of 2007.

1.5 The Evolution of UMTS

The need for 3G long-term evolution studies was stated at the end of 2004 within the 3GPP, in order to maintain the competitive position of UMTS-based technologies for the future. It was therefore decided to launch feasibility on overall system architecture and access network evolution, with an objective to finalize core specifications by mid-2007.

1.5.1 1st Evolution Driver: The Move towards Data Applications

As described above, initial 2G systems were built for circuit-switched applications – voice being the main one, and low-rate circuit-switched data being supported as well. As the World Wide Web became a reality, 2G systems evolved towards packet data. This was not a complete system redesign, as packet data network architecture was defined as an add-on to existing networks – meaning increased costs for deployment and operation.

Within Evolved UMTS networks, all communication streams will be considered as packet data, allowing converged and simplified network architectures.

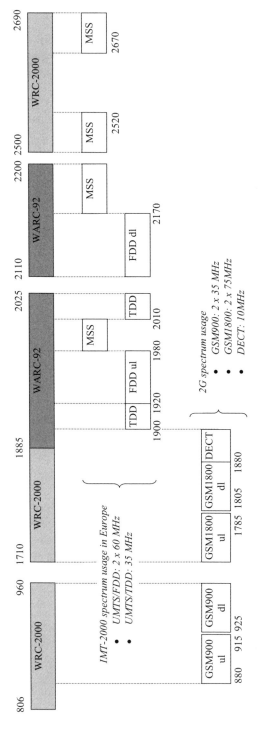

Figure 1.12 Overall cellular spectrum overview.

Introduction

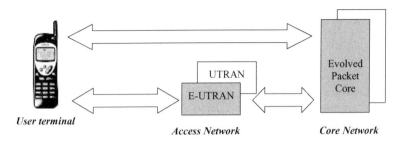

Figure 1.13 Towards 'Evolved UMTS' networks.

1.5.2 2nd Evolution Driver: Enhanced Radio Interface Capabilities

Over the past few years, fixed Internet access capabilities have been improving, from the 56-Kb/s V90 modem-based access to 10-Mb/s ADSL and 100-Mb/s fiber access, enabling new services and much better user experience. As a consequence, it is not surprising that wireless communication systems are also moving towards increased capabilities and performances, so that the Quality of Service for existing and new services is kept acceptable when used over the radio interface.

1.5.3 What Will Change Within the Network?

The evolution of UMTS consists of two parts (Figure 1.13):

- The Core Network evolution – as a consequence of the global move towards packet data applications. The Evolved Packet Core network (or EPC) aims to provide a converged framework for packet-based real-time and nonreal-time services.
- The Access Network – existing UMTS access network, also named as UTRAN (Universal Terrestrial Radio Access Network) is evolving towards E-UTRAN (for Evolved-UTRAN) so as to offer high-data-rate, low-latency and packet-optimized radio access technology. This implies a new radio interface, new protocols and also new access network architecture, as further described in this book.

Some vocabulary

The *E-UTRAN* Access Network (for Evolved UTRAN) is often referred to as LTE (for Long Term Evolution) in 3GPP reports and specifications, which was the initial name given to the 3G/UTRAN evolution study item.

Similarly, on the Core Network side, the evolution towards Evolved Packet Core, or *EPC*, is referred to as SAE (for System Architecture Evolution).

The Evolved UMTS, which is basically the concatenation of the E-UTRAN access network and EPC Core Network, is referred to as *EPS* (Evolved Packet System) in 3GPP standard documents.

1.5.4 What is Described in this Book?

The technical details of UMTS evolution are further described, through the following chapters:

- **Chapter 2** provides an overview of 'Evolved UMTS'. The major blocks and concepts are described, as an introduction to the in-depth view presented in the following chapters.
- **Chapter 3** is dedicated to the physical radio layer, as being a disruption with existing 2G (TD/FDMA) or 3G (CDMA) systems.
- **Chapter 4** is dedicated to the network architecture. Major blocks are presented – Access and Core – as well as the main protocol layers and Transport Network.
- **Chapter 5** aims at presenting the steps and procedures being implemented throughout the life of the terminal (from power-on to mobility in ACTIVE mode).
- **Chapter 6** describes the main services built on top of 'Evolved UMTS', focusing on how the service is implemented over the architecture.

All through this book, lots of references are provided to standards and specifications, so that the reader willing to dig into a specific area can refer to the right document.

1.6 Links and Documents

1.6.1 Useful Web Sites

Here is a list of Web sites which may be consulted for further information:

- **http://www.3gpp.org/** This is the homepage of the 3GPP consortium. Lots of information can be found here, from details on technical working groups to technical specifications. Access to technical documents is free of charge.
- **http://www.3gpp2.org/** This is the homepage of the 3GPP2 consortium. Access to 3GPP2 specifications is free of charge.
- **http://www.rfc-editor.org/** The RFC Editor is the publisher of the IETF RFC documents and is responsible for the final editorial review. All the IETF RFC mentioned in this book can be found here, free of charge.
- **http://www.openmobilealliance.org/** This link points to the home page of the OMA (Open Mobile Alliance). Specifications of services presented in Chapter 6 of this book can be accessed from this link, free of charge.
- **http://www.itu.int/home/imt.html** This address points to the IMT-2000 page hosted by the ITU. General information about IMT-2000 can be found here. In most cases, access to ITU documents is subject to charging.

1.6.2 Evolved UMTS Specifications

(i) The 3GPP Standard Releases
As for many standard committees, 3GPP work is organized into releases. This allows the different groups to work on future evolutions and still allow corrections and bug tracking on

Introduction

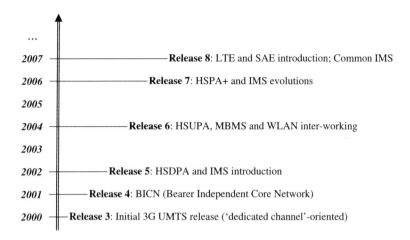

Figure 1.14 3GPP standard releases.

previous stable versions. Each manufacturer (either on the network or on the terminal side) has the flexibility to implement any version it prefers. In principle, specification progress is backward-compatible, meaning that a given version includes all the basic and optional features which were defined in all previous versions (except in some rare cases of 'option pruning', where useless features are withdrawn from the standard).

Figure 1.14 represents all 3G standard releases supported by the 3GPP. The picture is limited to 3G and post-3G evolutions and does not include GERAN versions of the standard.

Each version actually contains a great number of new features. The picture only shows the main ones. E-UTRAN and complete Evolved Packet Core definition are actually part of Release 8 of the 3GPP standard. The completion of all protocols and messages is planned for the end of 2007.

(ii) The Main Documents
This section describes all the main 3GPP specifications related to Evolved UMTS, all available from the 3GPP homepage.

The following documents are maintained by the SA2 3GPP group. They specify overall network architecture (including Access and Core parts) as well as general procedures such as network attachment, session setup and mobility:

- 23.401, 'GPRS Enhancements for E-UTRAN Access'.
- 23.402, Architecture Enhancements for Non-3GPP Accesses'.

These documents are maintained by the RAN1 3GPP group. They specify the physical layer of the E-UTRAN radio interface:

- 36.201, 'LTE Physical Layer: General Description'.
- 36.211, 'Physical Channels and Modulation'.
- 36.212, 'Multiplexing and Channel Coding'.
- 36.213, 'Physical Layer Procedures'.
- 36.214, 'Physical Layer Measurements'.

These documents are maintained by the RAN2 3GPP group. They specify the layer 2 and E-UTRAN control protocols supported by the E-UTRAN radio interface:

- 36.300, 'E-UTRAN Overall Description: Stage 2'.
- 36.302, 'E-UTRAN Services Provided by the Physical Layer'.
- 36.304, 'User Equipment (UE) Procedures in Idle Mode'.
- 36.306, 'User Equipment (UE) Radio Access Capabilities'.
- 36.321, 'Medium Access Control (MAC) Protocol Specification'.
- 36.322, 'Radio Link Control (RLC) Protocol Specification'.
- 36.323, 'Packet Data Convergence Protocol (PDCP) Specification'.
- 36.331, 'Radio Resource Control (RRC) Protocol Specification'.

These documents are maintained by the RAN3 3GPP group. They specify the interfaces and procedures supported by interfaces between network nodes of E-UTRAN:

- 36.401, 'E-UTRAN Architecture Description'.
- 36.410, 'S1 General Aspects and Principles'.
- 36.411, 'S1 Layer 1'.
- 36.412, 'S1 Signalling Transport'.
- 36.413, 'S1 Protocol Specification'.
- 36.414, 'S1 Data Transport'.
- 36.420, 'X2 General Aspects and Principles'.
- 36.421, 'X2 Layer 1'.
- 36.422, 'X2 Signalling Transport'.
- 36.423, 'X2 Protocol Specification'.
- 36.424, 'X2 Data Transport'.

In addition to those specifications, 3GPP also produces a number of Technical Reports, which record working assumptions and agreements until actual specifications are made available:

- 24.801, '3GPP System Architecture Evolution: CT WG1 Aspects'.
- 29.804, 'CT WG3 Aspect of 3GPP System Architecture Evolution'.
- 29.803, '3GPP System Architecture Evolution: CT WG4 Aspects'.
- 32.816, 'Telecommunication Management; Study on Management of LTE and SAE'.
- 32.820, 'Telecommunication Management: Study on Charging Aspects of 3GPP System Evolution'.
- 33.821, 'Rationale and Track of Security Decisions in LTE/SAE'.

2

Evolved UMTS Overview

The objective of this chapter is to present an overview of the Evolved UMTS system, as an introduction to the more in-depth view provided by the subsequent chapters.

In the two first sections, the main requirements and basic concepts are looked at, from the radio interface and overall network architecture perspectives. Then, the overall architecture of Evolved UMTS is described, highlighting common views and disruptions with existing 3G UMTS systems, including a brief description of the terminal logical architecture and SIM card aspects.

The IP Multimedia, as well as policy and charging features, is also described in this chapter. Although being more related to the application level, those two domains are part of the wireless network architecture. This chapter also describes how these features relate to the existing 3G nodes, and how they progress in the scope of Evolved UMTS.

2.1 The Access Network Requirements

As stated in the previous chapter, one of the main objectives of Evolved UMTS is to design a high-performance radio interface. The rest of this section describes in more detail the performance criteria and the associated requirements of this new Radio Access Network. These include:

- Radio interface throughput.
- Data transmission latency.
- Terminal state transition requirements.
- Mobility requirements.
- Flexibility in spectrum usage.
- Mobility requirements between systems.

In this section, as well as in the rest of this book, E-UTRAN (Evolved Universal Terrestrial Radio Access Network) is referred to as the access network of Evolved UMTS (or EPS for

---

Evolved Packet System (EPS) P. Lescuyer and T. Lucidarme
Copyright © 2008 John Wiley & Sons, Ltd.

Evolved Packet System), which not only includes the radio interface, but also the network nodes and terrestrial interfaces supporting the radio-related features.

2.1.1 Radio Interface Throughput

The radio interface of E-UTRAN shall be able to support an instantaneous downlink (from network to terminal) peak data rate of 100 Mb/s within a 20 MHz downlink spectrum allocation and an instantaneous uplink (from terminal to network) peak data rate of 50 Mb/s within a 20 MHz uplink spectrum allocation. This corresponds to a spectrum efficiency of 5 bits/s/Hz for the downlink, and 2.5 bits/s/Hz for the uplink.

It is interesting to compare those requirements with the initial objectives of UMTS, as defined in 1998. At that time, 3G systems were expected to provide the following throughput:

- 144 Kb/s in rural outdoor radio environments.
- 384 Kb/s in urban or suburban outdoor radio environments.
- 2048 Kb/s in indoor or low range outdoor radio environment.
- More than 2 Mb/s in urban or low-range outdoor radio environments – this later was added when HSDPA (High Speed Downlink Packet Access) was introduced in UMTS specifications.

If we consider that UMTS is actually providing a theoretical maximum of 2 Mb/s within a 5 MHz spectrum allocation for dedicated channels and a theoretical maximum of 14.4 Mb/s within a 5 MHz spectrum allocation when using HSDPA shared-channel techniques, the corresponding spectrum efficiencies are, respectively, 0.4 and 2.9 bits/s/Hz for downlink transmission.

2.1.2 Data Transmission Latency

The new E-UTRAN specifications shall enable a user data latency of less than 5 ms in a non-loaded condition (meaning that the system is only supporting a single data stream) (Figure 2.1). This value corresponds to the maximum transfer delay between the E-UTRAN access gateway and the user terminal, including the transfer delay between nodes within the Access Network and the transmission over the radio interface.

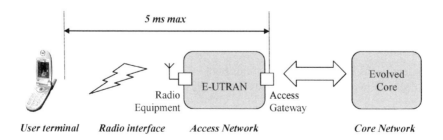

Figure 2.1 E-UTRAN latency requirements.

This very constraining requirement is the guarantee that the E-UTRAN Access Network will be suitable for the most demanding existing and upcoming real-time services.

For information, and for an audio or video conversation, the maximum one-way transfer delay that human perception can tolerate is 400 ms, which corresponds to more than one syllable (one syllable is around 250 ms long); 150 ms is the preferred value under normal operation, as a commonly agreed value and as reflected by IMT-2000 performance and Quality of Service requirements. Above the 400 ms transfer delay, the quality of experience is generally considered as nonacceptable for a conversational service.

It has to be noted that the 150 ms figure is an end-to-end value, including not only the Access Network, but also data transmission within the Core Network and a public phone network. This indicates that the contribution of the evolved Access Network in terms of transfer delay is set to a very small value when compared with the overall time budget.

2.1.3 Terminal State Transition

This requirement is often referred to as 'control plane latency' in the 3GPP specifications. This requirement is actually related to what is better known as 'efficient Always-On support'.

In traditional circuit-switched cellular networks, as for classical telephony, it was generally understood that the end-user terminal could only be into two different states:

- The IDLE mode – a state in which the mobile is reachable (e.g. for the case of user-terminated call setup) but no communication sessions are active.
- The ACTIVE mode – corresponding to the active phase of a session. In this mode, data are actively transmitted to and from the user terminal, as in a voice call or circuit-switched data session.

This duality is very much in line with the classical 'dial-in' communication mode, inherited from the traditional voice technology and modem-based circuit-switched data sessions.

From the emergence of new fixed Internet access technologies (such as ADSL) and the development of new packet services characterized by bursts of activity separated by silent periods of time, the need for a new terminal mode rose. The Standby mode was an answer to this need, filling the gap between the IDLE and the ACTIVE modes. This state was initially defined in the scope of 2G GPRS, and re-used from the first version of 3G UMTS specifications. In such a state, the user terminal is silent (very limited data exchange is allowed anyway) but always connected. In addition, this state is defined in such a way that it allows a very short period of time for transition to the ACTIVE mode in case end-user session activity is resumed.

The transition between the terminal states is under the Access Network's responsibility, as this mechanism is closely related to radio activity and radio resource usage. For that reason, specific requirements have been introduced in E-UTRAN for terminal state transitions.

The support for efficient mechanisms to support always-connected terminals is a key requirement for services like 'Presence' (in which a group of people share each other's availability information in real time), 'Instant Messaging' (in which people exchange a short set of information at any time) and 'Push-To-Talk' (which is the voice version of Instant Messaging).

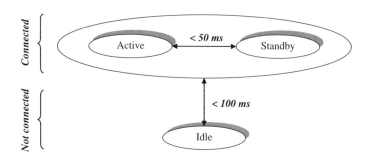

Figure 2.2 Terminal state transition requirements.

As illustrated by Figure 2.2, the requirements related to E-UTRAN operation mandate that:

- The transition from IDLE to ACTIVE state shall be less than 100 ms (this time does not include the time needed for paging procedures, in the case of mobile-terminated sessions). Obviously, this does not correspond to a full call setup, but only to the period of time needed for the network and the terminal to move from the IDLE state to an ACTIVE mode of operation in which a dedicated resource is allocated for the transmission of signalling or user information.
- The transition between ACTIVE and STANDBY states shall be less than 50 ms (also excluding possible paging procedures, if needed).

2.1.4 Mobility

In terms of mobility, E-UTRAN requires that user connectivity across the cellular network shall be maintained at speeds from 120 to 350 km/h (or even up to 500 km/h, depending on the frequency band).

2.1.5 Spectrum Flexibility

One of the major operational constraints of the UMTS network is about spectrum flexibility. As in current UMTS, the spectrum allocation possible schemes are very limited.

In FDD (Frequency Division Duplex) mode, not less than 5 MHz of spectrum can be allocated to a given cell. This value is determined by the radio interface characteristics: a CDMA direct spread system based on a chip rate of 3.84 Mchips/s.

In TDD (Frequency Division Duplex) mode, the situation is a bit different. Like the FDD option, TDD is also a CDMA direct spread system; however, two chip rates have been defined by the 3GPP:

- The low chip rate option (1.28 Mchips/s) corresponding to a channel occupation of 1.6 MHz.
- The high chip rate option (3.84 Mchips/s) corresponding to a channel occupation of 5 MHz.

Evolved UMTS Overview

As an improvement to this situation, the evolved Access Network shall be able to operate in spectrum allocations of different sizes, including 1.25, 2.5, 5, 10, 15 and 20 MHz, for both uplink and downlink. The intention is to allow flexible deployment and operation of spectrum accordingly to the expected traffic and service requirements.

This flexibility will also be beneficial to the frequency attribution process held by local regulators. Evolved UMTS introduction will be made easier in regions where large chunks of spectrum are scarce. Besides, it will help to make the spectrum license price more affordable and allow Evolved UMTS deployment to operators not willing to pay a heavy price for spectrum that they don't plan to use.

In addition, operation in paired and unpaired spectrums shall be supported by Evolved UTRAN, so that the system supports asymmetric services (such as content streaming, or broadcast/multicast services) in the most efficient way.

2.1.6 Co-existence and Inter-Working with Existing UMTS

From the beginning of UMTS standardization, inter-working between new and old technologies has always been a subject of concern. For that reason, lots of effort has been dedicated to ensuring smooth and efficient transition between UMTS and GSM for terminals in ACTIVE or IDLE mode.

It is not surprising that such a constraint is part of Evolved UTRAN foundations. Thus, interruption time for handover between E-UTRAN and UTRAN/GERAN shall be less than 300 ms for real-time services and 500 ms for nonreal-time services.

Reference documents about wireless network requirements

ITU-R and ITU-T specifications:

- M.1079, 'Performance and Quality of Service Requirements for IMT-2000 Access Networks'
- G.114 'One-Way Transmission Time'
- G.174 'Transmission Performance Objectives for Terrestrial Digital Wireless Systems'

3GPP specifications:

- 22.105 'Services and Service Capabilities'

2.2 Evolved UMTS Concepts

This section describes the main concepts driving the structure of Evolved UMTS networks, through the following sub-sections:

- The move towards 'packet only' architecture.
- The evolution towards a fully shared radio interface.
- The opening towards other access types.

2.2.1 A Packet-Only Architecture

In order to understand the major trends of evolved 3G architecture, it is necessary to look at the main steps of wireless network evolution, starting with 2G networks.

(i) 2G Initial Architecture
2G GSM cellular networks were initially designed for voice and circuit-switched services. For that reason, the architecture of such networks was comparatively simple and comprises two main parts:

- The Access Network part, which includes the radio interface as well as the network nodes and interfaces supporting radio-related functions. In initial 2G GSM systems, the radio interface was specifically designed and optimized for voice or low bit rate circuit data transmission.
- The CS part – or circuit-switched core network domain providing circuit services support (this includes call setup, authentication and billing) and inter-working with classical PSTN (Public Switched Telephone Network).

(ii) 2G Packet Evolution
With the emergence of IP and Web services, 2G GSM networks eventually evolved to efficiently support packet data transmission:

- The Access Network part was partly redesigned to support packet transmission and shared resource allocation schemes – as for GPRS and EDGE evolutions.
- A new Core Network domain (PS for Packet Switched) was added, in parallel to the CS domain. This new domain has the same role as the CS domain, meaning support for packet transmission (including authentication and billing) as well as inter-working with public or private Internet (or IP) networks.

For illustration, the Figure 2.3 describes a simplified view of the dual Core Network Domain. The CS domain is composed of a MSC/VLR (Mobile Switching Center/Visitor Location Register) responsible for end-to-end call setup and in charge of maintaining user location information (such information is typically used to page a user terminal in order to establish user-terminated communication sessions). The GMSC (Gateway MSC) is a specific type of MSC, as being the gateway switch responsible for PSTN inter-working.

The PS domain is composed of the SGSN (Serving GPRS Support Node), which basically plays the role of a MSC/VLR for the packet domain, and the GGSN (Gateway GPRS Support Node), which is equivalent to the GMSC for inter-working with external packet networks.

PS and CS domains may possibly be linked together in order to maintain consistent user location information between the two domains and therefore reduce the amount of radio and network signalling.

In addition to the domain-specific nodes, the Core Network also contains the HLR (Home Location Register), accessed by both the CS and PS domains. The HLR is a key part of the network architecture, containing all information related to user subscription.

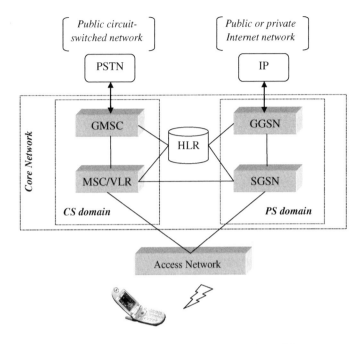

Figure 2.3 The 2G dual Core Network model.

From a network operation point of view, the addition of a new Core Network domain is not a simple move. As shown above, the deployment of a brand new PS domain architecture means new network nodes, and increased cost for network management and engineering.

The network operation also becomes more complex, as the user terminal location needs to be known by both domains (meaning a duplication of processes and information) and new procedures and interfaces need to be defined between network nodes.

(iii) 3G IMS Evolution

From a system overview, initial 3G UMTS network architecture was more or less the same as the 2G, as it included both circuit and packet Core Networks. Eventually, a new domain was added on top of the PS domain: the IMS (IP Multimedia Subsystem).

The main objective of IMS was to allow the creation of standard and interoperable IP services (like 'Push-To-Talk', 'Presence' or 'Instant Messaging') in a consistent way across 3GPP wireless networks. The interoperability of IMS-based services comes from the fact that IMS is based on flexible protocols like SIP (Session Initiation Protocol) developed by the IETF (Internet Engineering Task Force – an international community dedicated to the evolution of the Internet).

In addition, the IMS standard offers VoIP (Voice over IP) support and provides interworking with classical PSTN through signalling and media gateways.

As presented in Figure 2.4, the CS domain was still part of the 3G Core Network architecture, along with the PS/IMS structure. The main reason for keeping the CS domain

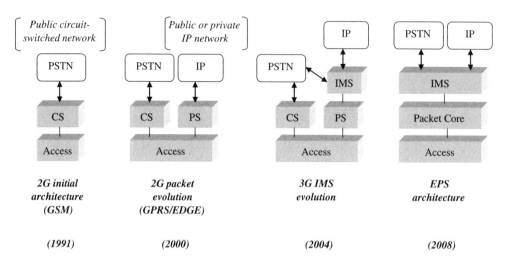

Figure 2.4 Evolution of network architecture (from circuit to packet).

was motivated by the need to support the – still dominant – circuit-switched voice services and H324M-based video-telephony support.

Although IMS was presented as an interesting step towards service integration, legacy network operators refrained from widely deploying and using it as a common platform for all services (including voice, real-time and nonreal-time services) because of the lack of support for voice services' seamless mobility between existing CS-based networks and IMS.

(iv) EPS Architecture
Evolved UMTS networks have a clear objective to integrate all applications over a simplified and common architecture. The main components of EPS architecture are the following:

- A packet-optimized Access Network which can efficiently support IP-based nonreal-time services as well as circuit-like services requiring constant delay and constant bit rate transmission.
- A simplified Core Network, composed of only one packet domain, supporting all PS services (possibly IMS-based) and inter-working capabilities towards traditional PSTN.

The CS domain is no longer present, as all applications (including the most real-time-constraining ones) are supported over the PS domain. This obviously requires specific gateway nodes – part of the IMS architecture – so that IP traffic is converted to PSTN circuit-switched-based transport.

As a consequence of this network simplification, specific efforts have been produced in the scope of Evolved UMTS standardization activity in order to maintain voice call continuity between old and new systems. More details about the evolved 3G Core and Access architecture are provided further in this chapter.

Evolved UMTS Overview

> *Reference documents about 3GPP network architecture*
>
> 3GPP technical specifications:
>
> - 23.002, 'Network Architecture'

2.2.2 A Shared Radio Interface

As Core Network architecture is moving towards packet or 'all-IP' architecture, it becomes critical that the Access Network provides efficient radio transmission schemes for packet data.

When using dedicated resources, as in early UMTS implementations, it is required to allocate to each user a fixed amount of resource which may be used well under its capacity for certain periods of time.

Evolved UTRAN relies on a fully shared radio resource allocation scheme, which allows maximizing resource usage by combining all radio bearers on a sort of shared high bit rate radio pipe. All services are therefore supported over shared radio resource, including background or interactive services like Web browsing as well as real-time-constraining services like voice or streaming applications. Dedicated and shared resources are compared in Figure 2.5.

This differs from initial UTRAN radio interface, which allowed the possibility of allocating within the same cell both dedicated radio resources (typically for CS domain guaranteed bit rate services) as well as high-speed shared channels like HSDPA.

This evolution towards a fully shared radio access looks like a simplification, similarly to what Wifi (IEEE 802.11) or WiMAX (IEEE 802.16) wireless Ethernet standard proposes today. However, such an access scheme requires a fine and specific radio resource management scheme, in order to ensure that all real-time services can be supported according to their requirements in terms of bit rate and transfer delay.

2.2.3 Other Access Technologies

From early 2000, lots of high-speed and high-performance radio interfaces have been developed for the use of WLAN or Wireless local networking. This includes the well known

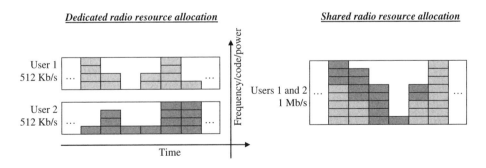

Figure 2.5 Dedicated versus shared resource allocation.

Table 2.1 IEEE radio access technologies.

Radio technology	Peak bit rate over the air
802.11b	11 Mb/s
802.11g	54 Mb/s
802.11n	200 Mb/s
802.16	70 Mb/s

Wifi family (802.11b, 802.11g as well as the latest 802.11n multiple antenna evolution) and the new 802.16 WiMAX.

As presented in Table 2.1, those technologies can provide data rates which are actually quite close to what UTRAN and E-UTRAN propose. However, these technologies are far from full network systems and solutions. IEEE specifications are focused on the radio interface data link level and do not intend to specify higher levels, such as network architecture and interfaces, subscriber management, services, network Quality of Service policing, etc.

Because these radio technologies are (or will become) increasingly popular, 3GPP architecture had to consider them as potential alternative access technologies, taking advantage of a widespread high radio chipset and devices combined with a complete packet-networking framework which 3GPP provides.

Inter-working with WLAN access technologies has been a subject of interest since the beginning of UMTS, and there is no doubt that Evolved UMTS will also follow that path in providing support to seamless mobility between heterogeneous access networks.

2.3 Overall Evolved UMTS Architecture

Figure 2.6 describes the overall network architecture, not only including the Evolved Packet Core and Access Network, but also other blocks, in order to show the relationship between them. For simplification, the picture only shows the signalling interfaces. In some cases, both user data and signalling are supported by the interface (like the S1, S2 or 3G PS Gi interfaces) but, in some other cases, the interfaces are dedicated to the Control plane, and only support signalling (like the S6 and S7 interfaces).

The new blocks specific to Evolved UMTS evolution, also known as the Evolved Packet System (EPS), are the Evolved Packet Core (or EPC) and the Evolved UTRAN (or E-UTRAN).

Other blocks from the classical UMTS architecture are also displayed, such as the UTRAN (the UMTS Access Network), the PS and the CS Core Networks, respectively, connected to the public (or any private) IP and Telephone Networks. The IMS (IP Multimedia Subsystem) is located on top of the Packet Core blocks and provide access to both public or private IP networks, and the public telephone network via Media Gateway network entities. The HSS, managing user subscription information is shown as a central node, providing services to all Core Network blocks of 3G and evolved 3G architecture. The picture does not represent the nodes involved in the support of charging function. The architecture and interfaces related to charging are described in a specific section, further in this chapter.

Evolved UMTS Overview

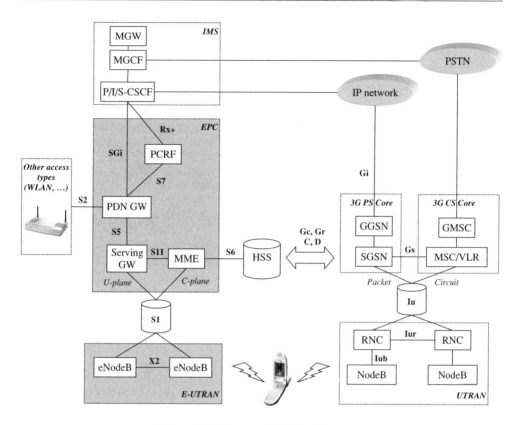

Figure 2.6 The overall EPS architecture.

For further information about dynamic interactions between the different network nodes during network life (e.g. procedures like user registration, session setup, etc.), please refer to Chapter 5.

The next sections of this chapter provide a more in-depth view of the evolved Access and Core parts (respectively E-UTRAN and EPC) as well as the IMS subsystem and corresponding interfaces. However, for further information about the detailed node features and the protocols being used between nodes, please refer to Chapter 4.

2.3.1 E-UTRAN: The Evolved Access Network

Coming back to the first releases of the UMTS standard, the UTRAN architecture was initially very much aligned with 2G/GSM Access Network concepts. As described in Figure 2.7, the UTRAN network is composed of radio equipment (known as NodeB) in charge of transmission and reception over the radio interface, and a specific node – the RNC (Radio Network Controller) – in charge of NodeB configuration and radio resource allocation. The general architecture follows the good old 2G/GSM 'star' model, meaning that a single controller (the RNC) may possibly control a large number – the typical number in commercial networks is about several hundreds – of radio Base Stations (the NodeB) over the Iub interface.

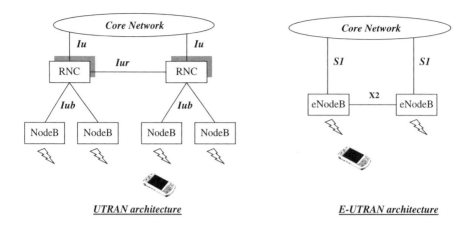

Figure 2.7 UTRAN and Evolved UTRAN architectures.

In addition, an inter-RNC Iur interface was defined to allow UTRAN call anchoring at the RNC level and macro-diversity between different NodeB controlled by different RNCs. Macro-diversity was a consequence of CDMA-based UTRAN physical layers, as a means to reduce radio interference and preserve network capacity. The initial UTRAN architecture resulted in a simplified NodeB implementation, and a relatively complex, sensitive, high-capacity and feature-rich RNC design. In this model, the RNC had to support resource and traffic management features as well as a significant part of the radio protocols. Compared with UTRAN, the E-UTRAN OFDM-based structure is quite simple. It is only composed of one network element: the eNodeB (for 'evolved NodeB').

The 3G RNC (Radio Network Controller) inherited from the 2G BSC (Base Station Controller) has disappeared from E-UTRAN and the eNodeB is directly connected to the Core Network using the S1 interface. As a consequence, the features supported by the RNC have been distributed between the eNodeB or the Core Network MME or Serving Gateway entities.

The standard does not provide much detail about the architecture of the eNodeB. It is only defined as the network node responsible for radio transmission and reception in one or more cells to the terminals (more details are, however, provided in Chapter 3 about the block model of a Base Station). The 'Node' term comes from the fact that the Base Station can be implemented either as single-cell equipment providing coverage and services in one cell only, or as a multi-cell node, each cell covering a given geographical sector. Possible eNodeB models are shown in Figure 2.8.

A new interface (X2) has been defined between eNodeB, working in a meshed way (meaning that all NodeBs may possibly be linked together). The main purpose of this interface is to minimize packet loss due to user mobility. As the terminal moves across the access network, unsent or unacknowledged packets stored in the old eNodeB queues can be forwarded or tunnelled to the new eNodeB thanks to the X2 interface.

From a high-level perspective, the new E-UTRAN architecture is actually moving towards WLAN network structures and Wifi or WiMAX Base Stations' functional definition. eNodeB – as WLAN access points – support all Layer 1 and Layer 2 features

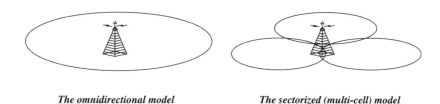

The omnidirectional model   *The sectorized (multi-cell) model*

Figure 2.8 Possible eNodeB models.

associated to the E-UTRAN OFDM physical interface, and they are directly connected to network routers. There is no more intermediate controlling node (as the 2G/BSC or 3G/RNC was).

This has the merit of a simpler network architecture (fewer nodes of different types, which means simplified network operation) and allows better performance over the radio interface. As described in Chapter 4, the termination of Layer 2 protocols in eNodeB rather than in the RNC helps to decrease data-transmission latency by saving the delay incurred by the transmission of packet repetitions over the Iub interface.

From a functional perspective, the eNodeB supports a set of legacy features, all related to physical layer procedures for transmission and reception over the radio interface:

- Modulation and de-modulation.
- Channel coding and de-coding.

Besides, the eNodeB includes additional features, coming from the fact that there are no more Base Station controllers in the E-UTRAN architecture. Those features, which are further described in Chapter 4, include the following:

- Radio Resource Control: this relates to the allocation, modification and release of resources for the transmission over the radio interface between the user terminal and the eNodeB.
- Radio Mobility management: this refers to a measurement processing and handover decision.
- Radio interface full Layer 2 protocol: in the OSI 'Data Link' way, the layer 2 purpose is to ensure transfer of data between network entities. This implies detection and possibly correction of errors that may occur in the physical layer.

2.3.2 EPC: The Evolved Packet Core Network

The EPC (Evolved Packet Core) is composed of several functional entities:

- The MME (Mobility Management Entity).
- The Serving Gateway.
- The PDN Gateway (Packet Data Network).
- The PCRF (Policy and Charging Rules Function) – the role of this node is further described in this chapter, within the section about 'Policy and Charging'.

The rest of this section gives some more details on EPC nodes' role, as well as EPC architecture interaction in the following cases:

- Non-roaming architecture – the simplest EPC architecture case.
- Roaming architecture – how the EPC works in a roaming situation in the 'home routed traffic' and 'local breakout' cases.
- Non-3GPP access case – how the EPC interacts with non-3GPP access technologies like WLAN, in trusted and non-trusted modes.
- 2G/3G mobility architecture – how the EPC interacts with 2G/3G PS Core nodes in cases of user mobility.

(i) The MME (Mobility Management Entity)
The MME is in charge of all the Control plane functions related to subscriber and session management. From that perspective, the MME supports the following:

- Security procedures – this relates to end-user authentication as well as initiation and negotiation of ciphering and integrity protection algorithms.
- Terminal-to-network session handling – this relates to all the signalling procedures used to set up Packet Data context and negotiate associated parameters like the Quality of Service.
- Idle terminal location management – this relates to the tracking area update process (described more in Chapter 5) used in order for the network to be able to join terminals in case of incoming sessions.

The MME is linked through the S6 interface to the HSS which supports the database containing all the user subscription information.

(ii) The Serving GW (Serving Gateway)
From a functional perspective, the Serving GW is the termination point of the packet data interface towards E-UTRAN. When terminals move across eNodeB in E-UTRAN, the Serving GW serves as a local mobility anchor, meaning that packets are routed through this point for intra E-UTRAN mobility and mobility with other 3GPP technologies, such as 2G/GSM and 3G/UMTS.

(iii) The PDN GW (Packet Data Network Gateway)
Similarly to the Serving GW, the PDN gateway is the termination point of the packet data interface towards the Packet Data Network. As an anchor point for sessions towards the external Packet Data Networks, the PDN GW also supports Policy Enforcement features (which apply operator-defined rules for resource allocation and usage) as well as packet filtering (like deep packet inspection for virus signature detection) and evolved charging support (like per URL charging).

(iv) The Non-Roaming Architecture
The standard does not require any physical architecture for EPC node implementation. For example, it may be possible that the Serving and PDN Gateways are supported by a single node, or that the MME and Serving Gateway are implemented in one physical node. These

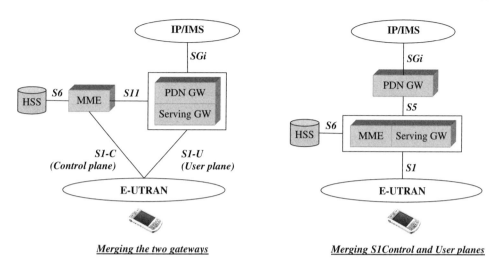

Figure 2.9 Different options for merged EPC physical node implementation.

two options are described in Figure 2.9; for simplicity, only the EPC nodes are shown, all belonging to the same network and operator (in other words, 'the non-roaming case').

This model shows the interface S1 which transports both signalling (the S1-C part) and User plane data (the S1-U part) between E-UTRAN and EPC as well as interface SGi – the equivalent of the 2G and 3G Packet Core Gi – towards public or private external IP networks, or the IMS domain.

The right-hand option of the picture (with combined MME and Serving Gateway) may be seen as a simpler solution to operate a Packet Core network, since it does not require implementing and operating the S11 signalling interface between the MME and Serving GW. However, implementing separated MME and User plane functions allows further flexibility in terms of deployment, allowing independent scalability between the signalling load on the Control plane and traffic handling on the User plane.

In any kind of implementation, the S5 interface has been defined in such a way as to allow connectivity to multiple Packet Data Networks, providing the same flexibility as the Gn interface defined between SGSN and GGSN 2G/3G nodes – the S5 interface is actually an evolved version of the Gn interface based on the GTP (GPRS Tunnelling Protocol). The typical use case of such a configuration would be, for example, to separate the access to an Intranet from the public IP network access, as in Figure 2.10. In this case, access to each of the networks is done through a specific PDN GW, which allows implementing specific policy-enforcement rules, or packet-filtering algorithms, depending on the type of packet network being addressed.

From a functional perspective, the association of MME and Serving GW features is actually very close to a combination of 2G/3G SGSN. Similarly, the PDN GW logical node can also be considered as a kind of 2G/3G GGSN, in its role to support policy-enforcement and charging rules, inter-working with external packet networks, as well as terminal IP address allocation.

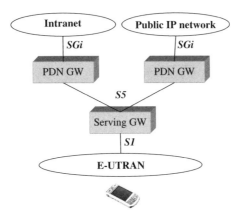

Figure 2.10 An example of S5 connectivity to multiple packet networks.

It is interesting to note that the separation of SGSN User and Control parts in EPS, combined with the fact that S5 is optional (as in combined Serving and PDN GW), looks very similar to one of the latest improvements brought to UMTS Packet Core and known as 'Direct Tunnel optimization'. This enhancement is described in Figure 2.11.

This initial objective of this optimization, which followed the introduction of HSDPA/HSUPA high-speed radio techniques and IMS evolutions, is to allow a gain on network scalability and better cope with increased user data volumes, due to the fact that packet data are routed directly from the UTRAN edge node to the GGSN Internet gateway.

In this evolution, the SGSN has the possibility to set up a direct tunnel from the RNC to the GGSN, instead of the two-tunnel approach from RNC to SGSN, and from SGSN to GGSN (represented using dotted lines).

At the end, the 3G/UMTS Iu interface is clearly split into User and Control parts (the same way as the S1 interface) and the Gn becomes a pure Control plane interface equivalent to S11 (and actually based on the same GTP protocol).

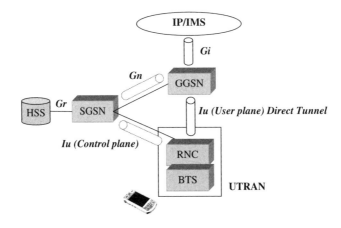

Figure 2.11 3G/UMTS 'Direct Tunnel' optimization – one step towards EPC.

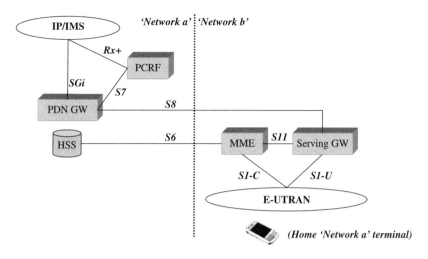

Figure 2.12 The EPC roaming architecture – home-routed traffic.

(v) The Roaming Architecture

Figure 2.12 presents the EPC architecture support for roaming cases. In this example, a user has subscribed to 'Network a' but is currently under the coverage of the visited 'Network b'. This kind of situation may happen while the user is travelling to another country, or in the case in which a national roaming agreement has been set up between operators, so as to decrease the investment effort for national coverage. In such a roaming situation, part of the session is handled by the visited network. This includes E-UTRAN access network support, session-signalling handling by the MME, and User plane routing through the local Serving GW nodes. Thanks to local MME and Serving GW, the visited network is then able to build and send charging tickets to the subscriber home operator, corresponding to the amount of data transferred and the Quality of Service allocated.

However, since the terminal user has no subscription with the visited network, the MME needs to be linked to the HSS of the user home network, at least to retrieve the user-specific security credential needed for authentication and ciphering. In the roaming architecture, the session path goes through the home PDN GW over the S8 interface, so as to apply policy and charging rules in the home network corresponding to the user-subscription parameters.

The S8 interface introduced in this model supports both signalling and data transfer between the visited Serving GW and the home PDN GW. Its definition is actually based on the Gp interface defined in the 2G and 3G Packet Core roaming architecture between the visited SGSN and the home GGSN.

Briefly, in such a model, the visited network provides the access connectivity (which also involves the basic session signalling procedures supported by the visited MME, with the support of the home HSS), whereas the home network still provides the access to external networks, possibly including IMS-based services.

In this model, the call is still anchored to the home PDN GW, hence the 'home routed traffic' denomination. The user packet routing in such a scheme may, however, be quite inefficient in

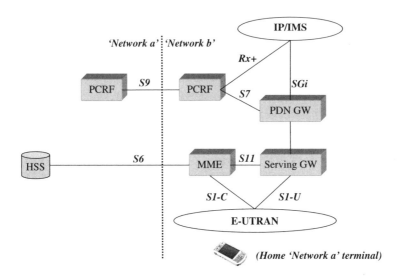

Figure 2.13 The EPC roaming architecture – local breakout.

terms of cost and network resources, as the home PDN GW and visited Serving GW may be very far from each other. This is the reason why the 3GPP standard also allows the possibility of the user traffic to be routed via a visited PDN GW, as an optimization. This may be very beneficial in the example of public Internet access – as routing the traffic to the home network does not add any value to the end-user – and even more in the case of an IMS session established between a roaming user and a subscriber of the visited network. In this last case, local traffic routing avoids a complete round trip of user data through the home network packet gateways.

Figure 2.13 describes possible network architecture in the case where the traffic is routed locally – or the 'local breakout' case. Both Gateways are part of the visited network. The visited PCRF retrieves Quality of Service policy and charging control information from the Home PCRF via the new S9 interface.

(vi) The Non-3GPP Access Architecture
Figure 2.14 represents the network architecture providing IP connectivity to the Evolved Packet Core using non-3GPP type of access. This architecture is independent from the access technology, which could be Wifi, WiMAX or any other kind of access type. This picture applies to the trusted WLAN access, corresponding to the situations where the WLAN network is controlled by the operator itself or by another entity (local operator or service provider) which can be trusted due to the existence of mutual agreements.

As described below, some new network nodes and interfaces are needed to support non-3GPP access types. In contrast, on the terminal side, no changes are required except some slight software adaptations. This comes from the fact that AAA (Authentication, Authorization and Accounting) mechanisms for mutual authentication and access control are based on known IETF protocols but make use of 3GPP UICC stored credentials.

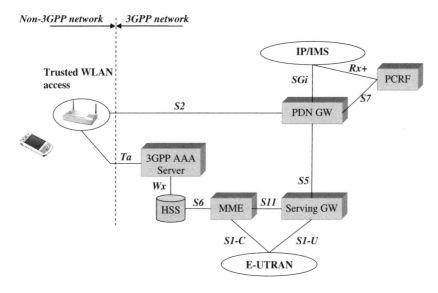

Figure 2.14 The EPC architecture for trusted WLAN access.

The 3GPP AAA Server's role is to act as an inter-working unit between the 3GPP world and IETF standard-driven WLAN networks from the security perspective. Its purpose is to allow end-to-end authentication with WLAN terminals using 3GPP credentials. For that reason, the 3GPP AAA Server has an access to the HSS through the Wx interface, so as to retrieve user-related subscription information and 3GPP authentication vectors.

From the 3GPP AAA Server, the Ta interface has been defined with the trusted access network, aiming at transporting authentication, authorization and charging-related information in a secure manner.

From the User plane perspective, the user data are transmitted from the WLAN network to the PDN GW through the new S2 interface. As in legacy EPC architecture, the PDN GW still serves as an anchor point for the user traffic.

In such a model, the Serving GW and MME nodes are not needed anymore. Terminal location management is under the responsibility of the WLAN Access as well as the packet session signalling and does not need any support from 3GPP EPC nodes (aside from the provision of 3GPP security credentials). In the example of a 802.11 Wifi access point, user association (the process by which a Wifi terminal connects to an access point), security features as well as radio protocols are handled by the access point itself.

In addition to the trusted model, the standard defines another model, for the situations where the WLAN network is nontrusted. This model is described in Figure 2.15. As an example, this may correspond to a business entity deploying a WLAN for its internal use and willing to offer 3GPP connectivity to some of its customers. In such a case, the WLAN-3GPP interconnection looks a bit different due to additional mechanisms to maintain legacy 3GPP infrastructure security and integrity.

This model introduced a ePDG node (for evolved Packet Data Gateway) which concentrates all the traffic issued or directed to the WLAN network. Its main role is to establish a

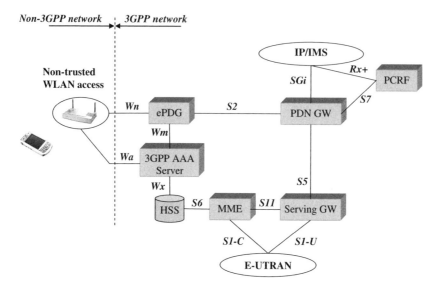

Figure 2.15 The EPC architecture for non-trusted WLAN access.

secure tunnel for user data transmission with the terminal using IPSec and filter unauthorized traffic.

In this model, the new Wm interface is introduced for the purpose of exchanging user-related information from the 3GPP AAA Server to the ePDG. This will allow the ePDG to enable proper user data tunnelling and encryption to the terminal.

The model described above is actually very similar to the architecture defined for non-3GPP access to 3G/UMTS packet core, defined in the following specification:

- 23.234, '3GPP System to WLAN Interworking; System Description'.

The Wa (and its trusted mode Ta equivalent), Wx, Wm and Wn interfaces have actually been inherited from this model and are functionally similar to their 3G/UMTS equivalent.

(viii) The 2G/3G Mobility Architecture
Figure 2.16 represents the EPC interactions with 2G/3G Packet Core nodes. This model is applicable to terminal mobility cases between E-UTRAN cells on one side and 3G/UTRAN or 2G/GERAN on the other side while a packet data context or a packet session is active. GERAN (GPRS EDGE RAN) refers to the 2G/GSM access network, as in 3GPP terminology. The 2G/3G mobility architecture model introduces two interfaces: S3 and S4.

The role of S3 is to support user and bearer information exchange between the SGSN and the MME, as the terminal is moving from one access type to the other. For example, user context, as well as negotiated Quality of Service and bearers' information, is exchanged between the two nodes, so that the new serving MME or SGSN gets all the necessary information associated with the on-going session. The S3 interface is based on the Gn interface designed for 2G/3G Packet Core architecture to support mobility between SGSN nodes.

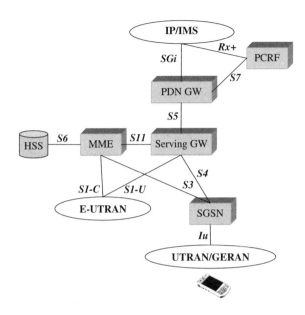

Figure 2.16 The EPC architecture for inter-working with 2G/3G access.

The S4 interface is associated with the session User plane. It supports packet data transfer between the SGSN and the Serving GW which plays the role of a mobility anchor point within the EPC architecture. From the SGSN perspective, the Serving GW plays a role which is very similar to a GGSN node. This is the reason why S4 is based on the Gn interface, defined between the 2G/3G Packet Core SGSN and GGSN nodes.

Reference documents about EPC architecture

3GPP technical specifications:

- 23.401, '3GPP System Architecture Evolution: GPRS Enhancements for LTE Access'
- 23.402, '3GPP System Architecture Evolution: Architecture Enhancements for Non-3GPP Access'

2.3.3 The HSS

The HSS (Home Subscriber Server) is the concatenation of the HLR (Home Location Register) and the AuC (Authentication Center) – two functions being already present in pre-IMS 2G/GSM and 3G/UMTS networks. The HLR part of the HSS is in charge of storing and updating when necessary the database containing all the user subscription information, including (list is nonexhaustive):

- User identification and addressing – this corresponds to the IMSI (International Mobile Subscriber Identity) and MSISDN (Mobile Subscriber ISDN Number) or mobile telephone number.

- User profile information – this includes service subscription states and user-subscribed Quality of Service information (such as maximum allowed bit rate or allowed traffic class).

The AuC part of the HSS is in charge of generating security information from user identity keys. This security information is provided to the HLR and further communicated to other entities in the network. Security information is mainly used for:

- Mutual network-terminal authentication.
- Radio path ciphering and integrity protection, to ensure data and signalling transmitted between the network and the terminal is neither eavesdropped nor altered.

Introduced from the very beginning of GSM network standardization, HLR and AuC boxes were eventually joined together in a single HSS node as IMS was defined by the 3GPP.

In its extended role, the HSS of Evolved UMTS networks integrates both HLR and AuC features, including classical MAP features (for the support of circuit and packet-based sessions), IMS-related functions, and all necessary functions related to the new Evolved Packet Core.

This is illustrated by Figure 2.17, showing HSS support to the evolved Packet Core through the S6 interface in addition to connections to 3G non-IMS Core Network nodes such as the GMSC and MSC (through the C and D interfaces) for the CS domain and the SGSN and GGSN (through the Gr and Gc interfaces) for the PS domain, but also to the CSCF IMS nodes via the Cx interface.

As an illustration of the functions supported by the HSS, there are actually three main cases in which the HSS is actively involved:

- At user registration – the HSS is interrogated by the corresponding Core Network node as the user attempts to register to the network in order to check the user subscription rights. This can be done by either the MSC, the SGSN or the MME, depending on the type of network and registration being requested;
- In the case of terminal location update – as the terminal changes location areas, the HSS is kept updated and maintains a reference of the last known area;

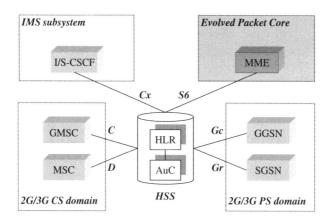

Figure 2.17 HSS structure and external interfaces.

Table 2.2 Subset of user information data stored in the HSS.

General data		
IMSI	International Mobile Subscriber Identity	Permanent
MSISDN	Mobile Subscriber ISDN Number	Permanent
Authentication vector	Security information for user authentication, data confidentiality and data integrity	Temporary
...	...	...
GPRS-related data		
SGSN Address	Address of the SGSN the user is currently registered to	Temporary
Subscribed QoS	Specifies the Quality of Service attributes the user has subscribed to	Permanent
...	...	...
IMS-related data		
Authorized networks	List of visited network identifiers the user is allowed for roaming	Permanent
Private user identity	In the form of 'user@realm'	Permanent
Public user identities	List of user identities in the form of 'sip:user@domain'	Permanent
Registration status	State of registration of a user IMS identity	Temporary
S-CSCF name	Identifies the S-CSCF the subscriber is registered to	Temporary
...	...	...

- In the case of user-terminated session request – the HSS is interrogated and provides a reference of the Core Network node corresponding to the current user location.

For further details, please refer to Chapter 5, about various steps and procedures applied by living networks.

The standard does not specify how the HSS database is structured, but what kind of information shall or should be stored. For illustration purposes, Table 2.2 describes part of the user subscription information handled by the HSS. From a functional perspective, the HSS information can be split into different categories: the general information (mainly containing identifiers or security parameters), the GPRS-related information, the IMS-related parameters, as well as other categories not listed in this table, such as data related to supplementary service or location services. For each of the information pieces, the standard specifies whether it is part of the permanent data (meaning it can only be changed by administration means) or the temporary data (meaning the data may possibly change resulting from normal system operation).

> *Reference documents about HSS data organization*
>
> 3GPP technical specifications:
>
> - 23.008, 'Organization of Subscriber Data'
> - 23.016, 'Subscriber Data Management: Stage 2'

2.4 The IMS Subsystem

The IMS (IP Multimedia Subsystem) is a generic platform offering IP-based multimedia services. IMS provides functions and common procedures for session control, bearer control, policy and charging. From a network architecture perspective, IMS has to be considered as an access-type agnostic overlay to the Packet Core.

Of course, IMS can be deployed on top of UTRAN access as well as over the 2G GPRS/EDGE network. Because of its clear orientation towards efficient packet transmission support, Evolved UTRAN will add a lot of value to IMS and allow more quality-demanding services to be supported in a more efficient way.

The rest of the section is not to provide a full description of the IMS subsystem. The objective here is to describe the main network entities, as an introduction to the detailed service description which can be found in Chapter 6.

2.4.1 The Session Control Function

The CSCF (Call Session Control Function) plays a key role in IMS architecture, for establishing, terminating or modifying IMS sessions. The CSCF is actually a specific type of SIP (Session Initiation Protocol) server. SIP is the base protocol developed by the IETF for the management of IP-based multimedia sessions and services; it is described in the following document:

- RFC3261, 'SIP: Session Initiation Protocol'.

A more detailed description of the SIP protocol in the scope of IMS session control is provided in Chapter 4.

The CSCF can play three different roles in the IMS network: the Proxy (P-SCSF), the Interrogating (I-CSCH) or the Serving (S-CSCF) one.

The P-CSCF is the first contact point reached by user terminal SIP messaging within the IMS subsystem. In the case of roaming, the P-CSCF belongs to the visited network, and forwards user-originated SIP requests towards the relevant I-CSCF or S-CSCF. As a SIP session controlling entity belonging to the visited network, the P-CSCF actually supports two main functions:

- The control of allocated bearers within the visited network.
- The generation of charging records, so that the visited network can charge the user home network according to the allocation and usage of resources for this part of the session path.

Additionally, the P-CSCF is responsible for SIP header compression, in order to minimize the bandwidth requirement for the transmission of SIP signalling over the visited Access Network. More details about this feature are provided in Chapter 4.

The I-CSCF is an entry point for all connections related to a user belonging to its network. The I-CSCFs main task is to identify the relevant S-SCSF (based on HSS interrogation) for a user performing SIP registration, and to forward the registration request accordingly. In addition, the I-CSCF may also serve as a THIG (Topology Hiding Inter-network Gateway) in case an operator has some specific security requirements and is willing to hide the configuration and topology of its network from the outside. In this case, the I-CSCF applies a specific process to the incoming and outgoing SIP signalling. This point is further described in the IMS protocol section of Chapter 4.

The S-CSCF is the SIP server to which the user will be registered, and actually provides access to Application Servers to the end-user. As in SIP terminology, the S-CSCF play three types of roles:

- The 'registrar' (the entity to which the SIP user is registered).
- The 'proxy server' (when SIP service requests are forwarded to another SIP server).
- The 'user agent' (when SIP request terminates at the P-CSCF).

The S-CSCF performs the session control services for the UE. It maintains a session state as needed by the network operator for support of the services. Besides, the S-CSCF is in charge of CDR generation (Charging Data Record) so that the end-user can be charged according to the requested service, Quality of Service or amount of resource being allocated to the session.

As for the MME and UPE PS Core functional split, the standard does not mandate for a specific hardware architecture for CSCF nodes. Thus, it is possible that some network implementers aggregate the P-CSCF and I-CSCF logical nodes into one physical network entity. Similarly, the I-CSCF and S-CSCF may also be integrated into one single physical box. This may be the right approach for early low-capacity IMS deployments, leaving open the possibility of scaling the network while reducing the number of inter-node messages.

Figure 2.18 describes the IMS architecture. In the case of a roaming subscriber accessing a home server from a visited network, the P-CSCF belongs to the visited network 'A', whereas the I-CSCF and S-CSCF to which the user will register belong to the home network 'B'. As

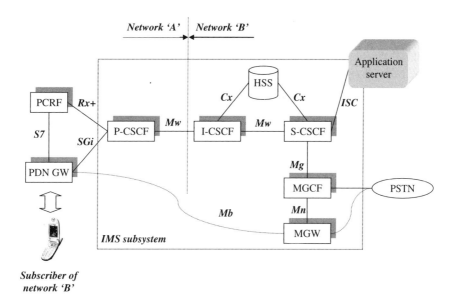

Figure 2.18 The IMS subsystem architecture.

regards to the roaming architectures presented above in this chapter, this case corresponds to the 'local breakout' where traffic is routed through the local PDN GW. The figure only describes the signalling path between network entities. The data bearer follows a different path and is not processed by CSCF nodes. As described in IETF SIP specifications, a SIP server (like the CSCF nodes) only cares about session signalling.

The figure shows, however, one example of a data bearer path, using dotted lines over the Mb interface, in the case of a mobile terminal to fixed-line terminal voice call. In this example, the Mb interface actually represents the User plane of the Gi interface coming from the PDN GW node. Part of the bearer path, up to the MGW (Media Gateway), is carried over IP. The call setup signalling procedures are SIP-based, up to the MGCF (Media Gateway Control Function) which is in charge of signalling translation.

Figure 2.18 also describes the main interfaces defined by standard documents for the communication between network nodes. The Mw as well as the ISC (IP multimedia Subsystem Service Control) interfaces carry SIP signalling exchanged between CSCF session control servers or between the SCSCF and the Application Servers. The CSCF nodes are connected to the HSS via the Cx interface, which provides services for user data handling and user authentication. Like the Mw interface, the Cx interface is using another IETF-originated protocol known as Diameter, and specified in the following document:

- RFC3588, 'Diameter Base Protocol'.

The Diameter protocol is one of the standard IETF protocols for network access and IP mobility AAA (Authentication, Authorization and Accounting). In the scope of IMS, the Diameter protocol provided by the IETF has been extended to support IMS and UMTS-specific commands and corresponding information elements.

2.4.2 The Media Gateway Nodes

One of the IMS objectives is to be able to interact with legacy PSTN (Public Switched Telephone Network) for voice services. Due to the fact that IMS handles voice as IP traffic, this requires some adaptations for signalling and voice traffic, performed by the MGCF and MGW nodes.

TheMGCF (Media Gateway Control Function) supports three main functions: call control protocol conversion, MGW (Media Gateway) control and I-CSCF identification. The SIP signalling used to set up the session is forwarded by the S-CSCF to the MGCF over the Mg interface. This SIP signalling is then translated into ISUP (ISDN User Part) signalling by the MGCF before transmission to the PSTN network, and vice versa. In addition, the MGCF is in charge of selecting the relevant I-CSCF when a mobile-terminated call request is arriving from the PSTN. This is performed on the basis of the analysis of the called party routing number provided by the PSTN.

The MGW (Media Gateway) is responsible for media conversion, bearer control and payload processing (e.g. codec, echo canceller, conference bridge) under the control of the MGCF. The Mn interface between the MGW and the MGCF uses a H248 protocol. This protocol is a sort of toolbox which contains all the necessary and generic features to describe, create, modify and delete media streams within a Media Gateway. The H248 standard, published by the ITU, is the international standard for media gateway control. An identical (and public) specification has also been released by the IETF under the following reference:

- RFC3525, 'Gateway Control Protocol Version 1'.

> **Reference documents about the IMS subsystem**
>
> 3GPP technical specifications:
>
> - 22.228, 'Service requirements for the IP Multimedia Subsystem (IMS): Stage 1'
> - 23.228, 'IP Multimedia Subsystem (IMS): Stage 2'

2.5 Policy Control and Charging

Policy control is described in 3GPP specifications as being part of the Packet Core network architecture. Actually, this feature interacts not only with Packet Core nodes, but also with SIP servers belonging to the IMS subsystem, such as the P-CSCF.

This section describes the main concepts and added value of Policy Control, as well as the evolution of this mechanism (together with the charging feature), from the early Release 5 UMTS implementation up to the Release 7 enhancements and Evolved Packet Core architecture.

2.5.1 Policy Control in UMTS

Policy control has not been invented with UMTS. Lots of attention has been paid to this in IETF groups, as part of the activity related to the evolution an enhancement of IP networks. Practically speaking, what could be a fair definition of 'Policy Control'? One possible answer can be found, for example, in IETF document RFC2753, 'Framework for Policy-Based Admission Control', stating that 'Policy Control is the application of rules to determine resource access and usage'.

In early UMTS implementation (including the Release 5 of UMTS 3GPP specifications), Policy Control was user-terminal driven. Depending on the requested service (Web browsing, streaming, Push-To-Talk), the user terminal was requesting a PDP context (or Packet Data Protocol context) with Quality of Service attributes being set accordingly to the type of the service (real-time versus nonreal-time service, maximum and mean bit rate, delay requirements, etc.).

The requested QoS parameters were eventually checked by the SGSN based on user subscription limitation stored in the HSS. Using a set of CDR (Charging Data Record) defined by the standard and generated by network elements such as the SGSN and GGSN, the operator had the possibility of charging the end customer either on time, volume or on allocated Quality of Service. However, it was not possible to apply differentiated charging rules for the different service data flows which could possibly be aggregated within a single PDP context, as the end-user has actually no constraints for opening a new PDP context for each new type of application being used.

As IMS really began to emerge from the standard, and considering the future of IP-based applications (including the upcoming VoIP transition), the 3GPP community decided to define a new architecture for more flexible policy control and charging mechanisms.

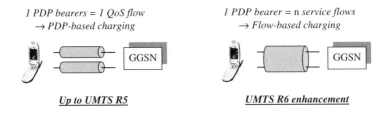

Figure 2.19 The evolution of Policy Control and Charging.

(i) The 'Flow Based Control' Enhancement

Figure 2.19 illustrates this major evolution. As in the pre-Release 6 UMTS standard, one PDP context corresponds to one set of negotiated Quality of Service attributes. Charging is applied to each PDP context as a whole. Thanks to Release 6 evolution, the network has the possibility to identify the different SDF (Service Data Flows) being aggregated within a single PDP context. This gives the possibility to the network of controlling (meaning allowing or blocking) each of the flows and charging the end-user having a much better accuracy.

Each of those elementary flows, also known as SDF (Service Data Flow), is defined in 3GPP specifications as a 5-tuple (source IP address, destination IP address, source port, destination port, protocol used above IP). This definition allows identifying each of the information flows from the mass of IP packets sent and received by the terminal, for example:

- A Web-browsing session towards server 'A'.
- Another Web-browsing session towards server 'B'.
- A streaming session from server 'Y'.
- A SIP-signalling flow associated to an IMS service.
- ...

Figure 2.20 illustrates the new network elements introduced in the R6 standard to allow flow-based policy control and charging. For that purpose, two new network elements have been introduced:

- The Policy Decision Function (PDF).
- The Charging Rules Function (CRF).

The **PDF** is the network entity where the policy decisions are made. As the IMS session is being set up, SIP signalling containing media requirements are exchanged between the terminal and the P-CSCF. At some time in the session establishment process, the PDF receives those requirements from the P-CSCF and makes decisions based on network operator rules, such as:

- Allowing or rejecting the media request.
- Using new or existing PDP context for an incoming media request.
- Checking the allocation of new resources against the maximum authorized.

Evolved UMTS Overview

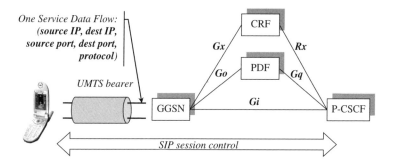

Figure 2.20 UMTS R6 Policy Control architecture.

The **GGSN** is in charge of enforcing policy decisions received from the PDF over the Go interface. The policy rules are either 'pushed' by the PDF, e.g. as new media are added to an existing session, or 'requested' by the GGSN itself, when the establishment of a new PDP context is requested by the terminal. The policy-enforcement process performed by the GGSN takes the form of a 'gating' process. Each packet received by the GGSN in the upstream or downstream direction is classified (meaning associated with one of the existing Service Data Flow) and checked against filters being defined by the PDF for the corresponding Service Data Flow (SDF).

The **CRF**s role is to provide operator-defined charging rules applicable to each service data flow. The CRF selects the relevant charging rules based on information provided by the P-CSCF, such as Application Identifier, Type of Stream (audio, video, etc.), Application Data Rate, etc.

Charging rules are then provided by the CRF to the GGSN in the form of a packet filter similar to the 5-tuple gate definition above. Using the charging rules, the GGSN is able to count packets for each of the service data flows and generates corresponding charging records.

The two new CRF and PDF network nodes interact with the GGSN and the P-CSCF using specific interfaces named in the figure. As for many IMS-related interfaces, Gx, Go, Gq and Rx interfaces make use of already existing IETF protocols. Go is based on COPS, which is specified in the following IETF document:

- RFC2748, 'The COPS (Common Open Policy Service) Protocol'.

The COPS protocol proposes generic policy control for packet networks and is based on a simple client/server model. In the COPS terminology, two entities are defined:

- The PDP (Policy Decision Point), which is the policy server making the decision – in the UMTS case, this role is supported by the PDF.
- The PEP (Policy Enforcement Point), which is the policy client, responsible for enforcing the policy decisions – in the UMTS case, this role is supported by the GGSN.

The other policy control and charging interfaces (Gq, Gx and Rx) are all based on an extended version of the IETF Diameter protocol, similarly to the Cx interface mentioned in the HSS section above.

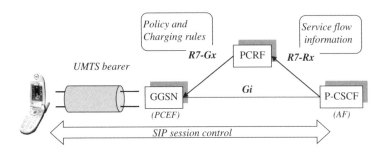

Figure 2.21 The UMTS R7 evolution.

(ii) The 3GPP R7 UMTS Enhancement
The major improvement brought by the Release 7 of 3GPP in terms of policy and charging is the definition of a new converged architecture, so as to allow the optimization of interactions between these two functions. The R7 evolution involves a new network node PCRF (Policy and Charging Rules Function) which is actually a concatenation of PDF and CRF. As a result, evolved versions of the R6 interfaces have been defined, as illustrated in Figure 2.21.

This model is actually not specific to UMTS or UTRAN access networks, as it was defined for all types of IP access, including 3GPP access types and also WLAN and fixed IP broadband access. In the generic policy and charging control 3GPP model, the PCEF (Policy and Charging Enforcement Function) is the generic name for the functional entity which supports service data flow detection, policy enforcement and flow-based charging. In the case of WLAN non-3GPP access, the PCEF is implemented by the PDG (Packet Data Gateway).

Similarly, the AF (Application Function) represents the network element which supports applications that require dynamic policy and/or charging control. In the IMS model, the AF is implemented by the P-CSCF.

The new **R7-Rx** interface combines both the former Rx and Gq with message definition – both based on the Diameter IETF protocol – were actually very close. The P-CSCF can therefore provide service-dynamic information to the PCRF using a single procedure. The R7-Gx is defined in a backward-compatible way, in order to ease introduction in already deployed networks. This means that a R7 PCRF can interact with R6 P-CSCF using former Rx and Gq interfaces. Similarly, a R7 P-CSCF can also interact with R6 PDF and CRF nodes.

The new **R7-Gx** interface supports Gx and Go capabilities, so that policy decisions and charging rules are provided from the PCRF to the GGSN using a single message. As R6 Gx and Go are not based on the same protocols (Gx is Diameter-based whereas Go relies on COPS), the choice was made to use Gx as a basis and to enhance it with all necessary features to allow service-based local policy.

Reference documents about policy control

3GPP technical specifications:

- 23.125, 'Overall High Level Functionality and Architecture Impacts of Flow Based Charging: Stage 2'
- 23.203, 'Policy and Charging Control Architecture'

- 23.207, 'End-to-End Quality of Service (QoS) Concept and Architecture'
- 29.211, 'Rx Interface and Rx/Gx Signalling Flows'

IETF documents:

- RFC2748, 'The COPS (Common Open Policy Service) Protocol'
- RFC2753, 'Framework for Policy-Based Admission Control'
- RFC3588, 'Diameter Base Protocol'

2.5.2 Evolved UMTS Policy Control

Not surprisingly, the model for policy and charging architecture in EPS networks is aligned with the latest 3GPP UMTS Release 7 evolution (Figure 2.22):

- The new S7 interface introduced in EPC is based on its R7-Gx.
- The PDN gateway plays the role of the PCEF, as an equivalent of the GGSN for the policy and charging control functions.

2.5.3 The Charging Architecture

From the network operator point of view, charging is one of the most critical features. Network subscriber charging is not only the major source of revenue, but also an area in which an operator can innovate and differentiate from its competitors by creating cost-attractive services and solutions while not jeopardizing the whole network profitability.

In legacy 2G or 3G circuit-switched based networks, charging was quite an easy task. Any granted user service request involved the allocation of a fixed amount of resource for a given time. Because circuit switched technology means guaranteed bandwidth and delay, the charging rules are generally simply based on the allocated resource size and use time.

When using packet applications and packet transmission, the picture is a bit different. The end-user may be inactive for long periods of time [e.g. during the silent phase of a PoC (Push-to-talk Over Cellular) session or during the time needed to read a Web page freshly

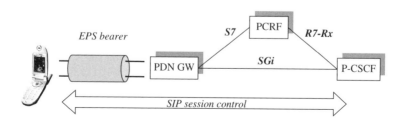

Figure 2.22 The EPS policy and charging architecture.

downloaded] and, during those inactivity phases, the resources may be used for another purpose – this is one of the main benefits of packet-based networks. Therefore, it may be seen as quite unfair to only charge the end-user for connection time or service-use duration.

The standard does not specify the charging schemes, leaving to the operator the choice to charge the end-user based on any of the following (nonexhaustive list):

- Data volume.
- Session or connection time.
- Service type (Web, email, MMS, etc.).
- Allocated Quality of Service.
- Type of technology being used (2G, 3G, evolved 3G, WLAN, etc.).

For that purpose, the 3GPP standard proposes all the necessary features to allow a flexible charging scheme to be implemented by operators. The charging process is based on the collection of various events and information which are stored in a formatted record called the CDR (Charging Data Record). The format of the CDR is specified by the 3GPP standard and depends on each type of service:

- PS domain access (as in 2G and 3G networks).
- IMS service.
- Any type of service hosted by an AS (Application Server) under the control of the operator. This includes Multimedia Messaging Service (MMS), Push-to-talk over Cellular (PoC), Multimedia Broadcast and Multicast Service (MBMS).

Table 2.3 Examples of CDR.

PS domain CDR	
IMSI	International Mobile Subscriber Identity
IMEI	The IMEI (International Mobile station Equipment Identifier) of the user terminal
Charging ID	An identifier of the PDP context being used
Record opening time	Time stamp when PDP context is activated
Duration	Duration of this record
Traffic data volumes	The amount of data per Quality of Service type
RAT type	Type of Radio Access Technology being used by the terminal
...	...
IMS CDR	
Session ID	The SIP session ID, as defined in the SIP specification
Role of node	Indicates if the CDR is issued by a P, I or S-CSCF
Service starting time	The SIP session starting time
Service ending time	The SIP session closing time
SDP description	Description of the session, the list of media components being used
...	...

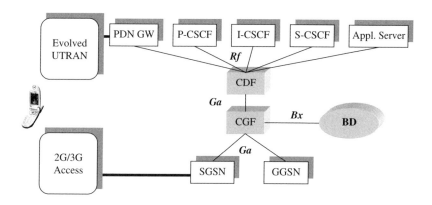

Figure 2.23 The architecture for Charging.

Table 2.3 is an example of CDR for the PS domain and IMS session.

Figure 2.23 presents the network elements involved in the charging process and their interaction with 2G, 3G and IMS network nodes. The role of the CDF (Charging Data Function) is to collect charging information from the different nodes through the Rf interface and build the corresponding CDR. The type of nodes being linked to the CDF is not limited. It includes the PDN Gateway from the EPC architecture, IMS nodes (such as the CSCF servers) but may also include all kinds of Application Servers like the BM-SC (Broadcast Multicast Service Centre) or the PoC server (Push-to-talk Over Cellular). Those two services are further described in Chapter 6.

The Rf interface is based on the Diameter IETF protocol, also used in many IMS interfaces. The Rf declination of Diameter makes use of extensions specific to the charging process. The CGF (Charging Gateway Function) is a gateway between the Core Network nodes and the BD (Billing Domain). Its main task is CDR collection through the Ga interface, CDR storage, CDR management (like CDR opening, closing, deleting) and secure transfer to the BD. The default CDR transfer method over the Bx interface proposed by the 3GPP standard is FTP (File Transfer Protocol), which is defined in the IETF specification RFC959. The Ga interface is based on a simple UDP/IP tunnelling protocol whose only purpose is to transfer the CDR.

For a given session, charging information is issued by different network nodes (SGSN and GGSN for a PS session; P, I or S-CSCF for an IMS session). This information is used by the CDF or CGF to build complete CDR, putting together the various pieces from the network elements. The redundant information (such as data traffic volumes or session start and stop Timestamp) is used to check the consistency between the views reported by the network elements.

The Roaming Case
Revenues generated from active subscribers in visited networks represent a significant part of operators' revenues. The roaming case requires specific mechanisms to be implemented between networks to ensure that a network operator gets the revenues corresponding to the resources being used in its network by other networks' subscribers.

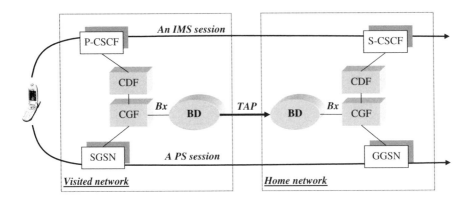

Figure 2.24 Charging in a roaming case.

Figure 2.24 presents the different network elements and interfaces involved in a roaming situation. In such a case, charging information needs to be collected by both visited and home networks. In the case of the IMS session, the visited network P-CSCF collects local information about the SIP session and the visited SGSN gathers information related to the PDP context being allocated (including allocated Quality of Service and data volumes). So does the home network through the S-CSCF and GGSN.

At some point, when the session is over, the visited network will send the bill to the subscriber home network using TAP (Transferred Account Procedure). TAP is a protocol defined by the GSM Association for interchange of billing data between different network operators, allowing them to bill each other for the use of networks and services through a standard process. The TAP format, originally defined for GSM circuit-switched services, has been enhanced to also cover packet-switched and IMS sessions. This includes specification of new TAP record attributes such as data volume counts, Timestamps, IP address, Access Point Number (APN – representing the GGSN the user has accessed to), requested and negotiated Quality of Service. The TAP record format is described by specifications maintained by the GSM association. It can be seen as a detailed charging bill submitted by the visited to the home network. As an illustration, here is a list of information contained in the TAP record:

- TAP record creation date and time.
- Charged units.
- Charge rate.
- Taxation information (including tax rate and type).
- Discount rate (when applicable).
- Local currency and exchange rates.
- Details about allocated PDP context.
- User data volumes.
- User connection time.
- ...

It is worth noting that even if the call is free in the visited network, e.g. in the case of a user-terminated session, the home network shall receive the TAP record. This may be needed in case the call or session transfer from the home to the visited network is subject to specific fees.

> *Reference documents about charging*
>
> 3GPP technical specifications:
>
> - 32.240, 'Charging Architecture and Principles'
> - 32.251, 'Packet Switched (PS) Domain Charging'
> - 32.260, 'IP Multimedia Subsystem (IMS) Charging'
> - 32.295, 'Charging Data Record (CDR) Transfer' (the Ga interface)
> - 32.297, 'Charging Data Record (CDR) File Format and Transfer' (the Bx interface)
> - 32.299, 'Diameter Charging Applications' (the Rf interface)
>
> The TAP record format, defined by the GSM Association:
>
> - TD.57, 'Transferred Account Procedure Data Record Format Specification Version Number 3'
>
> ITU-T Recommendation:
>
> - D.93, 'Charging and Accounting in the International Land Mobile Telephone Service'

2.6 The Terminal

2.6.1 The User Device Architecture

From the development of packet data services and support, mobile terminal integrate more and more features, as described in Figure 2.25. Because of the reduced size, supporting all applications on a single device may be too limiting to the end-user, for the following reasons:

- Limited man-to-machine interface: most of the time, only a small keypad is available or 'direction pad', limiting the possibilities for interaction with the device.
- Limited end-user experience caused by the poor audio quality or reduced display size.
- Limited storage capability.
- Limited battery life: supporting all features like radio transmission as well as colour display or headset powering is very consuming. All of this may quickly drain the batteries of a small device.

For these reasons, it may be of some interest to dissociate the provision of the service itself and the modem side of the service (this modem part relates to the pure data-transmission part of the terminal, including the ability to support high bit rate transmission and seamless mobility over a radio interface).

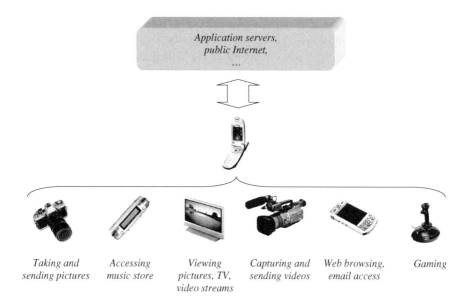

Figure 2.25 Mobile terminal and user services.

To answer this need, the standard has defined a flexible architecture for terminal realization. In this new model, all functions supported by a terminal are split between two elements: the MT (Mobile Terminal) and the TE (Terminal Equipment). The model, shown in Figure 2.26, presents the different parts of the mobile terminal (also called UE for 'User Equipment' – a term inherited from 3G/UMTS specifications). The terminal is described as a sum of a MT module and a TE device, or possibly multiple TE devices. In addition, the UE also integrates the user subscriber module (also known as the SIM card), further described into the next section. With such a model, several user devices (or 'TE' like a personal assistant or a camera) may have simultaneous access to different Application Servers, using the same radio equipment (the MT) and a single subscription (materialized by the SIM card).

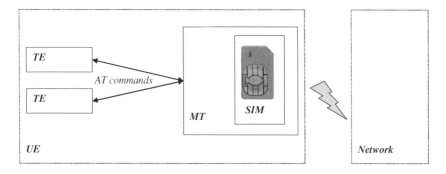

Figure 2.26 Mobile terminal architecture split.

The MT module contains all the functions related to the radio interface and wireless network access and data transmission in general. Therefore, it supports the following features:

- Transmission and reception of data and signalling over the radio interface.
- Authentication and registration to the UMTS network.
- Management (including creation, de-activation and modification) of PDP (Packet Data Protocol) contexts on request from a TE.
- Session control.
- Support of radio mobility functions, such as the handover.

The TE is the part the end-user has access to, as it supports all the functions related to user applications and interfaces. It contains the following features:

- Control of application-related hardware functions, such as speaker, microphones, video cameras, displays, etc.
- Support of user applications and services, such as email client, Web-browsing client, instant-messaging client, etc.
- Support of application-related protocol and session-management functions (for IMS-based applications, this includes protocol stacks like SIP, SDP and RTP).

The interface between MT and TE makes use of a simple command/response protocol known as AT commands (for 'ATtention'). This protocol interface allows the TE to control the MT, for instance to request the establishment of a PDP context for setting up an IMS session, or request the transmission of a SMS (Short Message), etc. The physical interface between the TE and the MT is not defined by the standard. It may use any wired interface, like Ethernet, USB or IEEE1394, or wireless interface, such as Bluetooth, IrDA infra red, 802.11 Wifi, etc.

Reference documents about the terminal interfaces

3GPP specifications:

- 27.007, 'AT Command Set for User Equipment (UE)'
- 27.901, 'Report on Terminal Interfaces'

2.6.2 Terminal Capabilities

In general, the standard defines mobile categories which correspond to different levels of capabilities (represented by a set of supported features) and performances (expressed in terms of maximum bit rate). This was the case in GPRS (the different classes of multi-slot terminals), as well as for UMTS HSDPA and HSUPA. The interest is to limit complexity for both terminal vendors (which can then choose which subset of the standard they will implement as well as the degree of performance) and network manufacturers (which can limit the list of possible terminal implementations and therefore reduce the effort in development and inter-operability testing). Similarly, the standard defines classes of E-UTRAN terminal, defined by a combination of reception and transmission bandwidth.

2.6.3 The Subscriber Module

The provision of services to end-users through 3GPP technology (including GSM, UMTS and Evolved UMTS) is dependent on a subscription to an operator. As a consequence, some user-specific information needs to be stored on the terminal side, such as subscriber identity and security credential, to allow the subscriber to authenticate to the network. For that purpose, the operators make use of a UICC card (Universal Integrated Circuit Card), also known as a SIM card (for Subscriber Identity Module) for GSM terminals or USIM (Universal SIM) for UMTS terminals. All those terms are often described as equivalent. Strictly speaking, the UICC only refers to the circuit card, its physical components, as well as the set of software and protocol for data exchange with an external device which is common to all UICC cards. In contrast, the SIM and USIM are referred to as the 'application' supported by the UICC, represented by the set of files, information and procedures specific to 3GPP networks.

The UICC has therefore become a generic multi-application platform which can support SIM and USIM applications, as well as ISIM (IMS SIM), which covers all information related to IMS functions and security procedures. In addition, the UICC can support any other kind of application for banking, ticketing or access control, e.g. based on the Java Card Technology. The availability of a valid UICC card is a pre-requisite for a terminal to register to the network and activate a service. However, the standard tolerates one exception: the emergency call. In the emergency situation, the network shall be able to accept calls to emergency numbers (such as 911 for North America or 112 in Europe) initiated by terminals not containing any UICC (or an invalid UICC).

The rest of this section provides some details about the hardware and software side of the UICC.

(i) The Hardware Part

From a high-level perspective, the UICC is nothing more than a piece of plastic supporting a microprocessor, ROM permanent memory containing the COS (Card Operating System), as well as erasable and dynamic memory. The physical and electrical characteristics of the UICC card are specified by a set of specifications issued by the ISO/IEC (International Organization for Standardization/the International Electrotechnical Commission). The main reference documents are:

- 7816-1, 'Physical Characteristics'.
- 7816-2, 'Dimensions and Location of the Contacts'.
- 7816-3, 'Electronic Signals and Transmission Protocols'.
- 7816-4, 'Interindustry Commands for Interchange'.

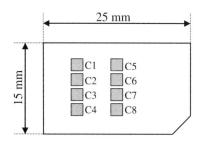

Figure 2.27 The UICC 'plug-in' format.

Table 2.4 UICC contact assignment (from ISO/IEC 7816-2).

(Voltage supply) VCC	C1	C5	GND (Ground)
(Reset signal) RST	C2	C6	VPP (Variable voltage for programming)
(Clock signal) CLK	C3	C7	I/O (Data input/output)
Not used	C4	C8	**Not used**

Figure 2.27 describes one of the possible formats of the UICC, also known as the 'plug-in'. Two other formats exist: ID-1, which is the format used for credit cards and 'mini-UICC' (12 × 15 mm), which was introduced in early 2004 for small terminals.

Table 2.4 describes the role of the external contacts supported by the UICC, as in the ISO/IEC 7816-2 standard. Over the years, the memory size of UICC modules has dramatically increased, from a few kilo-bytes to a few hundreds of mega-bytes, so that the main limitation of the UICC hardware is actually about the very limited data transfer rate (the current ISO/IEC 7816 standard transfer rate does not exceed 600 Kb/s). In the 1980s, when 2G/GSM systems were being defined, this was not felt as a limitation. However, the introduction of high-speed smart-card alternative technologies – like the MMC (Multimedia Card) or the SD (Secure Digital) card – able to cope with high-quality picture or video clip-based applications, makes the UICC slow data transfer rate a real limitation.

This is the reason why the ETSI Smart Card Platform group (SCP) has adopted a new technology for high-speed SIM card data transfer. This evolution was agreed at the end of 2006 for introduction in the Release 7 of the 3GPP standard and is therefore not specific to Evolved UMTS networks. The new interface makes use of the two unused UICC contacts (C4 and C8). The data transfer specification is based on the USB 'Inter-Chip' (IC) interface. This modified version of the well known USB (Universal Serial Bus) specification has been specially designed for short-distance communication between chips and is actually a minor adaptation of the technology used on personal computers. The adaptations refer to, for example, the interface voltage supply, which is no longer limited to 3.3 V in order to cope with other possible chip voltage classes, and the physical interface which does not require standard USB connectors and twisted-pair cables. Table 2.5 describes the UICC contact assignments when the USB-IC interface is active.

On the physical level of the USB interface, data are received and transferred using differential DC voltage levels between the D+ and the D− lines. As in the Inter-Chip USB standard supplement, the maximum data transfer rate is about 12 Mb/s, which corresponds to the legacy USB 2.0 Full-speed mode of operation. In order to allow smooth introduction, the UICC USB interface has been specified by the ETSI SCP group in a backward-compatible way. From a

Table 2.5 UICC contact assignment for USB interface.

(Voltage supply) IC-VDD	C1	C5	GND (Ground)
Not used	C2	C6	**Not used**
Not used	C3	C7	**Not used**
(D+ data line) IC-DP	C4	C8	IC-DM (D− data line)

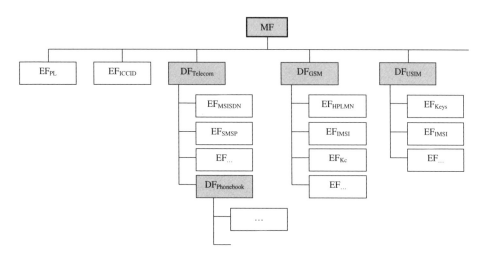

Figure 2.28 The USIM card file system.

practical perspective, this means that USB-capable terminal devices and UICC cards shall be able to interact with a non-USB-capable peer entity using the legacy ISO/IEC 7816 standard.

The ISO specifications listed at the top of this section are dedicated to the physical and electrical aspects of the UICC. However, they also specify how information is transmitted to and from the UICC as well as the file system to use in order to store and retrieve data on the UICC card. This is a very important aspect of the subscriber module.

(ii) The UICC File System

All the data stored in the UICC are structured in a sort of file tree, as shown in Figure 2.28. Each file contains one or several pieces of information and can be read individually by the user terminal. There are three sorts of files:

- MF (Master File) is the entry point of the data structure.
- EF (Elementary File) is a set of bytes which actually represent the data stored.
- DF (Dedicated File) which allows functional grouping of files. A DF can contain a number of EF plus some other DF.

A UICC card can contain quite a large number of files. Figure 2.28 only illustrates, in the USIM application case, the highest levels of the tree structure. At the root of the tree, the UICC contains key general files, such as:

- EF_{PL} (*Preferred Language*), which defines the preferred language to be used on the terminal user interface.
- EF_{ICCID}, which contains the UICC unique identifier.

The $DF_{Telecom}$ directory is also located at the top of the tree. It contains general-purpose information, such as:

- EF_{MSISDN}, related to the MSISDN numbers (or telephone numbers) associated to the subscriber.
- EF_{SMSP}, which contains information relative to the Short Message Service, such as the SMS Center phone number, the message coding type, etc.
- $DF_{Phonebook}$, which is a sub-structure containing the user-defined phone book.

In addition, two other DF are also present at the file root level: the DF_{GSM}, containing all the information needed by a GSM terminal to access to a 2G network, and the DF_{USIM}, which is used by the terminal when accessing a 3G network. The UICC file structure is actually designed in a backward-compatible way so that even a USIM card can be used in a 2G terminal. When reading the UICC, the 2G terminal can retrieve all the information it would have found in a genuine 2G UICC, thanks to the DF_{GSM} structure. However, when such a card is introduced in a 3G device, only the DF_{USIM} is considered by the terminal, which explains the redundancies between the two DF.

For information, the following briefly describe the purpose of the EF present in the figure:

- EF_{HPLMN}, which defines the home PLMN (or operator) scanning period when the subscriber is roaming.
- EF_{IMSI} contains the IMSI (International Mobile Subscriber Identity) of the subscriber.
- EF_{Kc} contains the GSM ciphering keys.
- EF_{Keys} contains the 3G ciphering and integrity keys.

(iii) The Subscriber Module in Evolved UMTS
In order to allow backward compatibility, the EPS subscriber module will be based on the UICC for the hardware part and on the USIM application for data organization and file organization. This comes from the fact that EPS network standards rely on the assumption that it may be possible for a subscriber having a 3G/USIM card plugged into an EPS-capable terminal to access an EPS network. The additions to the existing data tree will be limited to the additions specific to EPS networks, such as the specific security key and algorithms. More details about EPS security aspects are provided in Chapter 5.

Reference documents about the UICC content and interface specification

ISO/IEC technical specifications:

- 7816-1, 'Physical Characteristics'
- 7816-2, 'Dimensions and Location of the Contacts'
- 7816-3, 'Electronic Signals and Transmission Protocols'
- 7816-4, 'Inter-Industry Commands for Interchange'

ETSI specifications:

- TS 102 221, 'UICC-Terminal Interface: Physical and Logical Characteristics'
- TS 102 600, 'UICC-Terminal Interface: Characteristics of the USB Interface'

3GPP specifications:

- 31.101, 'UICC-Terminal Interface: Physical and Logical Characteristics'
- 31.102, 'Characteristics of the Universal Subscriber Identity Module (USIM) Application'
- 31.103, 'Characteristics of the IP Multimedia Services Identity Module (ISIM) Application'

2.7 The Evolved UMTS Interfaces

This section intends to list all the interfaces defined in the scope of the EPS networks. From the multiplicity of logical nodes in the architecture result many interfaces. Some of them are brand new ones, whereas some others rely on existing 3G/UMTS ones – sometimes with different names. Due to the difficulty in showing everything in a single picture, Table 2.6 should serve as a dictionary for all the EPS interfaces named in this chapter and in the rest of the book.

2.8 Major Disruptions with 3G UTRAN-FDD Networks

The choice of OFDM technology on the radio interface and the various options chosen for the E-UTRAN radio interface have some significant and disruptive impacts on the way the overall network behaves. This section aims at describing the main impacts of some E-UTRAN technical choices made in the standard definition on network architectures and supported features, as regards to 3G/UTRAN FDD networks.

2.8.1 About Soft Handover

One of the main consequences of using a CDMA-based radio interface is about the need for soft handover, also known as macro-diversity. In CDMA, each transmission channel behaves as an interferer for the other channels. The consequence is that transmission power tuning is a key point to preserve CDMA system capacity. This becomes critical at cell edge or in poor coverage areas, where maintaining the radio link transmission quality is often a synonym to increased transmission power. This is where soft handover helps, allowing the information to be transmitted on different links – called the active set – and adding transmission diversity gain. As illustrated in Figure 2.29, soft handover is a mechanism by which a terminal maintains simultaneously several radio links in different cells for one single session or data flow. Information is then recombined from the received radio links, either on the network or on the terminal side, for the sake of transmitted power and associated interference. In 3G/UTRAN, soft handover is applied to all dedicated channel transmission, including HSUPA (the high-speed transmission technique for the uplink) also based on dedicated channels. However, soft handover does not apply to HSDPA data transmission, which makes use of a physical shared channel.

The consequence of soft handover on 3G/UTRAN architecture is that the Iur interface between RNC is mandatory in case the different BTS involved by the soft handover are not controlled by the same RNC. This is also a consequence of the fact that uplink and downlink soft handover was actually defined as a RNC level feature. In such a case, 3G/UTRAN standards distinguishes the SRNC (for Serving RNC) as being the RNC which controls the

Table 2.6 Summary of Evolved UMTS interfaces.

Name	Comments
S1	Between MME/Serving GW and eNodeB.
	Supports the user and control plane traffic between the E-UTRAN and EPC.
S2	Between PDN GW and non-3GPP access.
	Supports control and mobility procedures for non-3GPP access technologies.
S3	Between MME and 2G/3G SGSN.
	Supports user and bearer information exchange for inter-system
	mobility in idle or active state.
	Based on the 2G/3G Gn interface (GTP protocol).
S4	Between Serving GW and 2G/3G SGSN.
	Supports user plane data transfer for inter-system mobility.
	Based on the 2G/3G Gn interface (GTP protocol).
S5	Between Serving GW and PDN GW.
	Supports bearer management and user plane data tunnelling between the two gateways.
	Based on GTP.
S6	Between the Evolved Packet Core nodes and the HSS.
	Supports the procedures for user subscription data retrieval and location update.
	Based on Diameter.
S7	Between the PDN GW and the PCRF.
	Supports the procedures for Policy and Charging rule transfer from the PCRF to the EPC.
	This interface is based on the 3GPP R7 Gx definition.
S8	Between the visited Serving GW and home PDN GW.
	It supports packet and user plane transfer between the two gateways for the roaming cases.
	This interface is a variant of S5, based on 2G/3G Gp interface (GTP protocol).
S9	Between the visited PCRF and the home PCRF.
	This interface is a variant of S7 for the roaming cases.
S10	Between the MME nodes.
	This interface is used in case of inter-MME mobility (or relocation) to
	exchange session and user contexts between nodes.
	Based on the Control part of GTP protocol (GTP-C).
S11	Between the MME and Serving GW.
	Supports bearer management e.g. at user attachment or service request.
	Based on the Control part of GTP protocol (GTP-C).
X2	Between eNodeBs.
	Supports mobility and user plane tunnelling features.
	Based on the same user plane protocol as S1.
Rx	Between PCRF and P-CSCF.
	Used to provide service dynamic information to the PCRF.
	Based on an evolution of the IETF Diameter protocol.
SGi	Between the PDN Gateway and the packet data network.
	Based on the 2G/3G Gi interface.
Ga	Between CDF and CGF.
	Supports the transfer of CDR (Charging Data Record) from the CDF to the CGF.
	Based on UDP/IP tunnelling.
Bx	Between the CGF and the BD.
	Supports secured transfer of the CDR to the BD.
	Based on the FTP protocol.
Rf	Between an IMS network entity or an Application Server and the CDF.
	Supports charging information transfer to the CDF.
	Based on an evolution of the IETF Diameter protocol.

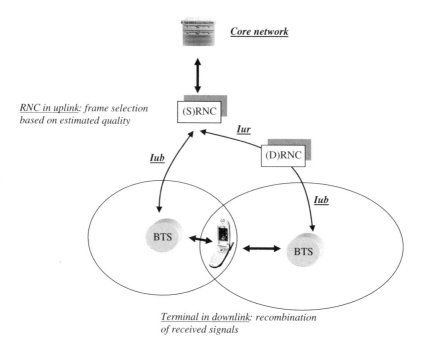

Figure 2.29 An example of UTRAN soft handover.

communication with the terminal from the DRNC (for Drift RNC) as being the RNC which only controls the BTS involved in a soft handover.

The cost for having an Iur interface is not negligible from an operational perspective. When an operator mixes equipments from different manufacturers in the access network, or decides to upgrade parts of it, lots of care has to be taken on interoperability between the RNC nodes being connected through the Iur. In addition, the Iur requires the operator to pay for supporting the inter-RNC connectivity and the traffic sent over the Iur. This traffic can, however, be reduced thanks to the 'relocation' mechanism which allows changing the call-controlling RNC to a more appropriate RNC. In any case, even if Iur is not needed in most of the soft handover cases (because all the cells involved in the active set are controlled by the same RNC), there is still the cost of multiple Iub link transmission to be supported between the SRNC and all the BTS.

The side effects of the Iur interface have, however, to be mitigated by the fact that a RNC can control a large number of cells, so that the Iur is actually only needed at the edge of the RNC area. The actual figure depends, of course, on the implementation and physical capacity of each RNC, as well as the traffic load. In any case, it is usually admitted that a RNC can control several hundreds of cells. CDMA macro-diversity therefore has significant impacts in terms network architectural definition and network operation. In E-UTRAN, from this perspective, the picture is very different. Soft handover is no longer needed or required, which is a significant change in terms of overall access network definition and operation.

As already mentioned, the E-UTRAN architecture introduces an X2 interface between eNodeB. However, this new interface is not an Iur equivalent, as there is no soft handover support in E-UTRAN. X2 is actually an optional interface, which provides a data-forwarding service in case of inter-eNodeB mobility, for the sake of lossless mobility. More details about X2 usage are provided in Chapter 5.

2.8.2 About Compressed Mode

One fundamental point of CDMA radio interface is that radio transmission is continuous, and simultaneous for both uplink and downlink transmission. In principle, this would not be a problem by itself, except for some specific radio mobility cases, which the CDMA radio interface is not able to work out without some additions. When moving around cells of the same frequency, the terminal CDMA receiver can quite easily monitor them – build its own list of best cell, so that the active set is managed at best, thanks to the soft handover mechanism. However, problems arise in the case of multi-frequency or multi-access technology deployment, as is more and more the case in countries where cellular communications have reached maturity. When continuously receiving radio frames, the CDMA terminal has no time to switch its receiver to another frequency, and monitor neighbouring cells, e.g. in 2G/GSM, of using another UMTS/FDD frequency band.

There may be multiple solutions to this issue. The most obvious one would be to build multiple receiver terminals, which is a very costly and power-consuming option. Another solution brought by the 3GPP standard was to change the CDMA radio frame structure in order to artificially create some holes so that the terminal has time to perform other frequency of even radio-system monitoring. This mechanism, widely deployed in UMTS networks, is known as Compressed Mode.

Figure 2.30 shows an example of Compressed Mode operation. In this case, periodic monitoring windows are created in uplink and/or downlink radio frames using Compressed Mode, so that the terminal can decode beacon information of a neighboring GSM cell. The 3GPP UTRAN standard documents define different methods for creating these transmission gaps, the main two being:

- **Spreading Factor reduction:** in this method, the same amount of data is sent in half the time thanks to the use of a reduced SF code. The decrease in spreading gain is then compensated by increased power during the time the code is used.

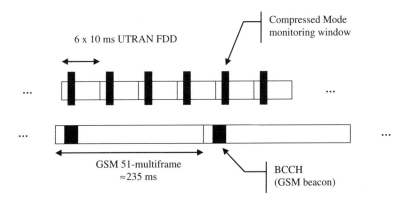

Figure 2.30 An example of UTRAN Compressed Mode operation.

- **Higher Layer Scheduling:** this method is only suitable for non-real time communications, as it relies on the MAC data scheduler to limit data transmission at some specific time.

There is no ideal method, and both of them have drawbacks. The Spreading Factor reduction method requires more transmission power and creates more interference when the compressed mode is active, which has some impact on network capacity. The second method only results in a slight increase in radio transmission latency, but it is not applicable to circuit services like voice.

As further described in Chapter 3, E-UTRAN OFDM-based radio interface relies on frequency and timeslot allocation and does not mandate continuous transmission and reception. This leaves the possibility for an E-UTRAN terminal to monitor neighboring beacons, frequencies and systems when it is not required to receive or transmit data on the radio interface. For that reason, Compressed Mode is no longer needed in E-UTRAN.

2.8.3 About Dedicated Channels

In its early definition, UMTS was mainly thought of in terms of a dedicated channel system, based on the 384-Kb/s max DCH Dedicated Transport Channel. Initial UMTS standard also proposed some limited possibilities for shared channel transmission on FACH (Forward link Access Channel) or on the DSCH (Downlink Shared Channel). Transmission on FACH was never intended for high bit rate, due to the lack of an efficient power control scheme and also because of the limitation of the uplink RACH (the RACH is the Random Access Channel used in combination with the FACH). Regarding DSCH, it was actually never developed and deployed as a commercial solution because of its high degree of complexity and the lack of real performance improvement.

HSDPA introduced a major change in UTRAN by allowing higher bit rate transmission on shared channel or shared group of physical resources, also known as the HS-DSCH transport

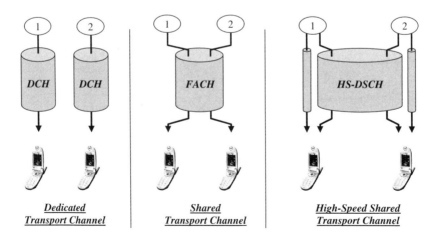

Figure 2.31 The three options for downlink data transmission in 3G/UTRAN.

channel. Thanks to new modulation scheme and fast HARQ packet repetition scheme, the radio performances were significantly better than before. However, HSDPA still relies on an associated dedicated channel used in downlink for power control commands. A dedicated physical channel is also present in uplink to carry HARQ indications and channel quality information.

In practice, all three transmission schemes (Figure 2.31) often co-exist with the same cell, set of cells, or geographical area. For circuit-based services requiring constant delay and bandwidth, transmission on DCH will still be applicable until HSDPA allows cost-efficient guaranteed bit rate solution. In addition, for low bit rate packet data service or MBMS (Multimedia Broadcast Service), transmission over the FACH will still be applied.

E-UTRAN data transmission on radio interface is based on shared channels, whatever the service type and requested Quality of Service. Although this puts more constraints on the radio scheduler (in the sense that the system needs to ensure that all data flows are transmitted with the requested Quality of Service), there is only one unique solution for user data transmission which is a major simplification for network design and operation. It is no more needed to partition radio interface physical resources in different and competing sets of shared or dedicated channels.

3

Physical Layer of E-UTRAN

3.1 Basic Concepts of Evolved 3G Radio Interface

From a radio and network point of view, E-UTRAN focuses on the UTRAN evolution and optimization, keeping in mind that UTRAN HSDPA and HSUPA will be highly competitive for years. E-UTRAN is a significant technology step, aiming to ensure the competitiveness continuation of the 3GPP family of technologies.

Among future challenges that E-UTRAN will have to face, one can list:

- The demand for higher data rate.
- The expectations of additional 3G spectrum allocations.
- A greater flexibility in the frequency allocation methods: E-UTRAN can operate in 1.25, 2.5, 5, 10, 15 and 20-MHz bandwidth in uplink or downlink, paired or unpaired spectrum. Coexistence with GERAN and UTRAN should be possible.
- The competition with unlicensed technologies like WiMAX.

Therefore, the objectives of the system will roughly be:

- Significantly increased peak data rates: up to 100 Mb/s in downlink and 50 Mb/s in uplink in a 20-MHz spectrum for a user throughput three or four times the user throughput of UTRAN.
- Increased cell edge bits rate.
- Improved spectrum efficiency (three to four times the UTRAN DL and two to three times the UL).
- Reduced latency: reduce the latency between RRC states change – from idle to active state – and for transmission over the network radio access (less than 5 ms).
- Scaleable bandwidth. (1.25/2.5/5/10/20 MHz).
- Reduced operation cost.
- Acceptable system and terminal complexity, cost and power consumption.

Evolved Packet System (EPS) P. Lescuyer and T. Lucidarme
Copyright © 2008 John Wiley & Sons, Ltd.

- Compatibility with previous releases and with other systems.
- Optimized for low mobile speed (<15 km/h) but supporting high mobile speed (up to 350 km/h).

Among all the modern concepts considered for future radio interfaces, two technologies seemed particularly promising to be considered for UTRAN evolution: OFDM (Orthogonal Frequency Division Multiplex) and MIMO (Multi Input Multi Output).

For most international experts, members of standardization bodies, it is granted now that these two technologies will be added to the technological E-UTRAN puzzle whose original 3G system is the very first foundation. They present indeed some advantages with respect to 2G CDMA in terms of flexibility of resources allocation for packet transmissions and data rate increase for a given complexity.

We will describe in a first part the main principles of these technologies and their variants before focusing in a second part on the E-UTRAN technologies as such.

3.2 OFDM (Orthogonal Frequency Division Multiplex)

The OFDM principles were elaborated in the early 1960s with the first multi-carrier systems, especially military systems but without massive impact because of the lacunas of the electronic circuits and signal processing available at this date.

This technology came back in the 1980s for the application to multipath channels. Such channels are characterized by a non-flat frequency response which includes deep holes known as 'selective fading'. The basic idea for OFDM is, as we will see, to spread the information on a lot of sub-carriers in order to create very narrow band channels, experimenting in each of them a frequency response that can be considered as uniform or 'flat'.

These multi-carrier modulations became practically interesting, since a completely numerical structure of a modulator was highlighted based on Fast Fourier Transform (FFT). Apart from the projects of future use of the OFDM for the applications of E-UTRAN radio communication, the OFDM is used particularly in the digital audio broadcast system (DAB, 'Digital Audio Broadcasting'), and digital video broadcast system (DVB, 'Digital Video Broadcasting'), High data rate local area networks, and DSL type wire line networks. In addition, this technology is used in the broadband wireless packet access of radio called WiMAX.

The OFDM technology is thus a well known technique which consists of multiplexing on frequency subcarriers some information to be transmitted on a channel of communication. Moreover, the subcarriers are orthogonal between them, owing to the fact that the minimal duration of information carried by each subcarrier is the reverse of the value of the band of modulation of the subcarrier (Nyquist criterion).

Figure 3.1 illustrates the general principles of a transmission carried out by using technology OFDM. A signal carrying information must be transmitted by a transmitter, to be received and interpreted by a receiver. The information carried by this signal includes a succession of binary characters.

Let X_n indicate a quantity of information in series to be transmitted for the n-user. Initially, a module transforms this flow series into several N parallel flows $X_{n,0}, X_{n,1} \ldots X_{n,N-1}$. On a purely illustrative basis, each one of these parallel flows can consist of a succession of binary characters of duration equal to T_u. A quick study of the modulation shows that the signal sent

Physical Layer of E-UTRAN

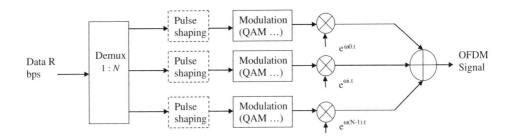

Figure 3.1 Generation principle of OFDM signals.

on the channel is the reverse FFT (or IFFT, 'Inverse Fast Fourier Transform') of the source signal. A reverse FFT is then carried out within the transmitter from the N input parallel flows $X_{n,0}, X_{n,1} \ldots X_{n,N-1}$.

At the end of the transmitting operation, an OFDM symbol S_m of T_u duration is obtained. This OFDM symbol represents a set of binary data coded on frequencies separated by the modulation band as indicated above. This modulation band obviously depends on the modulation chosen to create on each subcarrier a modulated symbol – we represent a classical QAM (Quadrature Amplitude Modulation) in Figure 3.1.

The OFDM symbol S_m is then transmitted by the transmitter on a channel of communication which can be a radio channel, for example.

Figure 3.2 underlines the N subcarriers case, in which the symbol OFDM is spread on subcarriers $S_m(i)$; $i \in [1, N]$.

From the reception side, the receiver, while listening to the channel, receives a symbol $\hat{S}_m$ corresponding then to the transmitted symbol OFDM S_m, putting aside disturbances, introduced by the channel or external interferences.

A Fast Fourier Transform or 'FFT' is then carried out on the portion of received signal corresponding to the symbol $\hat{s}_m$. Data elements $\hat{X}_{n,0}, \hat{X}_{n,1} \ldots \hat{X}_{n,N-1}$ are estimated starting

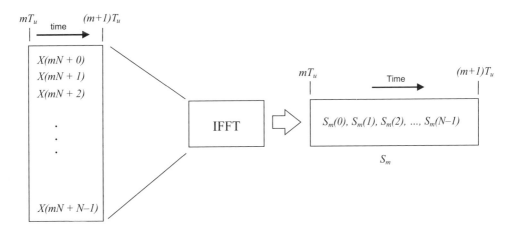

Figure 3.2 The OFDM symbol.

from this symbol, each data element corresponding to the information transmitted on a given frequency subcarrier. Then, a parallel/serial conversion allows obtaining an estimate $\hat{X}_n$ of the transmitted information X_n.

The pulse shaping filters of Figure 3.2 may then be used to window the temporal shape of the data with an aim of reducing their frequency spectrum spread out.

In addition, it is common to use a guard time interval in each symbol OFDM, aiming to reduce the Intersymbol Interference (ISI).

Indeed, when the communication channel on which the signal is transmitted includes multipaths, as it is generally the case for an urban radio channel, jamming replicas of the signal can be received at the receiver with respective delays corresponding to the various paths. It thus results in a certain overlapping between portions of signal relative to successive OFDM symbols, likely to make more difficult the estimate of information transmitted and thus to degrade the quality of the reception.

The guard time interval consists of increasing the duration of each useful symbol OFDM, by duplicating at the end of the symbol certain binary characters placed at the beginning of this symbol, or conversely.

Such addition of redundant information is named 'insertion of a cyclic prefix' to the symbol OFDM, which will be withdrawn in the receiver after convolution with the impulse response of the channel.

Obviously, this part of redundant information can be a prefix, i.e. consists of the duplication of a given number of the starting bits of a slot, or a suffix, i.e. the duplication of a given number of bits at the end of the slot, or even both a prefix and a suffix with duplication of bits of both types.

The receiver then benefits from the duplication of certain binary characters to improve the estimate of the useful information of each symbol.

This operation contributes to transform the convolution of the signal by the channel response into a cyclic convolution (property of the circulating matrix), making easy the demodulation of the symbol transmitted on each subcarrier. The T_s duration of the new extended transmitted symbol is then equal to $T_u + T_g$, where T_g indicates the duration of the guard time.

Mathematically, the basic idea of OFDM thus consists of dividing the band available into N sub-bands and transmitting in each one of them to an N times weaker rate than that which would be used in the total band. Subcarriers must be as close as possible, while preserving the orthogonality. We will see that the equalization then becomes extremely simple.

For instance, assuming f_0 is the bandwidth subcarrier,

$$S_n(t) = \sum_{i=1}^{N-1} X_{n,i}.\cos(2\pi n f_0 t + \phi_k) = Re\left(\sum_{i=1}^{N-1} c_{n,i}.e^{j.2\pi(n f_0 t + \phi_n)}\right), \quad (3.1)$$

in which we have, neglecting the index,

$$c_n = X_n e^{j.\phi_n}. \quad (3.2)$$

The X_n can be selected in order to exploit the band available in the best way possible. The signal $s(t)$ can be sampled at the frequency $2.N.f_0$ while respecting Shannon [the $2N$ samples of the OFDM symbol to be transmitted can thus be obtained by inverse FFT on the vector $(0, c1, \ldots, cN-1, 0, c*N-1, \ldots, c1*)^T$: this construction ensures simplifying that the inverse FFT is a real sequence but the inverse FFT can also be complex, as in E-UTRAN].

This very effective implementation is one of the keys of the success of modulations OFDM.

> ***To sum up: Advantages of OFDM for mobile radio***
>
> The advantages of the modulation OFDM essentially come from its performance in comparison with the simplicity of realization of the associated receiver, which incorporates only one device intended for carrying out the FFT of the received signal followed by a simplified equalizer correcting on each subcarrier the resulting flat fading (one constant complex gain per subcarrier).
>
> The long symbol time makes the signal resistant to multipaths and the guard interval limits the intersymbols' interferences.
>
> The orthogonality between subcarriers allows a huge spectral efficiency.
>
> No intra-cell interference cancellation system is required.
>
> Moreover the OFDM has other advantages, with respect to the good filling of the spectrum or flexible allowance of the frequencies.
>
> No one can today easily predict the evolutions allowed by the regulatory bodies dealing with the frequency spectrum in all the countries of the world. If the United States seems technology agnostic when compared to the rest of the world, it remains that certain military applications or television broadcasters occupy many portions of spectrum and do not intend to be dislodged that easily.
>
> However, it is particularly important to find some bands not too high in frequencies, to offer a reasonable range in cellular telephony for reasonable cost. It is one of the interests of the OFDM to be able to fill the small holes within a spectrum already partly allocated, while placing here and there the adequate number of subcarriers.
>
> Moreover, the final Fourier Transform can then be replaced by several Fourier Transform with less complexity granted the convexity of the log function [complexity of a fast transform of Fourier is in $N.\log.N$, but $n.\log(n) + m.\log(m) < (n+m).\log(n+m)$]. This also provides an advantage with respect to the spectrum scalability required for E-UTRAN.
>
> In case a feedback on the quality of the channel is available on the downlink, a base station can allocate to a user the better data rate on the better subcarriers in the signal-to-noise ratio sense, optimizing in mean the data rate for all users. This property is known as 'water-filing'.

A simplified global vision of an OFDM transmitter/receiver block scheme, for which we have arbitrarily chosen the QAM modulation, is shown in Figure 3.3.

The role of the 'subcarriers mapping' box will be discussed in the next section.

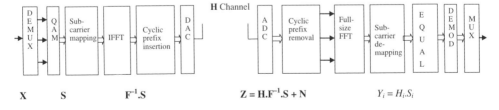

Figure 3.3 Simplified structure of a transceiver OFDM.

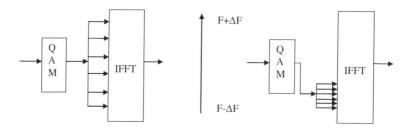

Figure 3.4 OFDMA methods of separating multiple users.

3.2.1 OFDMA Multiple Access

OFDMA is a scheme chosen for the downlink for E-UTRAN.

OFDMA is a reliable technology that has been already chosen for WiFi, WiMAX, ADSL/ADSL2+ or DVB technologies. It resists well to multipaths, allows a high spectral efficiency, especially with its MIMO compatibility, and a reduced complexity of implementation.

In OFDM systems, there are several processes of multiple accesses which can be used to distinguish the users. One of the simplest consists of choosing for a user given a unique law of choice of subcarrier frequencies. More precisely, each user is characterized by the choice of a set resource chosen in the frequency–time plane. The traffic multiplexing is performed by allocating to each user a pattern of frequency–time slots, depending on its data rate. We then speak of OFDMA. From a frequency point of view, according to the choice of mapping of symbols on subcarriers, the subcarriers allocated to one joint can be joint or separated. From a frequency diversity point of view, the separated scheme is obviously better.

Figure 3.4, still with an arbitrarily chosen QAM modulation, illustrates the two different ways the mapping box of the above figure can map the subcarriers for a given user. The scheme on the left provides much more diversity than the second one.

Figure 3.5 illustrates the resource distribution between user channels, common control channels and pilot symbols. Common control channels bring classically some information on the network, the cell, etc. Pilot symbols are useful to perform the identification of the channel response. Thanks to these known symbols, channel response can be interpolated both in time and frequency and simply equalized, as we will see in the following paragraph.

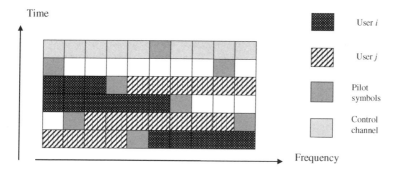

Figure 3.5 The time–frequency allocation pattern.

According to the speed authorized for the terminals, and the fading characteristics, a more or less dense pilot's pattern is needed, as the pilot position in the frequency domain is used to determine with enough accuracy the frequency selectivity of the channel and the pilot position in the time domain has to cope with high-speed mobiles.

It has been shown from a frequency and time channel sounding perspective that a good way of arranging these pilot symbols is to place them so that they form some diagonals in the time–frequency space.

(i) Pilot Symbols and Equalization

Once the cyclic prefix has been removed, the received signal can be written as

$$\mathbf{Z} = \mathbf{HF}^{-1}\mathbf{s} + \mathbf{n}, \tag{3.3}$$

where $\mathbf{s} = [s_0 \ldots s_{N-1}]^T$ are the transmitted OFDM symbols and $\mathbf{F}$ the matrix with Fourier coefficients: $f_{n,m} = e^{j2\pi nm/N}$

We have

$$\mathbf{H} = \begin{bmatrix} h_0 & 0 & \cdots & 0 & h_L & \cdots & h_2 & h_1 \\ h_1 & h_0 & \ddots & & & \ddots & \ddots & h_2 \\ \vdots & \ddots & \ddots & \ddots & & & \ddots & \vdots \\ h_L & \ddots & \ddots & \ddots & \ddots & & \ddots & h_L \\ 0 & \ddots & \ddots & \ddots & \ddots & \ddots & & 0 \\ \vdots & \ddots & \ddots & \ddots & \ddots & \ddots & \ddots & \vdots \\ \vdots & & \ddots & \ddots & \ddots & \ddots & \ddots & 0 \\ 0 & \cdots & & 0 & h_L & \cdots & h_1 & h_0 \end{bmatrix},$$

which is the matrix defining the impulse response of the channel, the form of which is due to the cyclic prefix addition and is well known as a circulating matrix whose properties allow to consider to write the corresponding formulas in the frequency domain: theoretically, this comes from the fact that the Fourier Transform of the convolution of the product is the product of the Fourier Transform.

$\mathbf{H}$ can be diagonalized and its eigenvectors is the FFT of the pattern.

Thus,

$$\mathbf{H} = \mathbf{F}^{-1}\mathbf{DF}, \quad \mathbf{D} = \text{diag}(H_0, \ldots, H_{N-1}) \text{ and } (H_0, \ldots, H_{N-1})$$
$$= FFT(h_0, \ldots, h_L, 0, \ldots, 0). \tag{3.4}$$

Then, the received signal is

$$\mathbf{Z} = \mathbf{F}^{-1}\mathbf{DFF}^{-1}\mathbf{s} + \mathbf{n} = \mathbf{F}^{-1}\mathbf{Ds} + \mathbf{n} \tag{3.5}$$

And, after FFT achievement in the receiver, this can be written as

$$\mathbf{Y} = \mathbf{FZ} = \mathbf{FF}^{-1}\mathbf{Ds} + \mathbf{Fn} = \mathbf{Ds} + \mathbf{Fn} \tag{3.6}$$

and is resuming in a pure simple scalar operation

$$Y_i = H_i s_i + n'_i, \quad i = 0, \ldots, N-1, \tag{3.7}$$

in which n' keeps the same Gaussian properties as n.

Therefore the equalization scheme is ultra-simplified, as it consists only of dividing the received signal after FFT Y_i by the complex value of the signal channel response interpolated on the pilot signals H_i:

$$S_i \approx Y_i/H_i. \tag{3.8}$$

3.2.2 MC-CDMA Multiple Access

Aside of OFDMA, some work has been done about the separation of the users on a CDMA basis, while keeping the advantages brought by the OFDM. This type of technology is often called MC-CDMA (Multi Carrier CDMA) (Figure 3.6). In this technique, like in classical CDMA, some spreading codes are used to multiplex the different users on a code-per-code basis before distributing the resulting summation on the OFDM subcarriers. This way of arranging the multiple access schemes is mentioned here in this section as it is a scheme that many actors of telecommunications, especially in Japan, advocated for in a fourth-generation context. However, it has not been chosen for E-UTRAN.

The receiver functions by reversing the stages of the emission: FFT initially then followed by a reverse spreading by orthogonal codes and/or pseudo-orthogonal ones (c_i).

3.2.3 Common Points between OFDM, CDMA, MC-CDMA, etc.

(i) The OFDM: Still a Spreading-Out Matter

Spread spectrum per direct sequence (DS-CDMA) is well known. It is established that the OFDM as well as the MC-CDMA is also obtained by spreading operators. Eventually, this leads to a unified representation of these processes of modulation or transformation of the signal in order to bring the two worlds closer.

For demonstration, we introduce a complex vector $\mathbf{X}(n.Ts) = [x_0(n.Ts) \ldots x_{N-1}(n.Ts)]^T$ containing a whole of N independent symbols of communications. These symbols must be transmitted in the same T_s time interval from a basic station or an access node to some terminals.

The DS-CDMA, OFDM, MC-CDMA solutions can be formalized as below, using a generic vector transformation:

$$\mathbf{S}(n.Ts) = \mathbf{O}.\mathbf{X}(n.Ts), \tag{3.9}$$

in which $\mathbf{O}$ is a matrix of spreading operator in the broad sense ($N.N$) for which each column represents a particular sequence of spreading. The k_{th} column of the $\mathbf{O}$ matrix is the code used

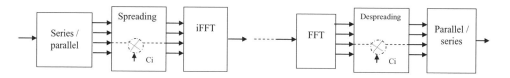

Figure 3.6 Transceiver MC-CDMA for the user m.

for the $x_k(n.Ts)$ symbol. The vectors $\mathbf{S}(n.Ts)$ represent the samples of the signal which will be received by the receiver after passing through the 'air' interface.

For the DS-CDMA (direct sequence), the operator **O** is quite simply resulting from the product of a diagonal matrix **D** whose elements come from a long scrambling code of gold type (case of the UMTS) by a square matrix **C** of Walsh-Hadamard codes.

Thus,

$$\mathbf{O} = \mathbf{D}.\mathbf{C}. \tag{3.10}$$

For the OFDM case, **O** is simply the matrix with Fourier coefficients: $f_{n,m} = e^{j2\pi nm/N}$, and $\mathbf{O} = \mathbf{F}$.

For the MC-CDMA, we thus have naturally, taking into account the linear character of the transmitter and the receiver,

$$\mathbf{O} = \mathbf{F}.\mathbf{D}.\mathbf{C}. \tag{3.11}$$

To reflect in a precise way the operations of demodulation, we need to introduce a model of channel, by supposing that fading remains constant over a T_s time; we then have an expression to model the channel:

$$H(n.Ts) = \sum_{k=0}^{L} \alpha_k(n.Ts).\delta(n.Ts - \tau_k), \tag{3.12}$$

in which the coefficients α_k are Gaussian-independent variables. Thanks to the insertion of a prefix for the OFDM (or even for the CDMA) and to its suppression in the receiver, the impulse response of the channel is finished and the receiver can only consider the symbols resulting from the vector $\mathbf{X}(\mathbf{n.Ts})$.

Using a matrix notation, the channel is thus noted as

$$\mathbf{H}(n.Ts) = \sum_{k=0}^{L} \alpha_k(n.Ts).\mathbf{J}^k, \tag{3.13}$$

where $\mathbf{J} = \begin{bmatrix} 0 & 1 \\ \mathbf{I}_{N-1.N-1} & 0 \end{bmatrix}$ is a **N.N** matrix.

The received signal is then

$$\mathbf{Z}(n.Ts) = \mathbf{H}(n.Ts).\mathbf{O}(n.Ts).\mathbf{X}(n.Ts) + \mathbf{B}(n.Ts), \tag{3.14}$$

where $N(n.Ts)$ is a white noise of power spectrum density N_0 such that $E(\mathbf{N}(n.Ts).\mathbf{N}^H(n.Ts)) = \sigma^2.\mathbf{I}$. In the gap time $[n.Ts, (n+1).Ts]$, we will simply write $\mathbf{Z} = \mathbf{H}.\mathbf{O}.\mathbf{X} + \mathbf{N}$.

From now, we can find the expression of a traditional receiver of MMSE type (Minimum Mean Squared Error), the goal of which is to isolate a dispreading (in the broad sense) matrix **W** which minimizes the error between the $X_{est} = \mathbf{W}^H.\mathbf{Z}$ estimate and X. $\mathbf{W}^H$ represents the Hermitian transform of the **W** matrix.

And, mathematically, we are looking for **W** such that

$$E[(\mathbf{W}^H \mathbf{Z} - \mathbf{X})^2] \tag{3.15}$$

is minimized. The well known solution of this equation is given by

$$\mathbf{W}_{\text{MMSE}} = (\mathbf{H}.\mathbf{O}.\mathbf{O}^H.\mathbf{H}^H + \sigma^2.\mathbf{I})^{-1}.\mathbf{H}.\mathbf{O}, \tag{3.16}$$

in which we have supposed that each communication symbol is independent, white and has a unit power, e.g. $E[\mathbf{X}.\mathbf{X}^H] = \mathbf{I}$.

In the case of the OFDM, one finds setting $\sigma = 0$ in Equation (3.16)

$$\mathbf{W}_{OFDM} = \mathbf{H}^{-H}.\mathbf{F}, \qquad (3.17)$$

which indicates, as already seen, that the receiver simply consists of a FFT followed by an equalization. For the CDMA, one finds the traditional Rake receiver

$$\mathbf{W}_{rake} = \mathbf{H}.\mathbf{D}.\mathbf{C}. \qquad (3.18)$$

3.2.4 Frequency Stability Considerations for OFDM Systems

Coming back to OFDM, an important constraint is the overall stability requirements of the system, whereas CDMA was more penalized by the number of fast calculations required by the 'chip-level processing' operations required for synchronization and demodulation.

It can easily be understood that if f_i and f_{i+1} are the closest subcarrier frequencies used in the OFDM system, the drift due to frequency instability of the electronics and the Doppler effects shall be estimated and compensated at the receiver side, in order to allow to keep the orthogonality of the subcarriers and the overall properties of the OFDM signals. The increase in the ICI (Inter-Carrier Interference) may reduce the performances of the network.

Therefore, an accurate frequency drift estimator is often necessary in OFDM receivers.

3.2.5 System Load in OFDMA Systems

In this section, still to compare with CDMA systems, we consider the downlink path of the FDD mode.

Correct performance of admission control algorithms and load balancing ones depends on the current expression of the load of the system. Contrary to the CDMA system, the OFDMA system load is not a direct function of the intra and inter-cell interferences, but a function of time–frequency–space resource assignments and power resource assignments. The power control, for instance, was of primary importance in WCDMA to mitigate the near–far effect deriving from the presence of the intra-cell interference, whereas in OFDM systems, like the E-UTRAN system, the users of a same cell are as orthogonal as possible with each other; thus, the only interference present is from the other cells (intercell). Though, still important, the role of the power control is reduced, compared to CDMA, to adapting the power to path loss and shadow fading fluctuations and reducing other cell interference.

Adaptations of parameters are the set of the assigned downlink subcarriers, coding, modulation, HARQ schemes including the spatial coding schemes and transmission power values.

The normalized consumed downlink time–frequency resources can be written as the ratio S_{DL}:

$$S_{DL} = \frac{S_{DL}}{S_{DL}^{max}}, \qquad (3.19)$$

where S_{DL} is the downlink time–frequency–space resources consumed by all the users and S_{DL}^{max} is the total time–frequency–space resource available on a downlink frame.

The normalized consumed downlink transmission power resources can be written as the ratio p_{DL}:

$$p_{DL} = \frac{P_{DL}}{P_{DL}^{max}}, \qquad (3.20)$$

where P_{DL} is the downlink power resources consumed by all the users and P_{DL}^{max} is the maximum power resource constraints.

The expression of the downlink load can then be defined as a nonlinear function of the two above quantities and can be power or frequency resources limited:

$$U_{DL} = \max(s_{DL}, p_{DL}). \qquad (3.21)$$

The downlink load can then be obtained by solving the optimization task:

$$u_{DL} = \min_{q,p,r}(U_{DL}), \text{ such that } p_{i,j,l}^{DL} \geq p_{i,j,l}^{QOS}, \qquad (3.22)$$

where q is the set of the coding and the modulation and spatial schemes assigned to the users in the downlink frame, p is the set of transmission power values assigned to the users in the downlink frame and r is the set of subcarriers time slots assigned to the users in the downlink frame. $p_{i,j,l}^{DL}$ is the transmission power for the service flow j of the user i on the downlink subcarrier l and $p_{i,j,l}^{QOS}$ is the minimum transmission power value that shall be assigned to the service flow j of the user i on the subcarrier l to satisfy the Quality of Service requirements for this service flow.

Optimization problem (1) can be solved thanks to fast suboptimal algorithms and the obtained estimation can be used in admission control algorithms or radio resource allocation algorithms. The packet scheduling aspect is detailed in Section 3.8.14.

3.2.6 SC-FDMA: The PAPR (Peak-Average-Power-Ratio) Problem

As OFDM is a multicarrier technique, and despite its benefits, it basically suffers from a number of drawbacks, including a need for an adaptive or coded scheme to overcome spectral nulls in the channel, and a high sensitivity to frequency offset and the high peak-average-power ratio (PAPR), which is a problem due to constructive addition of subcarriers on a random basis (Figure 3.7).

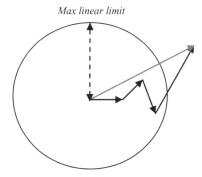

Figure 3.7 The PAPR problem on the Fresnel diagram.

The effect of a high PAPR on the transmitted OFDM symbols results in a spectral spreading (interferences between adjacent channels) and a higher BER (bit error rate) because of mistakes in the constellation.

On the downlink, this problem may be more easily overcome than in the uplink by using some high compression point PA (Power Amplifier) and sophisticated PAPR reduction mechanisms that are much more difficult to use in low processing power devices.

That is the reason why in the uplink, as it is crucial to reduce the cost of power amplifier directly linked to its compression point, some advanced scheme of OFDM called 'Spread OFDM' or SC-FDMA for 'Single Carrier FDMA' have been proposed.

This scheme basically consists of spreading the symbols of the uplink user on a group of subcarriers (contiguous or not) through the use of a unitary transformation F, 'unitary' meaning in the classical sense $F.F^{-1} = I$.

By doing this, the peak on average ratio of the signal is reduced to the PAPR of only one single carrier compared to the one resulting from association of the N subcarriers.

Concerning the F transformation, a Walsh-Hadamard transform (WHT) or a DFT transform provides some good choices. DFT or "Discrete Fourier Transform" is the generic name for the discrete version of the Fourier Transform, whose fast computing version is the FFT.

If a DFT is chosen, this technology evolution is sometimes called SC-SOFDM, where SC stands for 'single carrier'.

This option has been selected for the uplink side of E-UTRAN. Single carrier modulation should be followed by a frequency domain equalization and the complexity of SC-FDMA systems is the same as the overall complexity of OFDMA systems. A simplified SC-FDMA transceiver is shown in Figure 3.8.

In fact, the main difference with OFDM is that the signal is precoded before the IFFT in order to map on the subcarriers not the symbols themselves, but their spectrum components, introducing some small losses but helping to reduce this crucial PAPR level.

The DFT size M is a symbol block size and is much smaller than the IFFT of size N, which relates to the number of subcarriers available. Then, the resulting OFDM signal is over-sampled at a rate N/M compared to standard OFDM (see Figure 3.9). The signal becomes then M periodic and the negative effects of the phase constructive addition are averaged. The result is equivalent to an over-sampled single carrier ('S' in SC-FDMA stands for single) with a sin$(x)/x$ pulse shaping.

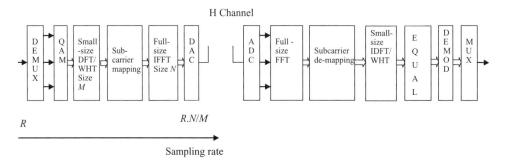

Figure 3.8 A simplified SC-FDMA transceiver.

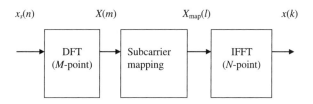

Figure 3.9 SC-FDMA transmit symbol.

In the time domain, when the process of subcarrier mapping is equidistant, the SC-FDMA is equivalent to a periodical distribution of the symbols over the equivalent time of the OFDM symbol with a user-dependent phase shift. Figure 3.10 provides a time domain interpretation.

As we explained in the OFDMA case, the allocation of the subcarriers can be made localized or distributed on the uplink band used by the terminal transmitters (see Figure 3.11).

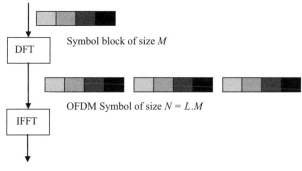

User-specific subcarrier allocation: distributed of localized

Figure 3.10 The SC-FDMA time domain interpretation.

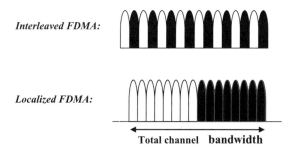

Figure 3.11 Subcarrier mapping modes.

Indeed, if we note $x_S(n)$ as the input symbols, $X(m)$ the signal after DFT, $X_{map}(l)$ the signal after mapping and $x(k)$ the signal after IFFT, and if we omit normalization factors, we can deduce the following for the two cases:

- Interleaved FDMA (IFDMA) case.
- Localized FDMA (LFDMA) case.

(i) Interleaved FDMA(IFDMA) Case

$$x(k) = \sum_{l=0}^{N-1} X_{map}(l) \cdot e^{\frac{2\pi jlk}{N}} \qquad (3.23)$$

and

$$x(k) = \sum_{m=0}^{M-1} X_{map}\left(m \cdot \frac{N}{M}\right) \cdot e^{\frac{2\pi jmkN}{N \cdot M}}. \qquad (3.24)$$

Therefore, as for the interleaved case,

$$X_{map}\left(m \cdot \frac{N}{M}\right) = X_{map}(m \cdot L) = X(m), \qquad (3.25)$$

we find

$$x(k) = \sum_{m=0}^{M-1} X(m) \cdot e^{\frac{2\pi jmk}{M}} = x_s(k), \qquad (3.26)$$

as $X(m)$ results from the direct transform of $xs(k)$, and, since $e^{\frac{2\pi jmk}{M}} = e^{\frac{2\pi jm(k+q \cdot M)}{M}}$, we find

$$x(k + q \cdot M) = \chi_s(k) \text{ for } k = 0 \ldots M-1 \text{ and for } q = 0 \ldots N/M-1, \qquad (3.27)$$

indicating that the users can be separated by assigning to each one a different phase shift.

Here, we have also to state that in SC-FDMA, the mapping of pilot symbol is not so straightforward compared to OFDMA, as one tone does not contain one symbol, but M.

Therefore, the symbol vector of length M at the input of the first DFT cannot contain both data and pilots, and the pilot symbols are necessarily time multiplexed.

Dealing with the PAPR, it is in this case determined by the number of users $M = N/L$ instead of the number of used subcarriers and will therefore, in most cases, be lower than for a conventional OFDMA system without additional spreading. In the case of $M = N$, of course, we find the classical OFDMA system. If the user data is spread over $M < N$ subcarriers, then the DFT of length M used for spreading does not directly cancel out with the length N IFFT of the OFDM modulator in general. Only when the spread symbols are mapped onto equidistant located subcarriers with a spacing $N/L = M$ can the DFT spreading and the OFDM modulation be removed in the transmitter structure.

(ii) Localized FDMA(LFDMA) Case

The expression of the time domain symbols on a given subcarrier is much more complex in this case, except for the first one of each of the L frames of M symbols.

We have

$$x(k) = \sum_{l=0}^{N-1} X_{map}(l) \cdot e^{\frac{2\pi jlk}{N}} \qquad (3.28)$$

and

$$x(k) = \sum_{m=0}^{M-1} X(m) \cdot e^{\frac{2\pi jm(L \cdot k + q)}{L \cdot M}}. \qquad (3.29)$$

Therefore, for $q=0$, we find that

$$x(k.L) = x_s(k) \qquad (3.30)$$

qnd, for $q \neq 0$, we find that

$$x(k.L + q) = \frac{1}{L}\left(1 - e^{j2\pi\frac{q}{L}}\right) \cdot \frac{1}{M} \cdot \sum_{p=0}^{M-1} \frac{x_p}{\left(1 - e^{j2\pi\left(\frac{k-p}{M} + \frac{q}{LM}\right)}\right)}. \qquad (3.31)$$

This expression indicates that, in the time domain, the localized FDMA signal has exact copies of input time symbols in the M-multiple sample positions. In-between values are the sum of all the time input symbols in the input block with different complex weighting.

(iii) Summary of Differences between OFDMA and SC-FDMA

SC-FDMA and OFDM have the same link-level performances. The data are sent in parallel on several subcarriers for OFDM, with a high PAPR result. The data are sent in a serial way on a single carrier for SC-FDMA, allowing a smaller PAPR.

For SC-CDMA, since the signals arrive at the eNodeB with substantial Inter-Symbol Interference and because SC-FDMA uses single carrier modulation, block equalization is performed at the receiver to cancel the effect of the radio channel over the received symbol and makes it nonsensitive to spectra nulls.

Whereas OFDMA is prone to Inter-Carrier Interference due to narrow subcarriers, SC-FDMA is not sensitive to frequency offset (Doppler) because of its single-carrier nature.

Practically, IFDMA is more desirable than LFDMA when choosing a subcarrier mapping method because it leads to a slightly reduced PAPR; however, so far, only the localized FDMA has been considered for the uplink solution of E-UTRAN.

3.2.7 Dimensioning an OFDM System

The purpose of this high-level study is now to look at the feasibility and basic dimensioning numbers of an OFDM radio system in some E-UTRAN bands. The values provided below are only given for example and are not directly related to the ones chosen for standardization purposes.

Key hypothesis are:

- Fc (carrier) around a few GHz.
- Small spectrum available <5 MHz.
- Max speed around 300 km/h.
- Propagation environment: urban.
- Services to accommodate: data up to 10 Mb/s.

In the following, we will choose $f_c = 2000$ MHz. We will suppose that:

- MAC efficiency layer is $R = 85\%$.
- Estimated modulation efficiency: efficiency = 3 bit/s/Hz (achievable trade-off).
- Propagation environment: urban, rural: delay spread = $T_d = 10$ μs.
- Application rate: 10 Mb/s.

(i) System Calculations
Required bandwidth:

- B = Mac-rate/eff = Data-rate/(R*eff) = 10/(0.85*3) = 3.9 MHz.

Thus, the minimum sampling frequency is 3.9 MHz and the sample duration is $T_s = 0.25$ μs.
Physical data:

- $V = 83$ m/s; $f_c = 2000$ MHz; $B_u = 5$ MHz could be available for future wireless multimedia services: $c = 3.10^8$ m/s; $T_d = 10$ μs typical.

Results:

- The Doppler spectrum is $B_d = 2^*f_d = 2^*f_c^*v/c = 2.2000.10^6.83/3.10^8 = 1106$ Hz.
- The coherency time is $T_c = 1/B_d = 0.9$ ms.
- The coherency bandwidth $B_c = 1/T_d = 100$ kHz, i.e. there is a lot of frequency diversity in the 5-MHz bandwidth. Therefore, something like 50 frequency pilots might be required.

(ii) Dimensioning Study
The guard time is supposed to null the ISI. Let D be the number of guard samples. Then, $D.T_s > T_d$. A classical choice (DAB system, HyperLan2 ...) is $D > 4.T_d/T_s$.

Let N be the number of useful samples. The channel is not supposed to vary a lot with respect to the OFDM symbol, which is dependent on the mobility of the terminal.

Assuming that D is of the order of T_d/T_s not to too lose to much useful data rate, then we have $D = 10/0.25 = 40$.

It is also generally assumed that $D/N = 1/4$; then, $N = 160$. The gap between subcarriers is $B_u/N = 5.10^6/160 = 31.25$ kHz.

It can be verified that:

- This inter-carrier spacing has usually to be much smaller than the coherency bandwidth in order to stay under a flat fading hypothesis.

- This inter-carrier spacing is much higher than the maximum Doppler frequency in order to counter the Doppler effect.
- The inter-carrier spacing has also to be compatible with the phase noise of the oscillators so that the noise created does not jam too much the neighbouring subcarriers.

The fact that the fading is slow or fast with our hypothesis really depends on the speed of the terminal. At 300 Km/h, we can then check that $(N+D).T_s = 200.25.10^{-8} = 0.05$ ms, which is far below the coherency time $T_c = 0.9$ ms of the channel.

Indeed, to remain under the flat fading hypothesis, it is better to keep the OFDM symbol duration far below the coherency time of the channel.

3.3 MIMO (Multiple Input Multiple Output)

3.3.1 Traditional Beamforming

Among traditional methods of doing antenna processing, one can mention transmit diversity, receive diversity or beamforming, which can be uplink and/or downlink (Figure 3.12).

None of these methods is handling multiple bit streams in the way that modern MIMO systems do.

With these traditional methods, the channel capacity is still bounded by the Shannon limit: $C = \log_2(1 + \text{SNR}.T.R)$ bits/s/Hs, in which T and R are the number of transmit and receive antennas, respectively.

Beamforming downlink or uplink still suffers some capacity limitation as compared to MIMO with multiple bit streams: channel capacity is still low and fading depth still large.

The paragraph below recalls the basics of beamforming principles. Such principles can still be applied on OFDM signals, although these are broadband signals. Indeed, narrowband

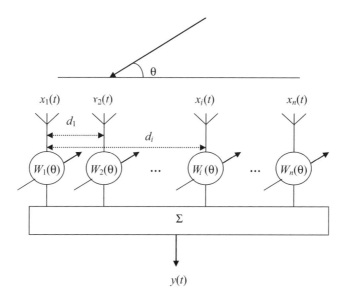

Figure 3.12 Rx beamforming.

beamforming can be implemented on individual subchannels. Notice that, since the broadband channels are frequency-selective, the spatial channel characteristics vary from subcarrier to subcarrier. By default, different beamforming vectors will be needed on different subchannels.

> **Adaptive Rx beamformer example**
>
> The goal of a network of antennas with a formation of beams is to recreate at the exit $y(t)$ of the network the conditions of optimal combination, that is to say in simple terms: to add the useful signal amplitudes in a coherent way (in phase), while noises, if they are non-coherent, can only be added in power.
>
> Thus, knowing that one receives N signals on N sensors coming from a $s(t)$ source, and knowing that each sensor receives the signal with a phase difference of $2.\pi.d.\cos(\theta)/\lambda$ compared to the neighbouring sensor spaced by D.
>
> Let us note d_i the distance from sensor i to a reference point expressed in a number of half-lengths of wave.
>
> We have $\mathbf{y}(t) = \mathbf{w_B}^H.\mathbf{x}(t)$ with $\mathbf{w_B} = [w_1(\omega), w_2(\omega) \cdots w_n(\omega)]^\#$ and $\omega \pi \cos(\theta)$ and $\mathbf{x}(t) = [x_1(t), x_2(t), \ldots, x_M(t)]^H$.
>
> By definition, the weights w of the network are selected in order to direct a beam in a specific direction θ, thus:
>
> $$w_i = (1/(n)^{1/2}).e^{-j.\pi.di.\cos(\theta)} = (1/(n)^{1/2}).e^{-j.di.\omega}.$$
>
> The output power is then $P_B = \mathbf{w_B}^H.\mathbf{R}.\mathbf{w_B}$, where $\mathbf{R} = <\mathbf{x}(t).\mathbf{x}^H(t)>$ is the $(n \times n)$ covariance matrix of the received signal.
>
> We can also find the expression of the weights w_B with other criterion. The weights defined above (w_B) present the disadvantage not to minimize the contributions of other signals in directions other than the useful signal's one; this is why we can prefer sometimes the estimator known as 'Capon' which minimizes $\mathbf{w}^H.\mathbf{R}.\mathbf{w}$ with $|\mathbf{w}^H.\mathbf{w_B}| = 1$ (the gain according to the 'useful' direction is constant).
>
> In this case, we get $\mathbf{w_c} = \mathbf{R}^{-1}.\mathbf{w_B}/(\mathbf{w_B}^H.\mathbf{R}^{-1}.\mathbf{w_B})$ and $Pc = 1/(\mathbf{w_B}^H.\mathbf{R}^{-1}.\mathbf{w_B})$.

The classical beamforming or SDMA (Signal Division Multiplex Access) is a way of increasing the cell capacity through the improvement of the data rate of the users especially located at the cell edge, e.g. the mobiles for which the available signal-to-noise ratio would not allow to have such a data rate with a MIMO multiplex scheme.

3.3.2 MIMO Channel and Capacity

This technology is based on the use of a multiplicity of antennas at the base station and/or in the mobile (which can be a portable computer). In the same way as for traditional 'beamforming', the principle of MIMO is to separate the users, depending on a different space signature experimented by each of them.

However, in some MIMO cases, the antennas radiate in generally different symbols and can be separated by a distance of about a few tens of wavelengths, sufficient to try out independent signals.

In contrast, traditional beamforming antennas are made up of radiating elements generally spaced by less than half of the wavelength to avoid the gratings lobes.

In this section, we consider a transmission channel between a transmitter equipped with T transmitting antennas and a receiver equipped with R receiving antennas. In a system with transmission space diversity, the number T is higher than 1, whereas the number R can be equal or higher than 1. The following analysis also includes the number of p physical multipaths available in the channel. This is the very general case of MIMO, and this way of considering the multipaths is valuable for the technologies for which the receivers are able to discriminate them, like the rake receiver in CDMA systems.

In what follows, a sequence of complex symbols $s_i(t)$ ($1 \leq i \leq n$) is transmitted. The signal $z_j(t)$ collected by the jth reception antenna ($1 \leq j \leq R$) can be written as

$$z_j(t) = \sum_{i=1}^{n}[s_i(t) \otimes h_{ij}(t)] + n(t), \qquad (3.32)$$

where $\otimes$ indicates the operation of convolution, and $n(t)$ indicates a supposed white and Gaussian noise. In a cellular system, the noise $n(t)$ contains contributions relative to other users of the system.

The impulse response $h_{ij}(t)$ of the channel of propagation between the ith transmitting antenna and the jth receiving antenna is classically estimated by the receiver thanks to known pilot sequences respectively emitted by T transmitting antennas.

It is generally modelled by a set of p paths taken into account by couple of antennas ($p \geq 1$), the kth path ($1 \leq k \leq p$) corresponding to a reception delay of τ_k and a complex received amplitude reception a_{ijk}. Each channel of propagation (ith transmitting antenna towards the jth receiving antenna) is associated to a vector of p amplitudes $H_{ij} = [h_{ij1} \quad h_{ij2} \quad \cdots \quad h_{ijp}]^T$ in which T is the transposition operator.

The received signal consists of a vector $\mathbf{Z} = [z_{11} \quad \cdots \quad z_{1p} \quad \cdots \quad z_{R1} \quad \cdots \quad z_{Rp}]^T$, where z_{jk} indicates the output of the matched filter relating to the antenna j, sampled with the delay τ_k. Thus, at a given time, the system of equations is obtained:

$$\mathbf{Z} = \mathbf{HS} + \mathbf{N}, \qquad (3.33)$$

where

$$\mathbf{H} = \begin{bmatrix} H_{11} & \cdots & H_{T1} \\ \vdots & \ddots & \vdots \\ H_{1R} & \cdots & H_{TR} \end{bmatrix} \qquad (3.34)$$

is a matrix representative of the multipath channel with $R.p$ lines and T columns, $\mathbf{S} = [s_1 \cdots s_n]^T$ is a vector containing n symbols transmitted at the considered time from the T transmitting antennas and N is a vector of noise of size $R.p$.

The system (2) is of a form very usually met in signal processing and can be easily solved by a traditional method of least squares estimation (MMSE, 'minimum mean squared error'), provided that the rank of the matrix $\mathbf{H}$ is at least equal to T. The solution of MMSE is written

$$\hat{\mathbf{S}} = (\mathbf{H}^*\mathbf{H})^{-1}\mathbf{H}^*\mathbf{Z}. \qquad (3.35)$$

By supposing that the antennas are not perfectly correlated, the row of the matrix **H** is generally equal to the smallest of T and $R.p$. The necessary and sufficient condition to solve the system by the MMSE method is then $R.p \geq T$.

Once this condition is checked, it is possible to solve the system according to the MMSE method or another method such as MLSE ('maximum likelihood sequence estimate'), which can also be applied if $R.p < T$, although less stable and sensitive to noise.

Generally speaking, the performances of the receiver depends on the conditioning of the matrix of the channel H, which depends on the number R of receiving antennas, on the number p of paths and on the decorrelation properties of the antennas.

In the case of OFDM systems, as pointed out before, there is an inherent resistance to multipath-generated intersymbols distortion, due to the length of modulation symbols. Natural multipath diversity can no more be used. The somehow artificial paths created by the MIMO antennas provide for OFDM systems a simpler (compared to CDMA rake receivers) and efficient way of getting some spatial diversity. Considering only OFDM systems in what follows, we will set $p = 1$.

Correlated antennas cause bad conditioning owing to the fact that matrix $\mathbf{H}^*\mathbf{H}$ then has eigenvalues close to zero, which disturb its inversion in the resolution according to Equation (3.25).

That is why the operator of MIMO system antennas arranges them so that they experiment enough decorrelation, by spacing them sufficiently and/or by making them radiate according to different polarizations, as, in practice, the receiver being generally mobile and of small size, it is common that the number R of receiving antennas is limited to 1 or lower than the number T of transmitting antennas (see Figure 3.13). This is why it is thought that in the future, mobiles will advantageously obtain several antennas with diversity of polarization, for example. In the known MIMO systems with, for instance, $T \geq 2$ and $R \geq 2$, one seeks to increase the rate of communications for a given emitted power, while transmitting different symbols $s_1, \ldots, s_n$ through the T transmitted antennas. These symbols can be mutually correlated, if they result from a space–time coding, or independent.

At the performance level, the fact of having several antennas within the terminals brings much profit compared to the only one antenna case (MISO system: Multiple Individual Input Output or SIMO system: Single Input Multiple Output), which were the well known ones in

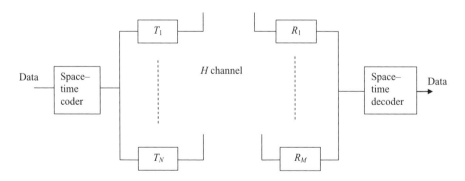

Figure 3.13 General MIMO system.

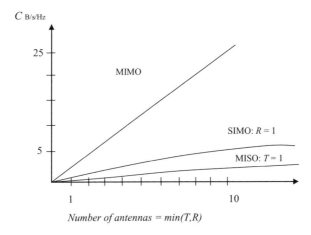

Figure 3.14 Performances of MIMO systems.

cellular radios for downlink and uplink, respectively (see Figure 3.14). It was shown in the mid-1990s that the capacity of a MIMO system was

$$C_{MIMO} = \sum_{i=1}^{\rho} \log 2\left(1 + \frac{P_T}{\sigma^2 \cdot T} \lambda_{H,i}^2\right), \tag{3.36}$$

with $\lambda_{H,i}$ the square roots of the eigenvalues of $\mathbf{H}^*.\mathbf{H}$, $\rho = \min(T,R)$, which can be compared with the famous Shannon equation for one channel:

$$C = \log 2\left(1 + \frac{1}{\sigma^2} P_T.|h|^2\right), \tag{3.37}$$

in which h is the complex gain of the channel and σ^2 is the noise variance. This can also be compared with the Equation (3.26) in the beamforming case.

Compared to a single antenna case or beamforming one, the capacity grows linearly with ρ, instead of logarithmically for classical transmission.

This phenomenon is properly the amazing one with MIMO systems, as it is basically stating that it is possible to increase the capacity of the system as a linear capacity of the number of antennas **without increasing the overall transmit power** of the system.

One can show with Rayleigh fading, uncorrelated antennas and a signal/noise ratio of 10 dB the curves in Figure 3.14.

Note on practical realization

In order to simplify, we assume that we are under noise-free conditions, so that the system in Equation (3.27) can now be written:

$$Z = \mathbf{H}.\mathbf{S}.$$

Provided that the MIMO system is not ill-conditioned, i.e. det(H) ≠ 0, and if **H** is a square matrix (**R** = **T**), the solution to Equation (3.27) is obtained in an elementary manner as:

$$S = H^{-1}.Z.$$

In practical applications, however, the number of channel outputs may be different from the channel inputs and an appropriate inverse, known as the Moore-Penrose pseudoinverse, needs to be calculated; instead of H^{-1}, we have:

$$H^+ = (H^*.H)^{-1}.H^*.$$

For a huge quantity of samples per column and under real-time processing conditions, calculation of the inverse or pseudo inverse matrix may require too many operations and computer load; therefore, methods have been developed to simplify such a task, like the QR decomposition.

We have **Z** = **H.S** and need to find **S**.

It can be shown in basic linear algebra that if **H** is a symmetric real matrix, it can be decomposed into the product of one orthogonal matrix **Q** (so that $Q.Q^T = I$) and one upper-triangular matrix **R**. Then, assuming **Q** and **R** are known:

$$Z = Q.R.S.$$

Let us note **A** = **R.S**; we then have:

A = **R.S** = $Q^{-1}.Z = Q^T.Z$ because **Q** is orthogonal and $Q^{-1} = Q^T$.

Therefore, calculating **A** is fast and only requires a few operations. The calculation of **S** is then straightforward, as **R** is upper-triangular and can be done by substitution by degrees.

The determination of **H** itself is the object of a well known identification method and can easily be done using the following steps:

- Find an orthonormal **U** base of a known input **S** (previous **Q** matrix is OK).
- Calculate $F_x = U.S$, the projection coefficient of **S** with respect to **U**.
- Calculate $F_y = U.Z$, the projection coefficient of **Z** with respect to **U**.

We find:
$Z = U^{-1}.F_y$ and

$$H.S = H.U^{-1}.F_x$$

$<H> = F_y.F_x^+$ providing the expected estimate of the MIMO channel.

3.3.3 A Simplified View of MIMO 2.2

A simplified vision of a MIMO system is shown in Figure 3.15. In practice, in order to be able to estimate the channel on each **R.T** 'natural' paths of the system of antennas, the receiver will have to benefit from different bits control transmitted on each transmitting antenna.

Once the matrix of transfer **H** is determined, it becomes possible by simple inversion to find the transmitted symbols.

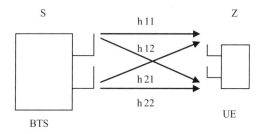

Figure 3.15 Simplified vision of a MIMO system.

Let us consider the simple example according to a system with two transmitting antennas and two receiving antennas. The **H** matrix is written as

$$H = \begin{bmatrix} h11 & h12 \\ h21 & h22 \end{bmatrix}. \tag{3.38}$$

Let us call **S** the vector of the emitted symbols and **Z** the vector of the received symbols. Then

$$\mathbf{Z} = \mathbf{H}.\mathbf{S} \tag{3.39}$$

and, from Equation (3.39),

$$\mathbf{S} = \mathbf{H}^{-1}.\mathbf{Z} \text{ (granted that } \mathbf{H}^{-1} \text{ exists)}, \tag{3.40}$$

which is the expected result.

There are obviously some cases for which the distribution (with equalized power) of symbols on the *T* antennas is not really useful in terms of data rate increase.

It is, for instance, the case when the channel only creates a few multiple paths, like the case when the mobile is in direct sight of the base station. In such conditions, it is then preferred to use another form of MIMO, named Alamouti's scheme.

3.3.4 The Harmonious Coupling between OFDM and MIMO

As previously stated, The OFDM systems being not able like the CDMA systems to use the multipaths delay diversity can exploit advantageously the spatial diversity brought by MIMO systems with a low additional complexity.

It has been shown previously that the OFM demodulation system was equivalent to the set of equations

$$Y_i = H_i S_i + n'_i, \quad i = 0, ..., N-1. \tag{3.41}$$

That is equivalent to *N* tone scalar channels to demodulate.

In the MIMO case, with *T* transmitting antennas and *R* receiving one, the system becomes as in Section (3.3.1):

$$Y_i = H_i S_i + N'_i, \quad i = 0, ..., N-1, \tag{3.42}$$

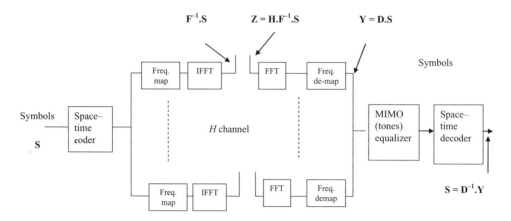

Figure 3.16 A MIMO–OFDM transceiver.

where Y_i and N_i' are vectors of R components and S_i is a vector of R components. H_i is the channel matrix of size **R.T**.

Therefore, granted the small size of the **H** matrix, only a small system (**R.T**) has to be inversed per tone. Moreover, space–time schemes can be implemented on a tone-per-tone basis.

Figure 3.16 generalizes the MIMO schemes with the notations used in the above formulas coming from Section 3.3.1. and neglecting the thermal noise.

3.3.5 MIMO: A Classification Attempt

According to a chronological approach of the apparition of the MIMO scheme, we can state the following:

- Time–space multiplex: this consists of operating a sequence of information without any space and time redundancy. The mathematical operation is a sort of space–time interleaving. The system then transmits T useful symbols in each symbol period. BLAST (Bell Labs Layered Space–Time) and its variants belong to such a category with its variants.
- Space Time Trellis Coded Modulation (STTCM): this a generalization for the MIMO case of what already existed for the SISO channel. The transmitter consists of a battery of shift registers and algebraic functions creating a codeword of T symbols simultaneously transmitted. The receiver is usually using a Viterbi algorithm to find the path with the best likelihood.
- Orthogonal Space Time Block Coded Modulation (OSTBCM): the complexity of the decoding of the previous STTCM modulation has limited its expansion, but while working on some way of reducing this complexity, Alamouti, whose scheme is provided below, found some ways to simplify the decoding by using an orthogonal structure of coding. Some real patterns exist and some complex ones as well. Real ones are restricted to 2, 4 and 8 spatial dimension. This type of coding optimizes the diversity gain (the slope of the BER vs C/I curve) but offers no coding gain (a C/I offset of the BER $=$ f(C/I) curve). Very often, the 'STTD' (Space Time Transmit Diversity) name is used to designate the Alamouti's family of solutions.

- Space–Time Coded Modulations with maximum diversity: this improves the former types of modulations in rotating the constellation points, optimizing the provided diversity at a higher price for the decoding.
- Linear Dispersion Coded Modulation (LDCM), Algebraic Coded Modulation and Code Concatenation can also be mentioned but the studies of their properties fall outside the scope of this book.

3.3.6 Some Classical Open Loop MIMO Schemes

Let us focus on the open loop MIMO scheme, whose advantages are easier implementation, less signaling overhead and nonsensitivity to speed.

In the sections above, we described the MIMO scheme which allows increasing the data rate while taking benefit from the multiplex allowed by the different spatial 'modes' proposed by the channel.

This mode is known as the BLAST (Bell Lucent Layered Space Time coding) mode and its variants. The simplest version of BLAST is known as V-BLAST (Vertical BLAST), whereas more complex variant exists like D-BLAST (Diagonal BLAST). The essential difference between D-BLAST and V-BLAST lies in the vector encoding process. In D-BLAST, redundancy between the substreams is introduced through the use of specialized intersubstream block coding. The D-BLAST code blocks are organized along an elegant diagonally layered coding structure in which code blocks are dispersed across diagonals in space–time. It is this coding that leads to D-BLAST's higher spectral efficiencies for a given number of transmitters and receivers. In V-BLAST, however, the vector encoding process is simply a demultiplex operation followed by independent bit-to-symbol mapping of each substream. No intersubstream coding, or coding of any kind, is required, though conventional coding of the individual substreams may be applied.

Other methods benefit from space diversity without increasing the data rate of the communication. It is, for example, the case of the diagram of the space–time diversity coding scheme (STTD, 'Space–Time Transmit Diversity') from Alamouti's work, already known within the framework of the cellular networks of the third generation with a single receiving antenna.

Classically, the way the current wireless cellular systems choose their MIMO schemes is part of link adaptation algorithms and can be done very dynamically, depending on the radio characteristic of the propagation channel.

(i) Rank of the Channel Correlation Matrix
One interesting notion is the notion of the rank of the space correlation matrix $\mathbf{K} = <\mathbf{H}.\mathbf{H}^*>$. This notion of rank can be seen in a purely spatial way. This notion is roughly equivalent to the notion of possible excitation energy modes inside a waveguide, in which, depending on its physical dimension, only some propagation modes can be generated or not.

The lower bound of the capacity of a MIMO system is obtained when there is only one single propagation path between the emitter and the receiver: Rank $(\mathbf{K}) = 1$.

Opposite to this, the upper bound is obtained when the system benefits from all the possible propagation modes offered by the channel. In this case, Rank $(\mathbf{K}) = \min(\mathbf{R}, \mathbf{T})$.

The 3G STTD diversity: The Alamouti coding scheme

A diversity scheme, also known as STTD (Space Time Transmit Diversity), produces for each physical channel two sequences, S_1 and S_2, starting from a unique entry sequence S.

S_1 and S_2 are radiated by two different antennas and carry the same symbol suits. The two sequences S_1 and S_2 are produced so that when processed simultaneously, the demodulations of the symbols are optimal. From a sequence of symbols s_1, s_2, s_3, s_4, one transmit on the first antenna: s_1, s_2, s_3, s_4 and on the second antenna: $-s_3{}^*, s_4{}^*, -s_2{}^*, s_1{}^*$ (see Figure 3.17).

The receiver will receive the following symbols on three successive time intervals:

	Antenna 0	Antenna 1
Time T	S_1	S_2
Time $t+3T$	$-S_2{}^*$	$S_1{}^*$

It is supposed that fading in the channel is constant from one symbol to another. Still, let us suppose that the impulse response of the channel between the first antenna (resp. the second) A_1 (resp. A_2) and the mobile at the moment T is characterized by the complex number:

$$h_1(t) = \alpha_1.\exp(j.\theta_1) = h_1 \text{ (resp. } h_2(t) = \alpha_2.\exp(j.\theta_2) = h_2).$$

The signals received by the receiver can then be written:

$$R(t) = h_1.S_1 + h_2.S_2 + n_1 \text{ and}$$

$$R(t) = -h_1.S_2{}^* + h_2.S_1{}^* + n_2,$$

in which n_1 and n_2 represent the thermal noise and the interferences received by the receiver at the two successive moments. As the receiver receives two distinct symbols permanently coming from two different antennas, it can therefore estimate the two answers h_1 and h_2 of the channel and transmit them to a diversity combiner which gives:

$$<S_1> = h_1{}^*.r_1 + h_2.r_2{}^* (<> \text{ means 'estimate' of) and}$$

$$<S_2> = h_2{}^*.r_1 - h_1.r_2{}^*.$$

Consequently, we obtain the estimate of the signals by the following relations:

$$<S_1> = (\alpha_1^2 + \alpha_2^2).S_1 + h_1{}^*.n_1 + h_2.n_2{}^* \text{ and}$$

$$<S_2> = (\alpha_1^2 + \alpha_2^2).S_2 - h_1.n_2{}^* + h_2{}^*.n_1{}^*.$$

These relations show that the final result is improved by the contributions in terms of signal to noise by the signals from the two sending antennas, in the same way as a receiver processing two ways of diversity in optimal recombination (Maximum Combining Ratio). The only difference consists of a rotation of the phases of the components of noises which do not influence the signal with resulting noise.

When the receiver includes several receiving antennas, the above equations remain the same but h_i are replaced for each given tone by a **R.T** matrix which is easy to invert.

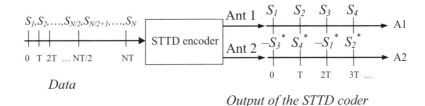

Figure 3.17 Alamouti's diversity scheme.

(ii) More on Orthogonal Space-Time Codes
The above-described Alamouti scheme, which is part of the OSTBCM family mentioned above, is characterized by its space–time code rate r, given by the ratio

$$r = \frac{K}{l}, \tag{3.43}$$

in which a number K of real or complex symbols is transmitted via N antennas over l symbol time intervals. For a M-ary modulation, we may then define the spectral efficiency of a space–time block code by the quantity

$$\eta = r \cdot \log_2(M). \tag{3.44}$$

From a general perspective, the patterns for such codes use symbols with the forms: $\{\pm S(n); \pm S(n)^*\}$.

For two antennas, as seen above, the scheme is:

- For real case: $S = \begin{bmatrix} S1 & S2 \\ -S2 & S1 \end{bmatrix}$.

- For complex case: $S = \begin{bmatrix} S1 & S2 \\ -S2^* & S1^* \end{bmatrix}$.

For the three-antenna case:

- For real case: $S = \begin{bmatrix} S1 & S2 & S3 \\ -S2 & S1 & -S4 \\ -S3 & S4 & S1 \\ -S4 & -S3 & S2 \end{bmatrix}$.

- For complex case: $S = \begin{bmatrix} S1 & S2 & S3 \\ -S2 & S1 & -S4 \\ -S3 & S4 & S1 \\ -S4 & -S3 & S2 \\ S1^* & S2^* & S3^* \\ -S2^* & S1^* & -S4^* \\ -S3^* & S4^* & S1^* \\ -S4^* & -S3^* & S2^* \end{bmatrix}$.

Here, the blocks of $K=4$ symbols are taken and transmitted in parallel using $N=3$ antennas over $l=8$ symbol periods. Therefore, the code rate is ½.

The principles of Alamouti's STTD diversity described above can be extended to the case when the receiver presents several antennas.

(iii) Open Loop Diversity Schemes Summary

To sum up, the MIMO systems may experiment with several types of scheme according to various optimization criteria:

- A mode (like BLAST type) which aims at increasing both uplink and downlink transmitted rates by using the additional degrees of freedom provided by the spatial modes available in the channel and then placing different symbols on each transmission antenna. In such a case, the richer the multipath profile, the higher the obtained rates.
- Some other modes [Alamouti's (STTD) or OSTBC type of variants] which target efficiency of transmission (less power for a given flow). These modes produce a diversity gain in emission if the channel does not bring it enough while placing in a more or less clever way the same symbols on different antennas.

Figure 3.18 provides a simplified vision of this two classical MIMO schemes.

Other types of diversity scheme – but not data-rate increasing – are also known, in which same symbols are transmitted on several transmitting antennas but with different codes and/or modulation MCS schemes (Modulation and Code Scheme). All of them have particular benefits.

3.3.7 Notions of Cyclic Delay Diversity (CDD)

While considering the open loop schemes for providing additional diversity to the radio channel, the following method has been selected for the E-UTRAN standards.

Cyclic delay diversity is part of the downlink of E-UTRAN. It is a simple approach that can be easily combined with a follower space coder stage, and which introduces a simple additional diversity without requiring any specificity inside the receiver.

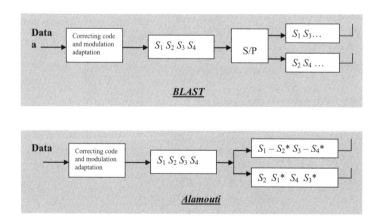

Figure 3.18 The classical MIMO schemes.

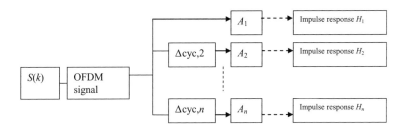

Figure 3.19 Cyclic transmit diversity scheme.

The different codewords result in a changed channel impulse response in the receiver as they insert some virtual echoes increasing frequency selectivity of the channel seen by this receiver.

Compared to a simple delay diversity scheme in which the signal is simply delayed from one antenna to another, thus being limited by the cyclic prefix time duration, the CDD provides a cyclical shift of the signal with no restrictions for the cyclic shift. The cyclic diversity scheme is given on Figure 3.19.

The OFDM initial signal to be transmitted can be written as

$$s(l) = \sum_{k=0}^{N-1} S(k) \cdot e^{\frac{2\pi j l k}{N}}, \qquad (3.45)$$

where N is the number of subcarriers available and $S(k)$ the symbol to transmit.

Then, the CCD signal can be written for the ith antenna port:

$$s((l-\delta_{cyc,i}) \bmod(N)) = \sum_{k=0}^{N-1} e^{\frac{-2\pi j k \cdot \delta_{cyc,i}}{N}} \cdot S(k) \cdot e^{\frac{2\pi j l k}{N}}. \qquad (3.46)$$

All these signals superimpose on the channels, thus creating a phase diversity effect that can be seen in the first term of the right member in the above equation.

It can be noticed that the signal can be shifted in the time domain before the OFDM transform, or in the frequency domain after the OFDM transform; in the scheme above, the shift can take any of the values between 1 and N.

On average, CDD does not increase or decrease the total number of errors received in an encoded OFDM system, but it changes the error distribution in a favourable way, yielding in increased coding gain.

3.3.8 MIMO Schemes and Link Adaptation

Generally speaking, the dynamic choice of the MIMO scheme is part of the global link adaptation process as well as the dynamic choice of the coding or modulation scheme, based on some feedback information linked to the quality or the nature of the channel. Therefore, there is a need for some adaptive switching algorithms between spatial multiplexing and diversity. The choice of a diversity mode or a spatial multiplexing mode (including classical beamforming, not shown in Figure 3.20) may be done, for instance, on

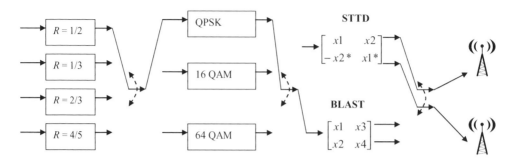

Figure 3.20 The global link adaptation process.

the statistical analysis of the signals measured by the terminals and periodically provided to the base stations.

Depending on the type of parameter to tune through the link adaptation process, several time constants may be used to ensure a good trade-off between algorithm complexity and measurement loads and performance optimization. As an example, for the downlink HARQ process in Section (3.8.13), the link adaptation may play very dynamically on the coding and modulation scheme for retransmission of the packets but may adapt the MIMO scheme with another time constant.

3.3.9 Improving MIMO with Some Feedback

We have described a simplified way of operation of these open loop systems, i.e. that the transmitter does not know the characteristics of the channel (H). It is clear that the performances of these systems can be significantly improved if the transmitter can benefit from a suitable feedback of the receiver to know the channel.

For instance, the mobile can estimate some weights to be applied by the transmitter to optimize the received signal. A channel quality indicator (CQI) is also useful to know for each of the possible subchannels.

For instance, similarly to UMTS FDD mode closed loop diversity, it becomes possible to allocate more or less power on given transmitting antennas, depending on whether they are privileged or not by the channel of propagation.

In general, simply said, it is better to increase the power of the antennas which are received best by the receiver. Respective phases can also be tuned.

These schemes allow reaching a near optimum capacity. They, however, require a huge feedback, sensitive to mobile speed, and therefore also present some drawbacks.

(i) The PARC Scheme

This scheme is a well known one and Figure 3.21 provides a general view of its implementation.

This is a closed loop scheme which is a BLAST variant. In this scheme, named PARC (Per Antenna Rate Control), a different coding and modulation scheme MCS is applied to each antenna stream. An individual CQI (Channel Quality Indicator) per channel is provided for each stream, so that individual rate adaptation is allowed.

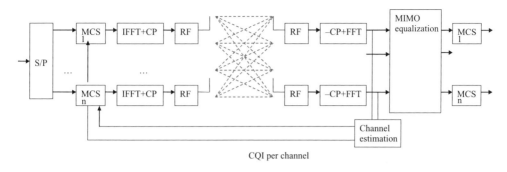

Figure 3.21 The PARC scheme.

(ii) Precoding

One concept proposed for E-UTRAN is known as the precoding method. This MIMO concept is a closed loop MIMO scheme where a set of precoding vectors (the codebook), arranged as columns in a precoding matrix, is applied at the base station to take advantage of the array gain and spatial interference suppression. The UE (the 3GPP name for mobile terminal) calculates one or several preferred precoding vectors based on the expected pattern of bits on the pilot channel, then finds the corresponding precoding vector index (PVI) described by a Q bit word in a look-up table with 2^Q entries and informs the base station about this. The base station collects the preferred index from all the users, performs the scheduling and assembles a precoding matrix with the columns corresponding to the precoding vectors of the scheduled users, to be used to precode the transmitted data.

The precoding vectors, roughly speaking, may be seen as phase shifts to be introduced on the signals transmitted from multiple antennas. This scheme is quite useful when the channel conditions on the forward link and the reverse link are quite different, which might be the case in FDD mode, where the downlink and uplink are several MHz apart.

We have

$$\mathbf{Z} = \mathbf{H}.\mathbf{W}.\mathbf{s} + \mathbf{N}, \qquad (3.47)$$

where $\mathbf{W}$ is the selected precoding matrix, $\mathbf{H}$ is the channel matrix which has to be known by the transmitter and $\mathbf{s}$ is the symbol vector.

The receiver will linearly compute an estimate of $\mathbf{s}$:

$$\mathbf{s}' = \mathbf{G}.\mathbf{H}.\mathbf{W}.\mathbf{s}. \qquad (3.48)$$

The choice of $\mathbf{W}$ is based on $\mathbf{H}$; thus, the transmitter has to be informed at least partly of $\mathbf{H}$ but limited feedback is possible, for instance, in:

- Restricting $\mathbf{W}$ to lie in a finite codebook and considering unitary precoder, i.e.

$$W^H W\ \mathbf{I} \qquad (3.49)$$

- Exploit coherence bandwidth of the channel.

Due to the channel frequency selectivity, these reported index and CQI are only valid for a certain limited frequency interval and, thus, there is a need to feed back PVI and CQI for all

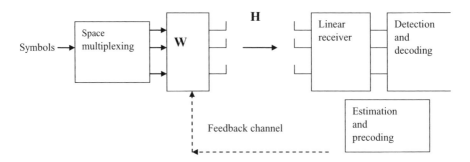

Figure 3.22 Closed-loop precoding concept.

the subchannels. However, the precoding scheme may be applied can be applied on a subchannel or on all the available bandwidths, as in Figure 3.23.

The optimum precoding matrix is the one which offers maximum capacity. The UE provides feedback on the uplink control channel regarding the preferred precoding matrix (precoding vector as a special case). Ideally, this information is made available per resource block or at least per group of resource blocks, since the optimum precoding matrix varies between resource blocks.

Note that instead of the terminal to feedback **W** matrix, also know as 'beamforming matrix', there exists some way of compressing the **W** matrix information, while using a codebook vector **u** which can provide the matrix through some classical Householder transform(s) $\mathbf{W} = \mathbf{I} - 2.\mathbf{u}.\mathbf{u}^H/\mathbf{u}^H.\mathbf{u}$. Such a method can be found in the last document provided in an annex dealing with technical documents.

These methods are based on the singular value decomposition theorem which operates as an operator of compression and decompression of the numbers contained in a matrix based on the number of its significant eigenvalues.

The codebook entries depend on the number of antenna elements in the base station, which must be known to the UE. Each precoding vector is associated to a group of other precoding vectors for which it has a spatial correlation less than some parameter ρ. Note that $\rho = 0$ corresponds to orthogonal precoding vectors and the precoding matrices then become (semi-)unitary. When a UE is scheduled for a certain subchannel using a certain precoding vector, this subchannel can only be reused using another precoding vector belonging to the same group, to assure low spatial intra-cell interference, i.e. another vector whose spatial correlation with the first one is less than ρ.

These rules have to be adopted by the scheduler. The grouping depends on the parameter ρ which can slowly be adapted by the base station according to the cell load. If a smaller ρ is selected, there are fewer options for the scheduler to re-use a subchannel, but the spatial intra-cell interference is lower. If a larger ρ is selected, more scheduling opportunities exist at the cost of increased intra-cell interference.

With such a scheme, a sub-band can be reused by different users (MU-MIMO for Multi-User MIMO). The MU-MIMO scheme can thus be seen as a spatial division multiple access (SDMA) scheme and it has large benefits in a macrocell where inter-cell interference is significant and the interference suppression by spatial precoding is useful, especially for cell edge users (Figure 3.24).

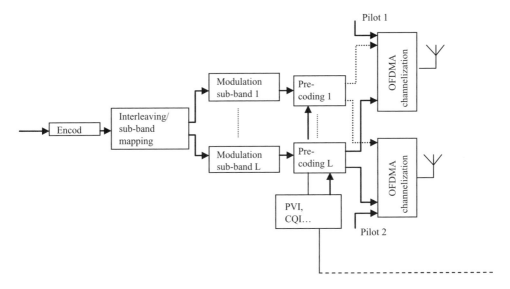

Figure 3.23 The precoding concept.

3.3.10 MU-MIMO, Virtual MIMO and Transmit Diversity

MU-MIMO consists of re-using the same physical resources in downlink or uplink for different users. This obviously requires adaptive algorithms to pair various users and switch between MU-MIMO mode and nonMU-MIMO modes. These algorithms are based on some feedback information about the degree of potential correlation the spatial path might experiment while re-using a given resource at a given power.

The result is an increasing spectrum statistical efficiency and a better average throughput in a cell.

Uplink MIMO schemes for E-UTRAN will differ from downlink MIMO schemes to take into account terminal complexity issues. For the uplink, MU-MIMO (Multi-User MIMO) can be used. Multiple-user terminals may transmit simultaneously on the same resource block. This is also referred to as spatial domain multiple access (SDMA). The scheme requires only one transmit antenna at the UE side, which is a big advantage

The realistic scheme to consider for the UE is the two-antenna case leading to a pragmatic 2×2 MIMO for which a lot of open or closed loop schemes can be considered.

It is interesting to notice that by comparison to a 2×2 MIMO with the two antenna supported by one mobile, a virtual MIMO can be defined as a mode for which the UE with two antenna is 'virtually' replaced by two UEs, each of them transmitting on a single antenna and using mutually orthogonal reference signals on their respective antenna. This scheme allows sharing for the two mobiles the same frequency and timing resource allocation. However, a pairing mechanism for the reference signals has to be planned between the two mobiles (see Figure 3.24).

To exploit the benefit of two or more transmit antennas but still keep the UE cost low, antenna subset selection can be used. In the beginning, this technique may be used, e.g. a UE will have two transmit antennas but only one transmit chain and amplifier. A switch will then choose the antenna that provides the best channel to the eNodeB.

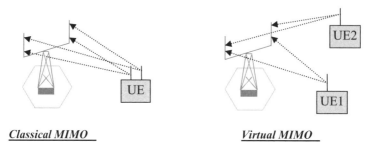

Figure 3.24 Classical vs virtual MIMO.

3.3.11 Towards a Generalized Downlink Scheme

In the rush to get the highest data rate possible, it can be thought that a generalized MIMO scheme will be envisaged in future standards for which each MIMO antenna port is not an antenna part of a base station configuration but an antenna part of a radio site, even if obviously the simpler case in which the transmit ports belong to a same NodeB is not precluded.

In this scheme, a centralized processor calculates a P-matrix which basically inverses the H-MIMO channel before the transmission. With the notation of Figure 3.22, we have:

$$\mathbf{P} = \mathbf{H}^*.(\mathbf{H}.\mathbf{H}^*)^{-1}. \tag{3.50}$$

(**P** is the Penrose inverse of the H channel) and

$$\mathbf{S}' = \mathbf{H}.\mathbf{P}.\mathbf{S} + n, \tag{3.51}$$

which should be close to **S**, in which 'n' is the thermal noise.

The calculation of **P** is effective if the number of users is less than the number of MIMO sites. If done all right, the co-channels interferences are totally removed.

Various types of spectrum efficiency can be achieved, depending on the level of synchronization among base stations, which may depend in some information exchanges on some interfaces between NodeBs partly shown in Figure 3.25 with dotted lines. The synchronization requirements are key to allow various types of spatial processing, from symbol synchronization for space–time coding or BLAST, to carrier phase processing for classical beamforming.

According to the performances and complexity wished at the system level, the scheme may requires noncoherent or coherent combining at the terminals level and limited or extensive cooperation between base stations.

Obviously, the knowledge of the channel feedback requires the processing of the feedback sent by all the terminals over all received channels.

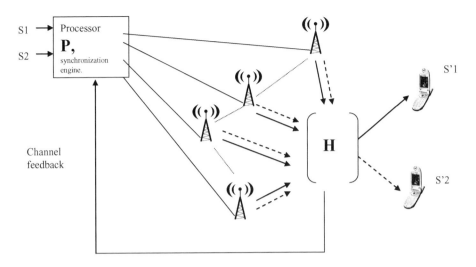

Figure 3.25 The generalized MIMO scheme.

In TDD mode, this estimation can be directly estimated on the uplink, as there is reciprocity of the downlink and the uplink channel conditions granted for the duration of the transmitted frames.

3.4 Architecture of the Base Station

3.4.1 The Block Scheme of the Base Station

Figure 3.26 presents an example of an E-UTRAN base station, able to support a large number of users. The architecture below is provided for illustrative purpose. This base station is connected to the backhaul network by means of physical interfaces like optical fibre, like STM1 of large data rate or even more classical E1/T1 2-Mb/s data link. It receives its user data information from the Serving Gateway through the S1 interface based on IP packet transport protocol.

(i) The Downlink Case
The downlink information coming from the MME/serving gateway to Node B reaches an IP routing module, typically called a Core Control Module (CCM). Then, each packet is routed to a modem – the Channel Element Module (CEM) – which is able to support all types of physical coding for any type of physical possible channels: pilots, common channels, dedicated channels, etc.

This module is able to treat N channels (flows) in parallel, performing for each the channel coding, the radio conditioning of the signal: OFDM modulation, cyclic prefix insertion, etc.

The processing of all the signals dedicated to one sector can happen in the channel module.

Each channel module then addresses the signals relative to a given sector to the routing module, whose output feeds a Transceiver Radio Module (TRM) connected physically to the relevant sector. According to some choices of architecture, a radio operator module can then address a sector at a given frequency or several sectors at several different frequencies.

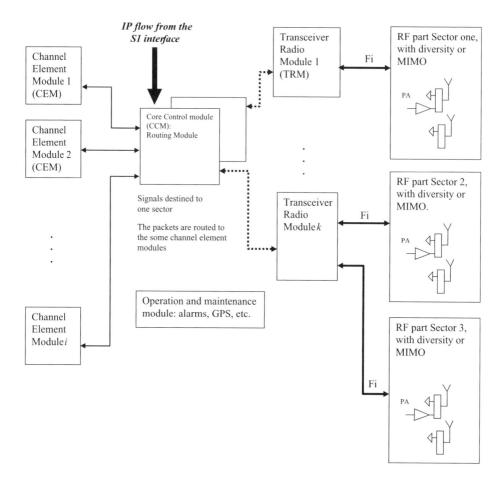

Figure 3.26 Architecture aspect for E-UTRAN base station.

The radio operator module in question is sometimes called a 'channelizer', whose primary goal is to over-sample the signals and shape them according to a chosen pulse filter, clip the total signals and modulate them at the frequency F_i of the radio carrier. It has to treat as many frequency bands as required by the operator.

In some cases, one may wish to share the power of one power amplifier (PA) on several sectors. It can then be done by distributing passively the output of the PA on each sector through microwave devices.

(ii) The Uplink Case

In uplink, each radio transceiver module receives the signals from the antennas of one or more sectors. For instance, one can associate a transceiver module to a given frequency band. The signals received in a sector are usually coming from several diversity antennas or from a MIMO or beamforming antenna array. Then, they are amplified, filtered and digitally converted by an analogical-to-digital converter.

They are then transmitted to the Core Control Module and routed to the Channel Element Modules, which include the respective modems performing the demodulation of the respective signals (cyclic prefix removal, FFT, etc.).

The signals demodulated by the channel module are then retransmitted to the IP network after demodulation by the routing module with the same transport protocol than the incident protocol.

(iii) The Distributed Base Station

Anticipating the evolution of the markets towards a product more easily installable (less weight, power consumption, etc.) and with lower engineering constraints, the base station manufacturers propose in their portfolio a distributed base station concept (see Figure 3.27), in which the TRM module and the RF part drawn in Figure 3.26 are physically localized in an outdoor packaging, named RRH (Remote Radio Head), and connected to the Core module by an optical fibre, or a RF link (dashed line on the above figure).

In that perspective, the Common Public Radio Interface (CPRI^{TM}), created in 2003, is an industry cooperation aimed at defining a publicly available specification for this internal interface of radio base stations between the Core module and the Radio Remote Head (RRH).

The CPRI consortium aims at defining a specification for Layer 1 (physical) and Layer 2 (transport) but keep the Layer 3 messages proprietary.

The data rate on the interface on one optic fibre is close to 1.2 Gb/s, allowing carrying of the Ethernet or HDLC signalling, operation and maintenance signalling and I/Q radio samples. The Core module can obviously feed numerous optic fibres like this one.

The RRH equipment can be daisy-chained, optimizing the overall installation for the best coverage of a given geographical zone.

3.4.2 The Analogue-to-Digital Conversion

The converters are one of the most important components in the base station. Their performances allows the conversion as close as possible of the antenna and to limit the number of RF expensive and frequency specific components.

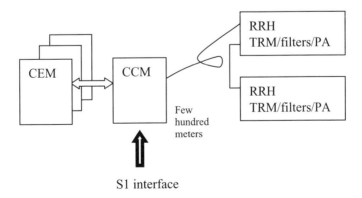

Figure 3.27 A distributed BTS scheme.

The key parameters of specification of the analogue-to-digital converter are: the sampling rate, the bandwidth, the signal-to-noise ratio and the dynamics of the signals.

During the design, two problems are particularly important to consider:

- The consumption of the energy in the terminals.
- The filtering of the harmonics of sampling (aliasing) before ADC converters.

With respect to the analogue-to-digital converter itself, the essential parameter is the signal-to-noise plus distortion ratio at the output side of the converter.

This one makes it possible to define the n-resolution of the converter in a number of effective bits by the relation

$$(\text{SNR})_{\text{dB}} = 6.02.n + 1.76. \tag{3.52}$$

'Today, the output performance of an ADC for radio-communication applications is in a range between 10^7 and 10^9 samples per second. For such a speed, the main limiting factor is clock time drift τ which creates an error on the sampled signal equal to half a quantification step $(1/2^{n+1})$. The maximum resolution n is thus

$$2^n = \frac{1}{\sqrt{3}.Fs.\tau}, \tag{3.53}$$

in which Fs is the input signal frequency. From the typical value of τ – around 0.5 ps – the maximum resolution can be deducted from the above equation.

The resolution equals 15 bits for 10 M-samples/s, then decrease with a slope of -1 bit/octave and reaches 7 bits at 2.5 G-samples/s. It is thus seen that this limits the performances of the ADC while, however, authorizing somehow the high-frequency acquisition of signals having an important dynamic.

The more common technique used in order to divide by 2 the sampling rate is to proceed as indicated in Figure 3.28, i.e. to carry out complex sampling by separating the received

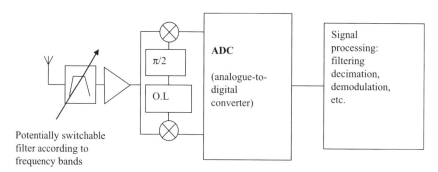

Low or null FI receiver

Figure 3.28 A digital receiver scheme.

analogue signal in two paths shifted by 90°, each one processing half of the original bandwidth. This obviously requires two converters.

Another promising technique consists of directly under-sampling the received signal at the useful bandwidth of the signal. The useful band is thus 'folded up' in a frequency band close to zero.

Certain constraints exist in the choice of the sampling rate in order to avoid a certain aliasing of the spectrum; however, once this carried out, one obtains a sufficiently low flow of bits able to be treated by DSPs (Digital Signal Processors).

3.4.3 Power Amplification (PA) Basics

This topic is really important in the design and practice of current and future wireless cellular networks. It is worth spending a section on this subject, as the PA arrangement and its associated cooling system represented approximately 50 % of the form factor of a classical base station. Therefore, some optimizations towards miniaturization, minimization of ecological impact, better install ability of the base stations required to improve the efficiency/ volume of the power amplifier.

The structure of the signals to be transmitted, such as phase–amplitude modulated signals, with large constellation implies to work far below the input power point for which the PA starts to compress the output signals and generates some that are spurious, as shown Figure 3.29.

The difference between this maximum allowed output power and the compression level of the PA is called the PA backoff.

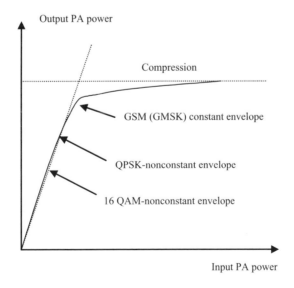

Figure 3.29 Compression of a power amplifier.

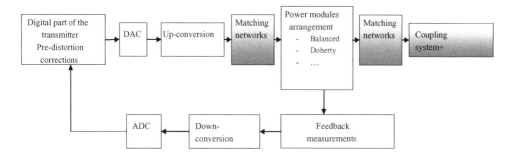

Figure 3.30 The PA regulation system plus antennas.

For example, the GMSK modulation chosen for the GSM is a frequency modulation with no envelope variation; therefore its backoff is null (peak on-average ratio is 0 dB).

HPSK modulation chosen for UMTS has at least 3.5 dB of PAR, whereas OFDM systems like 802.11a have more than 6 dB. The operating point of the PA on the Figure 3.29 curve is therefore chosen accordingly.

During the past few years, the PA efficiency has grown regularly but slowly. Figure 3.30 shows a state-of-the-art architecture of the transmission system in a base station.

This architecture is a common one. The grey boxes of the system are the frequency-selective parts. In the current design, this grey part limits the ability of a single design to address multiple frequency bands separated by a large gap of frequency.

Concerning the bandwidth that can be addressed, the limiting part is the DAC aperture time. Today, the joint progress of the semi-conductors (high-power, high-speed and more linear GaN transistors) and the regulation software of the transistor module of the PA, whose purpose is to perform the closed-loop compensation of the distortion in the whole chain, allow to foresee some improvement of the efficiency by using some architectures that can now be made possible, although their principles were known for a long time (Doherty, class S PA whose studies are beyond the scope of this book).

In a few years' time, it is expected to achieve 50 % power efficiency in a band of the order of 10-MHz bandwidth in cellular base station PA.

3.4.4 Cellular Antennas Basics

Antennas are the physical transducers that allow MIMO processing. This section gives some more details about cellular antennas.

Usually, in cellular systems, directive diagrams are often designed in the vertical plane (site) and broader ones in the horizontal plane (azimuth). To do so, the antenna is constituted as an array of radiating elements: dipoles, slots, microstrip patches, each having its own radiating D_{source}.

Moreover, this array would have its own radiating diagram $D_{isodiag}$ as if it was constituted with some isotropic sources. Radiation diagrams multiplication theorem leads to

$$D = D_{source} \cdot D_{isodiag}. \qquad (3.54)$$

Most often, the radiating element itself is weakly directive or almost isotropic; therefore, the directivity is provided by the array.

For instance, a classical result for an n sources array fed with a constant phase step δ is

$$R(\varphi) = \frac{1}{n} \cdot \frac{E}{E_0} = \frac{\sin\left(\frac{n.\phi}{2}\right)}{\sin\left(\frac{\phi}{2}\right)} \text{ with } \phi = \frac{2.\pi.\alpha}{\lambda}.\sin\varphi + \delta$$

To create a given diagram, one plays on the amplitude/phase weighting law of feeding the radiating elements (Figure 3.31).

Figure 3.32 shows a classical example in which the amplitude and phase law are chosen so that the diagram forms an inverse cosecant function which allows the mobile to receive more or less constant power, depending on the distance from the base station antenna. Assuming the mobile antenna gain is zero, the received power coming from the base station is

$$P_r = (A.4.\pi/\lambda^2).(P_e/4.\pi.R^2),$$

in which, in the first term, A is the effective area of the antenna, λ is the radiated wavelength and the second term is the isotropic power flow radiated at a distance R.

As $R = D/\sin\theta$, this expression becomes $P_r = [(A.P_e)/(\lambda^2.D^2)].\sin^2\theta$ showing the radiating diagram contained in A should vary as $1/\sin^2\theta$ to get a received power at the mobile level substantially constant and independent of R.

Practically, getting the complex amplitude/phase law to feed the radiating elements to form a given radiating diagram means solving an interpolation problem. The complex weights can be obtained digitally through signal processing, or analogically through microwave components like phasing or attenuating devices.

(i) Example of Possible Antennas for MIMO Applications

Figure 3.33 provides a simplified representation of a cross-polarization antenna for a base station able to radiate two orthogonal polarizations $\pm 45°$ with its feeding system. The number of radiating elements may be adapted to the vertical gain wished for the antenna. Only four of them are represented in Figure 3.33. For MIMO systems, the multi-polarization antennas are of special interest because two radioelectric antennas can be located in one physical 'box', avoiding the proliferation of spatially separated linearly polarized antennas.

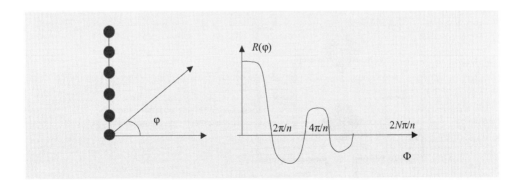

Figure 3.31 Form factor of a radiating array.

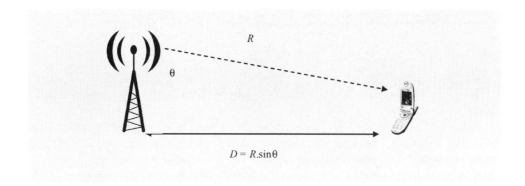

Figure 3.32 The inverse cosecant diagram.

Let us consider a design with microstrip radiating square patches. These elements can be modelled by two slots located at the edge, of one-half wavelengths long and radiating in phase. In each patch, two orthogonal resonating modes are produced, through a feeding circuit, so that it radiates a first radio signal with an electric field polarization of $+45°$ and a second signal with an electric field polarization of $-45°$.

Two antennas packaging like the one described below provide four antenna ports at the BTS side.

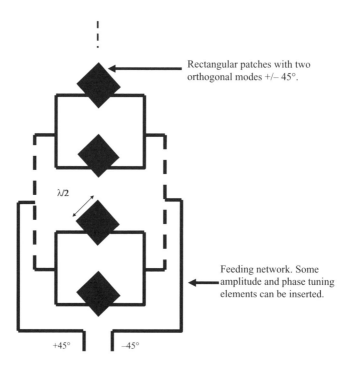

Figure 3.33 Example of cellular antennas with patches as radiating elements.

The microstrip patch antenna plays a big role in modern cellular antennas, as it allows reaching a high gain in restricted areas.

Figure 3.34 explains why the patch radiation is equivalent to the one generated by two horizontal slots of approximately λg long in the selected substrate. These slots are parallel to Oz in the xOy plane, spaced of $\lambda g/2$ and center-fed.

The radiating diagram in the yOz plane is equivalent to a slot one above a ground plane:

$$E_\varphi \frac{V_0}{\pi.R}.\sin\theta.\frac{\sin\left(\frac{\pi.w}{\lambda}.\cos\theta\right)}{\cos\theta},$$

where R is the distance from the antenna.

However, the radiated diagram in the xOy E plan is more directive, due to the vectorial combination of the two fields from the two slots:

$$E_\varphi = \frac{2.V_0}{\pi.R}.\cos\left(\frac{\pi.L}{\lambda}.\cos\right).$$

Finally, the width of the diagram in the two planes E and H is somehow the same and the directivity is approximately 6 dB.

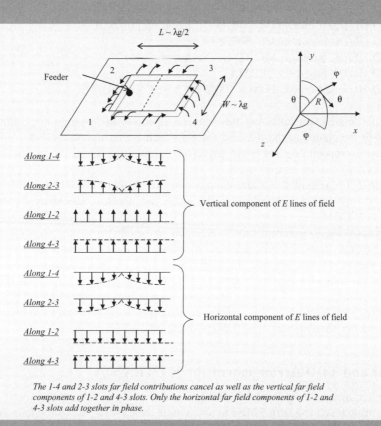

The 1-4 and 2-3 slots far field contributions cancel as well as the vertical far field components of 1-2 and 4-3 slots. Only the horizontal far field components of 1-2 and 4-3 slots add together in phase.

Figure 3.34 Principle of a radiating rectangular patch.

3.5 The E-UTRAN Physical Layer Standard

Although last 3GPP enhancements like HSDPA or HSUPA should provide a highly competitive access technology, 3GPP aims to define a longer time frame evolutionary path to ensure competitiveness. The new access technology is called E-UTRAN (Evolved UTRAN).

Requirements for this new technology include:

- Higher data rates.
- Reduced latency.
- Improved capacity and coverage.
- Reduced operational costs.
- Future additional spectrum allocations compatibility.
- Less than 5-MHz bandwidth compatibility.

Like in classical 3G/UTRAN, both FDD and TDD mode are still considered; both duplex schemes are still being imposed in most countries by the regulatory bodies.

At the early stage of E-UTRAN discussions in mid-2005, six new radio concepts were competing:

1. FDD UL based on SC-FDMA, FDD DL based on OFDMA.
2. FDD UL based on OFDMA, FDD DL based on OFDMA.
3. FDD UL/DL based on MC-WCDMA.
4. TDD UL/DL based on MC-TD-SCDMA.
5. TDD UL/DL based on OFDMA.
6. TDD UL based on SC-FDMA, TDD DL based on OFDMA.

We will not detail, in the following, the technical aspects of all the proposals; neither we will come back to the choice of the selected variant which is the first on the list.

All of these propositions had some attractiveness arguments and the classical game of industrial lobbies did the rest, as usual.

It has rapidly been decided that future radio interface would be based on OFDM, especially for receivers processing simplification purposes and spectral efficiency improvements compared to WCDMA.

Roughly speaking, a consensus rapidly emerged on the OFDMA technology choice for the downlink, whereas opinions were much more shared for the uplink choice.

Finally, the SC-FDMA reached the majority, mainly for its peak-to-average reduction property. In the following, we will focus on describing the standard implementation of this technical solution.

In the subsequent sections of this chapter, the details of E-UTRAN physical layer standard are described, for the Downlink and Uplink directions.

3.6 FDD and TDD Arrangement for E-UTRAN

Part of the requirements for 'evolved UTRAN' is the ability to cope with various spectrum allocations from much less than 5 MHz to much more than 5 MHz, accommodating future 3G spectrum allocations. Ultimately, the maximum achievable data rate available should be

Physical Layer of E-UTRAN

Figure 3.35 Possible TDD/FDD modes and interactions.

100 Mb/s in 20 MHz. Multicarrier technology should then allow a smooth migration from 1.25-MHz bandwidth to 20-MHz through 5, 10 and 15-MHz steps.

To increase this flexibility, the E-UTRAN air interface supports both frequency division duplex (FDD) and time division duplex (TDD) modes of operation (Figure 3.35). Moreover, most of the design parameters are common to TDD and FDD modes to reduce the complexity of the terminal.

3.6.1 A Word about Interferences in TDD Mode

TDD mode often requires synchronization between various transmitters and receivers in different cells. The reason why is discussed in the following and can easily be understood when looking at Figure 3.36. Two problem cases can be observed, in which either:

- Two non-synchronized close UEs (User Equipments – the 3GPP term for mobile terminal) jam each other in a near–far context. The strong signal from the UE1-TX (transmitter)

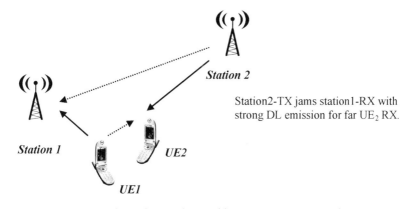

Figure 3.36 Interferences in TDD mode.

destined to far station 1 RX (receiver) can jam the close UE2-RX if the TX-UE1/RX-UE2 ranges overlap.
- Two nonsynchronized close base stations jam each other in a near–far context. The strong signal from the station2-TX (transmitter) destined to the far UE2 RX (receiver) can jam the far station1-RX if station2-TX/station1-RX ranges overlap.

It is now easier to understand the synchronization requirements developed in the following section.

3.6.2 Some Basic Physical Parameters

The size of various fields in the time domain is expressed as a number of time units $T_s = 1/(\Delta f \times N)$, where $\Delta f = 15\,\text{kHz}$ and $N = 2048$. In the frequency domain, the size is expressed as multiples of Δf. Physically, T_s represents somehow the achievable data rate period that could handle the system for a binary modulation.

(i) The Type 1 Frame Structure: Basic Numerology
A first type of frame structure is applicable to both FDD and TDD transmissions (Figure 3.37). Each radio frame is 10 ms long and consists of 20 slots of length $T_{sf} = 15360 \times T_s = 0.5\,\text{ms}$, numbered from 0 to 19. For FDD, all 20 slots are either available for downlink transmission or all 20 subframes are available for uplink transmissions. A subframe is defined as two consecutive slots where subframe i consists of slots $2i$ and $2i+1$.

For TDD, a subframe pair (0–1 subframe; 2–3 subframe; ...) is allocated to either downlink or uplink transmission. The first subframe pair in a radio frame is always allocated for downlink transmission.

For TDD only, and especially for coexistence operation with the UMTS LCR TDD, a second type of frame structure exists, in which each half frame of 5 ms is divided into seven subframes of 0.675 ms and three special fields (DwPTS,GP and UpPTS) (Figure 3.38).

The slot 0 and DwPTS are reserved for DL transmissions. The slot 1 and UpPTS are reserved for UL transmissions. Each slot benefits from a time slot interval which can be used as a guard period during the transition between UL to DL and reciprocally.

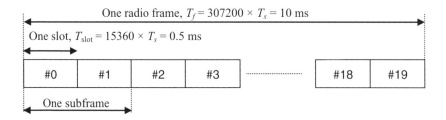

Figure 3.37 Type 1 frame structure.

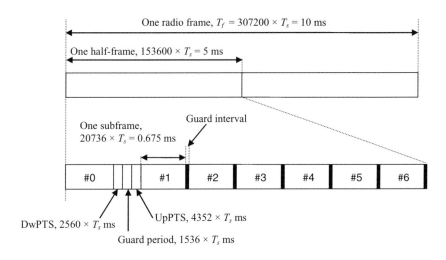

Figure 3.38 Type 2 frame structure.

3.6.3 TDD and Existing UTRAN Compatibility

The way the radio frames can be split into subframes and time slots can vary in order to be backward-compatible with the already existing 3G TDD systems like LCR-TDD (Low Chip Rate TDD, sometimes known as 'Chinese TDD') or HCR-TDD (High Chip Rate TDD) or current TDD.

For instance, for LCR-TDD compatibility, a 10-ms radio frame is cut in two radio subframes of 0.5 ms each, which are divided into seven slots of 0.675 ms each, compatible with the slot duration of LCR-TDD.

Compatibility with LCR-TDD can be achieved by inserting dynamically through the E-UTRA scheduler some idle symbols or subframes in the E-UTRA frame and applying some offset or delays between the LCR-TDD frame and the E-UTRAN frame.

Compatibility with HCR-TDD can be obtained in two ways:

- Having a E-UTRAN slot whose duration is a sub-multiple of the HCR-TDD one. In this case, the time slot is itself a subframe, whose duration is $0.01/15n$ s, where $n = \{1,2,3,\ldots\}$, and it may be configured as uplink or downlink.
- Having a E-UTRAN slot whose duration is chosen so that k.E-UTRAN slot duration is equal to p.HCR-TDD, as in Figure 3.39. Hence, the E-UTRA uplink and downlink may be aligned with the HCR-TDD uplink and downlink, provided that the HCR-TDD UL:DL time slot split is of the form $3.n/[3.(5-n)]$, where n is an integer. In this case, the E-UTRA UL:DL split is $4.n/[4.(5-n)]$. An example alignment of the HCR-TDD frame to the E-UTRAN frame is shown in Figure 3.39, in which $k=4$, $p=3$, showing a 6:9 UL/DL. The flexible frame structure of HCR-TDD allows existing HCR-TDD deployments to be migrated to a $3.n/[3.(5-n)]$ time slot split, so as to be ready for future E-UTRAN deployment in an adjacent carrier.

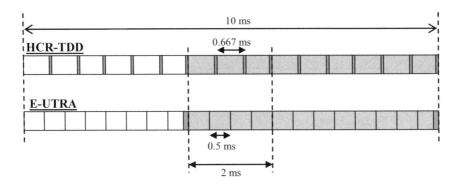

Figure 3.39 Compatibility between E-UTRA and HCR-TDD.

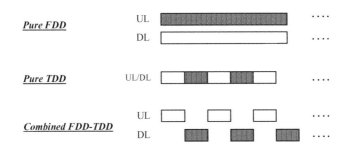

Figure 3.40 Combined FDD/TDD.

3.6.4 Combined FDD-TDD Mode

One possibility offered by the E-UTRAN system is to allow a combined FDD-TDD mode of operation, which is described in Figure 3.40.

This mode might simplify some terminals in avoiding the duplexing function in FDD mode.

3.7 Downlink Scheme: OFDMA (FDD/TDD)

The basic downlink parameters are provided in Table 3.1. Several transmission bandwidths (BW) are possible, in order to allow the positioning of the E-UTRAN physical layer in various spectrum gaps. The value of the narrow band spectrum that modulates each subcarrier is 15 kHz (OFDM symbol duration = 66.67 μs); 7.5-kHz subcarrier spacing, more sensitive to frequency offset, is considered for MBMS-dedicated channels, which gives a larger OFDM symbol duration (133 μs). Granted the constant gap of 15 kHz between subcarriers, the FFT size is growing as a function of selected bandwidth.

Two cyclic prefix lengths are possible, depending on the delay dispersion characteristics of the cell. The longer cyclic prefix (16.67 μs) should then target multi-cell broadcast (MBMS) and very-large-cell scenarios, for instance, for rural and low data rate applications at a price of bandwidth efficiency.

Table 3.1 Parameters for downlink transmission scheme.

Transmission BW	1.25 MHz	2.5 MHz	5 MHz	10 MHz	15 MHz	20 MHz
Subframe duration	0.5 ms					
Subcarrier spacing	15 kHz					
Sampling frequency	1.92 MHz (1/2 × 3.84 MHz)	3.84 MHz	7.68 MHz (2 × 3.84 MHz)	15.36 MHz (4 × 3.84 MHz)	23.04 MHz (6 × 3.84 MHz)	30.72 MHz (8 × 3.84 MHz)
FFT size	128	256	512	1024	1536	2048
Number of occupied subcarriers	76	151	301	601	901	1201
Number of OFDM symbols per subframe (short/long CP)	7/6					
CP length (μs/samples) Short	(4.69/9) × 6, (5.21/10) × 1* (16.67/32)	(4.69/18) × 6, (5.21/20) × 1 (16.67/64)	(4.69/36) × 6, (5.21/40) × 1 (16.67/128)	(4.69/72) × 6, (5.21/80) × 1 (16.67/256)	(4.69/108) × 6, (5.21/120) × 1 (16.67/384)	(4.69/144) × 6, (5.21/160) × 1 (16.67/512)
Long						

In what follows, the MBSFN acronym stands for Mobile Broadcast Single Frequency Network and refers to mobile broadcasting network using a single band on which dedicated and broadcasted signals are sharing a single 'carrier'. This notion will be developed in Section 3.8.11.

The short CP (5.21 μs) accommodates less delay spread and is suitable for urban and high data rate application. The number of OFDM symbols per slot depends on the size of this cyclic prefix, which is configured by the upper layers.

The minimum downlink TTI (Time Transmit Interval) duration, i.e. the time interval corresponding somehow to the interleaving depth of the data, is corresponding to a subframe duration, i.e. 1 ms. The possibility to concatenate multiple subframes into longer TTIs, e.g. for improved support for lower data rates and Quality of Service optimization, should be considered. In this case, the TTI can either be a semi-static or a dynamic transport channel attribute. The TTI can be implicitly given by the NodeB in indicating the modulation, the coding scheme and the size of the transport blocks. In the case of semi-static TTI, the TTI is set through higher layer signaling.

3.7.1 Downlink Physical Channels and Signals

A downlink physical channel corresponds to a set of resource atoms carrying information *originating from higher layers* (Figure 3.41). The following downlink physical channels are defined and synthetic information is provided:

- Physical Downlink Shared Channel: PDSCH. As for the HSDPA system, the radio channel is allocated dynamically in an opportunistic way, somehow breaking the notion of a dedicated channel to a given user. This may be seen as the major difference, or improvement, of E-UTRAN compared to UMTS.
- Physical Downlink Control Channel: PDCCH. This is the channel used by the eNodeB to carry control information to the UE. It carries an ACK/NACK response to the uplink channel,

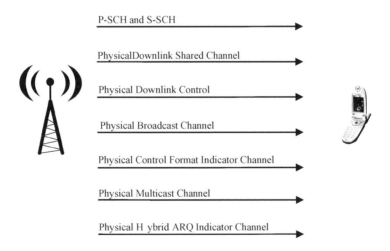

Figure 3.41 The physical downlink channels.

but also transport format allocation, UL scheduling grant and resource allocation information for the UE. In the future, there may be a need for multiple control signals, downlink scheduling control, uplink scheduling control and probably power control that needs to be dedicated for each mobile. A physical control channel is transmitted on an aggregation of one or several control channel elements, where a control channel element corresponds to a set of resource elements. Multiple PDCCHs can be transmitted in a subframe.

- Physical Broadcast Channel: PBCH. This one carries the paging and other control signals. The coded BCH transport block is mapped to four subframes (subframes #0) within a 40-ms interval which is blindly detected, i.e. there is no explicit signalling. Each subframe is assumed to be self-decodable, i.e the BCH can be decoded from a single reception, assuming sufficiently good channel conditions.
- Physical Multicast Channel: PMCH, which carries the MCH.
- Physical Control Format Indicator Channel: PCFICH. It informs the UE about the number of OFDM symbols used for the PDCCHs and is transmitted in every subframe.
- Physical Hybrid ARQ Indicator Channel: PHICH. The PHICH carries the hybrid-ARQ ACK/NAK.

Additionally, some physical signals are defined:

- Reference signal.
- Synchronization signal. A primary synchronization channel (P-SCH) and a secondary synchronization (S-SCH) channel can be used for slot synchronization and cell identification.

The 3GPP standard uses the term 'signal' rather than 'channel', as the carried information is originated from the lowest layers of the system and not from the higher protocol layers.

3.7.2 Physical Signal Transmitter Architecture

The following general steps can be identified for downlink transmission on a physical channel. All the following are mainly true for the PDSCH and the MCH channels, with some specific aspects for the other physical channels:

- Scrambling of coded bits. This scrambling step is very classical in radio systems. It aims at limiting the impact of radio fading or interferences that would be localized on all the bits of a given codeword.
- Modulation of scrambled bits to generate complex-valued modulation symbols.
- Mapping of the complex-valued modulation symbols onto one or several transmission layers.
- Precoding of the complex-valued modulation symbols on space layers for transmission on the antenna ports. This space coding stage aims at applying to the various symbol flows to be sent by all the antenna ports a given coding like the one we described in Figure 3.14. This precoding stage can be adapted to transmit diversity or spatial multiplex.
- Mapping of complex-valued modulation symbols for each antenna port to resource elements.
- Generation of a complex-valued time-domain OFDM signal for each antenna port (Figure 3.42).

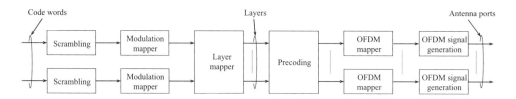

Figure 3.42 Overview of downlink baseband signal generation.

3.7.3 Downlink Data Multiplexing

The transmitted signal in each slot is described by a resource grid of $N_{RB}^{DL}N_{sc}^{RB}$ subcarriers and N_{symb}^{DL} OFDM symbols. The resource grid structure is illustrated in Figure 3.43. The quantity N_{RB}^{DL} depends on the downlink transmission bandwidth configured in the cell and shall fulfill $6 \leq N_{RB}^{DL} \leq 110$.

In the case of multi-antenna transmission, there is one resource grid defined per antenna port. An antenna port is defined by its associated reference signal. The set of antenna ports supported depends on the reference signal configuration in the cell:

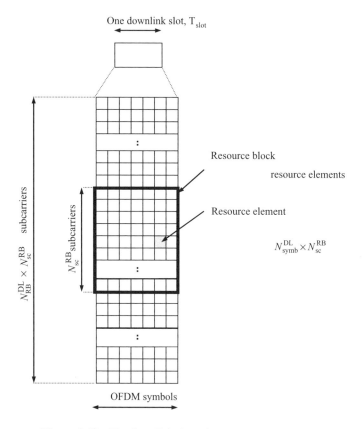

Figure 3.43 The downlink time–frequency resource grid.

- Cell-specific reference signals, associated with non-MBSFN transmission, support a configuration of one, two or four antenna ports, i.e. the index p shall fulfil $p=0$, $p \in \{0,1\}$ and $p \in \{0,1, 2, 3\}$, respectively.
- MBSFN reference signals, associated with MBSFN transmission, are transmitted on antenna port $p=4$.
- UE-specific reference signals, supported in frame structure type 2 only, are transmitted on antenna port $p=5$.

Each element in the resource grid for antenna port p is called a resource element and is uniquely identified by the index pair (k, l) in a slot where k and l are the indices in the frequency and time domains, respectively. Resource element (k, l) on antenna port p corresponds to the complex value $a_{k,l}^{(p)}$, where $k = 0, , N_{RB}^{DL} N_{sc}^{RB} - 1$ and $l = 0, , N_{symb}^{DL} - 1$. When there is no risk of confusion, the index p may be dropped. Quantities $a_{k,l}^{(p)}$ corresponding to resource elements not used for transmission of a physical channel or a physical signal in a slot shall be set to zero.

(i) Note on Resource Blocks
Physical and virtual resource blocks are defined.

N consecutive symbols can be organized into a block of M consecutive subcarriers called a RB (Resources Block). A block of subcarriers may also be composed of a nonconsecutive group of subcarriers but distributed, as discussed in the generalities chapter regarding OFDM. The notion of virtual resource block (VRB) is sometimes used. A VRB can be localized or distributed and can be mapped onto RBs.

In the literature, the notation of VRB is sometimes called a 'chunk'.

The size of the baseline physical frequency resource block, S_{RB}, is equal to M. Δf, where $M = 12$ and Δf is equal to the bandwidth spread on one subcarrier, i.e. 15 kHz, then we have $S_{RB} \sim 180$ kHz. This results in the segmentation of the transmit bandwidth shown in Table 3.2.

As also discussed in Section 4, the minimum size of the physical resources that can be allocated corresponds to the minimum TTI, i.e. one subframe of 1 ms. Therefore, the quantum of resources that can be allocated corresponds to two RBs, e.g. 14 OFDM symbols of 12 subcarriers (180 kHz). In addition, there is an unused DC subcarrier as it may be subject to a too high level of interference.

A physical resource block is defined as N_{symb}^{DL} consecutive OFDM symbols in the time domain and N_{sc}^{RB} consecutive subcarriers in the frequency domain, where N_{symb}^{DL} and N_{sc}^{RB} are given in Table 3.2. The physical resource block thus consists of $N_{symb}^{DL} \times N_{sc}^{RB}$ resource elements, i.e corresponding to one slot in the time domain and 180 kHz in the frequency domain.

Table 3.2 Resource block parameters.

Configuration		N_{sc}^{RB}	N_{symb}^{DL} Frame structure type 1	Frame structure type 2
Normal cyclic prefix	$\Delta f = 15$ kHz	12	7	9
Extended cyclic prefix	$\Delta f = 15$ kHz	12	6	8
	$\Delta f = 7.5$ kHz	24	3	4

Table 3.3 Physical resource block bandwidth and number of physical resource blocks dependent on bandwidth.

Bandwidth (MHz)	1.25	2.5	5.0	10.0	15.0	20.0
Physical resource block bandwidth (kHz)	375	375	375	375	375	375
Number of available physical resource blocks	3	6	12	24	36	48

The relationship between physical resource blocks and resource elements depends on N_{RB}^{DL} and the subframe number. The relationship between the physical resource block number n_{PRB} and the resource elements (k,l) in a slot is given by $n_{PRB} = \lfloor \frac{k}{N_{sc}^{RB}} \rfloor$, with the exception of subframe 0 in the case of N_{RB}^{DL} being an odd number, which are summarized in the 36.211 specification.

Another vision, depending on the spectrum available, can be found in Table 3.3, with another block frequency size (M = 25) and various bandwidths available.

It is the role of the scheduling function of the BTS to map the data for a given UE into a given RB at a given time. As already known in 2G or 3G systems, link adaptation (different modulation and coding scheme MCS) based on UE-reported CQI (Channel Quality Indicator) is also possible.

As a result of mapping VRBs to RBs, the transmit bandwidth is structured into a combination of localized or distributed transmissions which could be allowed to vary in a semi-static or dynamic (i.e. per subframe) way. The UE can be assigned multiple RBs by the scheduler. The information required by the UE to correctly identify its resource allocation must be made available to the UE by the scheduler.

(ii) Pilot symbols

As already explained in previous sections, pilot symbols are required for the following three purposes:

- Downlink-channel-quality measurements.
- Downlink channel estimation for coherent demodulation/detection at the UE.
- Cell search and initial acquisition.

The pilot or 'reference' symbols are arranged in the time–frequency domain so that they are time and frequency spaced, allowing correct interpolation of the channel. Depending on some radio conditions (not much time dispersion), only a first set of reference symbols might be required, spaced, for instance, by six subcarriers in the frequency domain and should be placed at the fifth and fourth OFDM symbols of each slot in case of normal and extended cyclic prefix. The obtained pattern is a rectangular one. If the radio conditions imply a second set of reference symbols, these one will be placed in a diagonal way, allowing the best time–frequency interpolation pattern for the radio channel estimation (Figure 3.44). The frequency domain positions of the reference symbols may also vary between consecutive subframes.

In the case of multiple transmit antennas for high-order MIMO or even beamforming, dedicated pilot symbols should be used in a given beam. Possible transmission of additional UE-specific downlink references should be considered.

Physical Layer of E-UTRAN

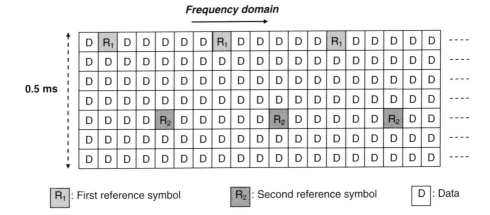

R_1 : First reference symbol R_2 : Second reference symbol   D : Data

Figure 3.44 Downlink reference signal structure (short CP).

Cell-specific reference signals shall be transmitted in all downlink subframes in a cell-supporting non-MBSFN transmission. In case the subframe is used for transmission with MBSFN, only the first two OFDM symbols in a subframe can be used for transmission of cell-specific reference symbols.

Cell-specific reference signals are transmitted on one or several of the antenna ports 0 to 3.

(iii) Reference Signals
Three types of downlink reference signals are defined:

- Cell-specific reference signals, associated with non-MBSFN transmission.
- MBSFN reference signals, associated with MBSFN transmission.
- UE-specific reference signals (supported in frame structure type 2 only). UE-specific reference signals are supported for single-antenna-port transmission of PDSCH in frame structure type 2 only and are transmitted on antenna port 5. The UE is informed by higher layers whether the UE-specific reference signal is present and is a valid phase reference for PDSCH demodulation or not.

There is one reference signal transmitted per downlink antenna port. We will develop mainly the cell-specific reference signals case.

The two-dimensional reference signal sequence $r_{m,n}$ is generated as the symbol-by-symbol product $r_{m,n} = r_{m,n}^{OS} \times r_{m,n}^{PRS}$ of a two-dimensional orthogonal sequence $r_{m,n}^{OS}$ and a two-dimensional pseudo-random sequence $r_{m,n}^{PRS}$. The two-dimensional sequence is a complex sequence, defined as $r_{m,n}^{OS} = [s_{m,n}]$, $n = 0, 1$ and $m = 0, 1, , N_r$, where N_r is the number of reference symbol positions in the OFDM symbol and $[S_{m,n}]$ is the entry at the mth row and the nth column of the matrix S_i, defined as'

$$S_i^T = \underbrace{\left[\bar{S}_i^T \ \ \bar{S}_i^T \ \ \cdots \ \ \bar{S}_i^T \right]}_{\left\lceil \frac{N_r}{3} \right\rceil \text{ repetitions}}, \quad i = 0, 1, 2, \qquad (3.55)$$

where

$$\bar{S}_0 = \begin{bmatrix} 1 & 1 \\ 1 & 1 \\ 1 & 1 \end{bmatrix}, \quad \bar{S}_1 = \begin{bmatrix} 1 & e^{j4\pi/3} \\ e^{j2\pi/3} & 1 \\ e^{j4\pi/3} & e^{j2\pi/3} \end{bmatrix}, \quad \bar{S}_2 = \begin{bmatrix} 1 & e^{j2\pi/3} \\ e^{j4\pi/3} & 1 \\ e^{j2\pi/3} & e^{j4\pi/3} \end{bmatrix} \quad (3.56)$$

for orthogonal sequence 0, 1 and 2, respectively.

There are $N_{os}=3$ different two-dimensional orthogonal sequences and $N_{PRS}=[170]$ different two-dimensional pseudo-random sequences. Each cell identity corresponds to a unique combination of one orthogonal sequence and one pseudo-random sequence, thus allowing for $N_{os} \times N_{PRS} = 510$ unique cell identities.

3.7.4 Scrambling

Each block of bits shall be scrambled prior to modulation. This is true for each type of downlink physical channel: PDSCH, PDCCH, PBCH, PCFICH and PHICH. This bit level downlink scrambling should be different for neighbours cells (except for MBSFN on the MCH) to ensure interference randomization between them and full processing gain of the channel code.

3.7.5 Modulation Scheme

Supported downlink data-modulation schemes are QPSK, 16 QAM and 64 QAM, depending on the physical channel considered. For instance, the three types of modulation are allowed for PDSCH and PMCH, whereas only QPSK is allowed for PBCH, PDCCH, PHICH and PCFICH.

An improving future option for modulation

Some improvements have been studied in the past decade in order to avoid the inefficiencies due to the cyclic prefix (CP). A new type of modulation has been introduced: OFDM/OQAM modulation, which does not require a CP. For this purpose, a pulse function modulating each subcarrier has been proposed, which must be very well accurately localized in the time domain, to limit the inter-symbol interference for transmissions over multipath channels. It is mathematically proven that when using complex-valued symbols, the prototype functions guaranteeing perfect orthogonality at a critical sampling rate cannot be well localized both in time and frequency. For instance, the unity function used in conventional OFDM has weak frequency localization properties and requires using a cyclic prefix between the symbols to limit inter-symbol interference.

OFDM/OQAM introduced a time offset between the real part and the imaginary part of the symbols. Orthogonality is then guaranteed only over real values. The OFDM/OQAM transmitted signal is expressed as

$$s(t) = \sum_n \sum_{m=0}^{M-1} a_{m,n} \underbrace{i^{m+n} e^{2i\pi m \nu_0 t} g(t-n\tau_0)}_{g_{m,n}(t)},$$

where $a_{m,n}$ denotes the real-valued information value (can be the real part or the imaginary part of the Offset complex QAM symbol) sent on the mth subcarrier at the nth symbol, M is the number of subcarriers, ν_0 is the inter-carrier spacing; it is the same as the classical OFDM system. τ_0 is the OFDM/OQAM symbol duration; it is equal to $T_u/2$ (T_u is the OFDM symbol duration), and g is the pulse function.

It is important to note that the OFDM/OQAM symbol rate is twice the classical OFDM symbol rate without cyclic prefix ($\tau_0 = N/2$); meanwhile, since the modulation used is a real one, the information amount sent by an OFDM/OQAM symbol is half the information amount sent by an OFDM symbol. Figure 3.45 depicts the signal generation chain of an OFDM/OQAM signal.

The modulator generates N real valued symbols, each τ_0, where $\tau_0 = T_u/2$. The real-valued symbols are then phase-shifted, multiplied by i^{m+n} before the IFFT. The main difference of OFDM/OQAM over conventional OFDM signal generation stays in the filtering by the prototype function g after the IFFT, instead of the cyclic prefix addition.

One good candidate for OFDM/OQAM pulse-shaping filtering is the IOTA (Isotropic Orthogonal Transform Algorithm) pulse obtained by orthogonalizing the Gaussian function in both time and frequency domains.

Another particularity of IOTA is the spectrum of the generated signal (Figure 3.46). Thanks to its good frequency localization, the resulting spectrum is steeper than for conventional OFDM.

With the OFDM/OQAM transmission scheme, we have CP = 0, which allows access to the full time gap for transmission and we have 15 OQAM symbols per subframe. However, the counterpart is the orthogonality of the subcarriers after filtering is no longer valid and the receiver needs more CPU to perform the demodulation. IOTA might be part of the future releases of E-UTRAN.

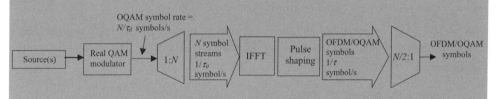

Figure 3.45 OFDM/OQAM signal generation chain.

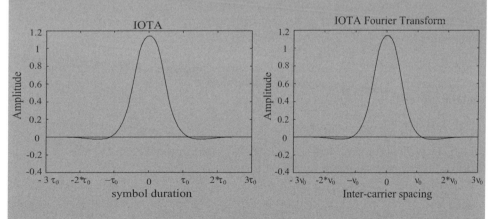

Figure 3.46 Temporal and spectral representation of the IOTA pulse filter.

3.7.6 Downlink Scheduling Information and Uplink Grant

Obviously, the UE has to be informed on how to process the downlink data transmission. This scheduling information may be more or less dynamically transmitted, depending on its type. Some may be known a priori as operator-dependent, like the specific group ID the UE belongs to or some resources assignment, whereas some are highly dynamic, like the ARQ in formations, transmitted for every TTI. Modulation scheme and payload sizes are also provided.

Moreover, the UE is also granted some uplink resources (localized or distributed) and transmission parameters (modulation scheme, payload size and MIMO-related information). The transport format may, however, be selected by the UE on its own.

Concerning the ARQ process that we will describe later, the hybrid ARQ response to uplink data transmission consists of a single ACK/NAK bit.

3.7.7 Channel Coding

Channel coding is based on UTRAN releasing six turbo-coding schemes. Other FEC (Forward Error Correction) schemes are also envisaged to cope with additional E-UTRAN requirements, like codes polynomial for lower rates or repetition coding for higher processing gain.

The main topics in the discussion about the coding scheme are:

- Extension of the maximum code block size.
- Support of code rates lower than 1/3.
- Removal of the tail.
- Reduction of decoder complexity in the UE.
- Improvement in power efficiency (lower Eb/N0).

While decoder complexity reduction and power efficiency improvement are still under discussion, other topics are already seen as acceptable options. The alternative coding schemes currently under investigation are:

- Duo-binary turbo codes.
- Inter-block permutation turbo code (IBPTC).
- Rate-compatible/quasi-cyclic LDPC code (RC/QCLDPC).
- Concatenated zigzag LDPC code.
- Turbo single parity check (SPC) low-density parity check (LDPC) code.
- Shortened turbo code by insertion of temporary bits.

Although some LDPC codes show either an improvement in power efficiency for high code rates or a potential complexity reduction of the decoder, the potential benefits may have not yet been proven to be significant enough for the selection of an alternative coding scheme.

3.7.8 OFDM Signal Generation

The OFDM symbols in a slot shall be transmitted in increasing order of l. The time-continuous signal $s_l^{(p)}(t)$ on antenna port p in OFDM symbol l in a downlink slot is defined by

$$s_l^{(p)}(t) = \sum_{k=-\lfloor N_{RB}^{DL} N_{sc}^{RB}/2 \rfloor}^{-1} a_{k^{(-)},l}^{(p)} \cdot e^{j2\pi k \Delta f(t-N_{CP,l}T_s)} + \sum_{k=1}^{\lceil N_{RB}^{DL} N_{sc}^{RB}/2 \rceil} a_{k^{(+)},l}^{(p)} \cdot e^{j2\pi k \Delta f(t-N_{CP,l}T_s)} \quad (3.57)$$

Table 3.4 OFDM parameters.

		Cyclic prefix length $N_{CP,l}$		
Configuration		Frame structure type 1	Frame structure type 2	Guard interval
Normal cyclic prefix	$\Delta f = 15\,\text{kHz}$	160 for $l=0$ 144 for $l=1,2,\ldots,6$	512 for $l=8$ in slot 0 224 otherwise	0 for slot 0 288 otherwise
Extended cyclic prefix	$\Delta f = 15\,\text{kHz}$	512 for $l=0,1,\ldots,5$	768 for $l=7$ in slot 0 512 otherwise	0 for slot 0 256 otherwise
	$\Delta f = 7.5\,\text{kHz}$	1024 for $l=0,1,2$	1280 for $l=3$ in slot 0 1024 otherwise	

for $0 \leq t < (N_{CP,l}+N) \times T_s$, where $k^{(-)} = k + \lfloor N_{RB}^{DL} N_{sc}^{RB}/2 \rfloor$ and $k^{(+)} = k + \lfloor N_{RB}^{DL} N_{sc}^{RB}/2 \rfloor - 1$. The variable N equals 2048 for $\Delta f = 15\,\text{kHz}$ subcarrier spacing and 4096 for $\Delta f = 7.5\,\text{kHz}$ subcarrier spacing.

Table 3.4 lists the value of $N_{CP,l}$ that shall be used for the two frame structures. Note that different OFDM symbols within a slot may have different cyclic prefix lengths. For frame structure type 2, note that the OFDM symbols do not fill all slots completely and the last part is left unused.

3.7.9 Downlink MIMO

The basic configuration (i.e. realistic) is two transmit antennas at the cell site and two receive ones at the UE, but higher-order MIMO (up to 4 × 4) should be possible. For simplicity, open loop MIMO should be preferred, as being able to cope with any UE speed. Simpler transmit diversity mechanisms can replace MIMO schemes if performances/complexity trade-off is in favor of that:

- Multiple codewords (including single codeword as a special case) that use the same time–frequency resource and are independently channel-coded with independent CRC should be investigated (see PARC MIMO scheme of Figure 3.12). Possible values for the maximum number of codewords per resource block transmitted by the NodeB are 1, 2, 3 or 4. Possible values for the maximum number of codewords that can be received by the UE are 1, 2, 3 or 4.
- The spatial division multiplexing of the modulation symbol streams for different UEs using the same time–frequency resource could be supported, which may be denoted as Spatial Division Multiple Access (SDMA) or multi-user (MU)-MIMO. Use of precoding as a means to convert the antenna domain MIMO signal processing into the beam domain processing should be investigated in the future.

Some transmit diversity techniques, whose detailed description is outside of the scope of this book, can be applied when the rank of the spatial correlation matrix is determined to be one; this is known as a rank adaptation mechanism, as explained in the MIMO generalities chapter. Possible candidates for the transmit diversity mode are:

- Block-code-based transmit diversity (STBC, SFBC).
- Time (or frequency)-switched transmit diversity (or antenna permutation).
- Cyclic delay diversity.
- Complex rotation matrix.
- Precoded transmission using selected precoding vector(s) (including selection transmit diversity); see Section 3.4.9.

Layer mapping and Precoding stage in the generic case.

The complex-valued modulation symbols for each of the codewords to be transmitted are mapped onto one or several layers. Complex-valued modulation symbols $d^{(q)}(0), \ldots, d^{(q)}(M^{(q)}_{symb}-1)$ for codeword q shall be mapped onto the layers $x(i) = [x^{(0)}(i) \ldots x^{(-1)}(i)]^T$.

The split between layer mapping and precoding steps allows to include all the antenna processing schemes in a single formulation.

(i) Layer Mapping and Precoding for Spatial Multiplexing

The layer mapping operation is an intermediary operation preparing the mapping of the symbol supported by the subcarriers at a given time and the mapping of these symbols on the respective antennas. Based on the measurements of the downlink reference signals of the different antennas, the mobile decides on the suitable rank to apply i.e. the number of layers and the suitable matrix.

For spatial multiplexing, the number of layers v is equal to the rank ρ of the transmission and the mapping shall be done according to Table 3.5.

Here, the notion of rank is associated with the number of Layers.

The precoding concepts have been partly discussed in Section 3.4.9. The precoder takes as input a block of vectors $x(i) = [x^{(0)}(i) \quad \cdots \quad x^{(-1)}(i)]^T$ from the layer mapping and generates a block of vectors $y(i) = [y^{(0)}(i) \quad \cdots \quad y^{(P-1)}(i)]^T$ to be mapped onto resources on each of the antenna ports, where $y^{(p)}(i)$ represents the signal for antenna port p. The number of antenna

Table 3.5 Codeword-to-layer mapping for spatial multiplexing.

Transmission rank	Number of codewords	Codeword-to-layer mapping
1	1	$x^{(0)}(i) = d^{(0)}(i)$
2	2	$x^{(0)}(i) = d^{(0)}(i)$ $x^{(1)}(i) = d^{(1)}(i)$
3	2	$d^{(0)}(i)$ is mapped to layer 0 $d^{(1)}(i)$ is mapped to layers 1 and 2
4	2	$d^{(0)}(i)$ is mapped to layers 0 and 1 $d^{(1)}(i)$ is mapped to layers 2 and 3

Physical Layer of E-UTRAN

Table 3.6 Cyclic delay diversity (CDD).

Number of antenna ports P	D(i)	Transmission rank ρ	δ No CDD	δ Small delay	δ Large delay (1,2)
1	$[1]$	1	—	—	—
2	$\begin{bmatrix} 1 & 0 \\ 0 & e^{-j2\pi \cdot i \cdot \delta_1} \end{bmatrix}$	1 2	0	$[1/\eta$ or $2/\eta]$	0 $[1/2]$
4	$\begin{bmatrix} 1 & 0 & 0 & 0 \\ 0 & e^{-j2\pi \cdot i \cdot \delta_1} & 0 & 0 \\ 0 & 0 & e^{-j2\pi \cdot i \cdot 2\delta_2} & 0 \\ 0 & 0 & 0 & e^{-j2\pi \cdot i \cdot 3\delta_3} \end{bmatrix}$	1 2 3 4	0		0 [0 or 1/4] [0 or 1/4] [1/4]

ports, P, is equal to or larger than the rank ρ of the transmission. Precoding shall be done according to $y(i) = D(i)W(i)x(i)$, where the Precoding matrix $W(i)$ is of size $P \times \nu$ and the quantity $D(i)$ is a diagonal matrix for the support of cyclic delay diversity, as described in the chapter dedicated to generalities.

The matrix $D(i)$ shall be selected from Table 3.5, where UE-specific values of δ are semi-statically configured in the UE and the Node B by higher layers. The quantity η in Table 3.6 is the smallest number from the set {128, 256, 512, 1024, 2048} such that $\eta \geq N_{BW}^{DL}$.

Note that, compared to the expression given in Section 3.4.5, in which the delay parameter was assimilated as a number of frequency subcarriers, the parameter δ is now given as a fraction of the entire spectrum N_{BW}^{DL}.

The current standard has planned that δ could handle a small delay value or a bigger one, depending potentially on the propagation conditions of the radio environment, which can be more or less static.

For spatial multiplexing, the values of $W(i)$ shall be selected from the codebook configured in the Node B and the UE. The configured codebook shall be equal to Table 3.7 or a subset thereof. Note that the number of layers is equal to the transmission rank ρ in the case of spatial multiplexing. The table lists the different possible values of $W(i)$ for a number of antennas of up to two.

The choice of the spatial code pattern can be done dynamically on a link adaptation basis, based on the PVI and the CQI indications.

It can be noted that this precoded matrix is somehow associated with a given vector from a predetermined codebook and can be obtained from a householder transform of such a vector, leading to an efficient compression of the amount of data to feedback. The codebook of such vectors is provided in the 36.211 specification.

Table 3.7 Codebook for spatial multiplexing.

Number of antenna ports P	Transmission rank ρ	Codebook					
1	1	$[1]$	—	—	—	—	—
2	1	$\begin{bmatrix}1\\0\end{bmatrix}$	$\begin{bmatrix}0\\1\end{bmatrix}$	$\frac{1}{\sqrt{2}}\begin{bmatrix}1\\1\end{bmatrix}$	$\frac{1}{\sqrt{2}}\begin{bmatrix}1\\-1\end{bmatrix}$	$\frac{1}{\sqrt{2}}\begin{bmatrix}1\\j\end{bmatrix}$	$\frac{1}{\sqrt{2}}\begin{bmatrix}1\\-j\end{bmatrix}$
	2	$\frac{1}{\sqrt{2}}\begin{bmatrix}1&0\\0&1\end{bmatrix}$	$\frac{1}{2}\begin{bmatrix}1&1\\1&-1\end{bmatrix}$	$\frac{1}{2}\begin{bmatrix}1&1\\j&-j\end{bmatrix}$	—	—	—

The beamforming case is a special case of the spatial multiplexing, in which there is a single codeword, a transparent layer mapping and a precoding beamforming vector applied at the precoding stage.

(ii) Layer Mapping and Precoding for Transmit Diversity

For transmit diversity, layer mapping shall be done according to Table 3.8. There is only one codeword and the transmission rank shall be equal to one. The number of layers υ is equal to the number of antenna ports P used for transmission.

The precoding operation for transmit diversity is defined for two and four-antenna ports. For transmit diversity, the precoding operation should not be understood under the adaptive sense of Section 3.4.9. The precoding matrix is a deterministic coding allowing to apply the Alamouti generalized form on the respective carriers symbols at a given time.

For transmission on antenna ports 0 and 1, the output $y(i) = [y^{(0)}(i) \; y^{(1)}(i)]^T$ of the precoding operation is defined by

$$\begin{bmatrix} y(2i) \\ y(2i+1) \end{bmatrix} = \begin{bmatrix} 1 & 0 & j & 0 \\ 0 & -1 & 0 & j \\ 0 & 1 & 0 & j \\ 1 & 0 & -j & 0 \end{bmatrix} \begin{bmatrix} Re(x(i)) \\ Im(x(i)) \end{bmatrix} \quad (3.58)$$

for $i = 0, 1, \cdots, \lfloor (M_{symb}-1)/2 \rfloor$ where M_{symb} is the size of the block of complex-valued modulation symbols.

As an exercise, the interested reader can check that successively applying on a symbol flow the operation of layer mapping and precoding leads to finding the Alamouti scheme described in Section 3.4.6.

Table 3.8 Codeword-to-layer mapping for transmit diversity.

Number of layers	Number of codewords	Codeword-to-layer mapping	
1	1	$x^{(0)}(i) = d(i)$	$i = 0, 1, \ldots, M_{symb} - 1$
2	1	$x^{(0)}(2i) = d(2i)$ $x^{(1)}(2i+1) = d(2i+1)$	$i = 0, 1, \ldots, \lfloor (M_{symb} - 1/2) \rfloor$

Such a scheme has been generalized for a number of antennas greater than two. For instance, for a transmission on four antenna ports, $p \in \{0, 1, 2, 3\}$, the output $y(i) = [y^{(0)}(i) \ y^{(1)}(i) \ y^{(2)}(i) \ y^{(3)}(i)]^T$ of the precoding operation is defined by

$$\begin{bmatrix} y^{(0)}(4i) \\ y^{(1)}(4i) \\ y^{(2)}(4i) \\ y^{(3)}(4i) \\ y^{(0)}(4i+1) \\ y^{(1)}(4i+1) \\ y^{(2)}(4i+1) \\ y^{(3)}(4i+1) \\ y^{(0)}(4i+2) \\ y^{(1)}(4i+2) \\ y^{(2)}(4i+2) \\ y^{(3)}(4i+2) \\ y^{(0)}(4i+3) \\ y^{(1)}(4i+3) \\ y^{(2)}(4i+3) \\ y^{(3)}(4i+3) \end{bmatrix} = \begin{bmatrix} 1 & 0 & 0 & 0 & j & 0 & 0 & 0 \\ 0 & -1 & 0 & 0 & 0 & j & 0 & 0 \\ 0 & 0 & 0 & 0 & 0 & 0 & 0 & 0 \\ 0 & 0 & 0 & 0 & 0 & 0 & 0 & 0 \\ 0 & 1 & 0 & 0 & 0 & j & 0 & 0 \\ 1 & 0 & 0 & 0 & -j & 0 & 0 & 0 \\ 0 & 0 & 0 & 0 & 0 & 0 & 0 & 0 \\ 0 & 0 & 0 & 0 & 0 & 0 & 0 & 0 \\ 0 & 0 & 0 & 0 & 0 & 0 & 0 & 0 \\ 0 & 0 & 0 & 0 & 0 & 0 & 0 & 0 \\ 0 & 0 & 1 & 0 & 0 & 0 & j & 0 \\ 0 & 0 & 0 & -1 & 0 & 0 & 0 & j \\ 0 & 0 & 0 & 0 & 0 & 0 & 0 & 0 \\ 0 & 0 & 0 & 0 & 0 & 0 & 0 & 0 \\ 0 & 0 & 0 & 1 & 0 & 0 & 0 & j \\ 0 & 0 & 1 & 0 & 0 & 0 & -j & 0 \end{bmatrix} \begin{bmatrix} Re(x^{(0)}(4i)) \\ Re(x^{(1)}(4i)) \\ Re(x^{(2)}(4i)) \\ Re(x^{(3)}(4i)) \\ Im(x^{(0)}(4i)) \\ Im(x^{(1)}(4i)) \\ Im(x^{(2)}(4i)) \\ Im(x^{(3)}(4i)) \end{bmatrix}$$

for $i = 0, 1, \ldots, \lfloor (M_{symb}-1)/4 \rfloor$.

3.7.10 Channels Layer Mapping, Precoding and Mapping to Resource Elements

(i) Layer Mapping and Precoding
Dealing with the layer mapping and precoding stage, all the following physical channels: PDSCH, PDCCH, PBCH, PCFICH and PHICH, can be arranged the same way as generically described in Section 3.7.9, i.e. being conditioned for either transmit diversity (Alamouti's Space Time Coding) or spatial multiplexing.

(ii) Generic Mapping to Resource Elements
For each of the antenna ports used for transmission of the physical channel, the block of complex-valued symbols $y^{(p)}(0), \ldots, y^{(p)}(M_s^{(p)}-1)$ shall be mapped in sequence, starting with $y^{(p)}(0)$, to virtual resource blocks assigned for transmission. The mapping to resource elements $(k,l,)$ on antenna port p not reserved for other purposes shall be in increasing order of first the index k and then the index l, starting with the first slot in a subframe.

(iii) Dealing with the PDSCH
The physical downlink-shared channel shall be processed and mapped to resource elements as described previously, with the following exceptions:

- The set of antenna ports used for transmission of the PDSCH is {0}, {0,1} or {0,1,2,3} if UE-specific reference signals are not transmitted.

- The antenna ports used for transmission of the PDSCH is {5} if UE-specific reference signals are transmitted.

(iv) Dealing with the PDCCH
The block of complex-valued symbols $y^{(p)}(0), \ldots, y^{(p)}(M_{symb}-1)$ for each antenna port used for transmission shall be permuted in groups of four symbols, resulting in a block of complex-valued symbols $Z^{(p)}(0), \ldots, Z^{(p)}(M_{symb}-1)$.

The block of complex-valued symbols $Z^{(p)}(0), \ldots, Z^{(p)}(M_{symb}-1)$ shall be cyclically shifted by $4N_{CSS}$ symbols, resulting in the sequence $w^{(p)}(0), \ldots, w^{(p)}(M_{symb}-1)$, where $w^{(p)}(i) = z^{(p)}(i + 4N_{CSS}) \bmod M_{symb}$.

The block of complex-valued symbols $w^{(p)}(0), \ldots, w^{(p)}(M_{symb}-1)$ shall be mapped in sequence starting with $w^{(p)}(0)$ to resource elements corresponding to the physical control channels. The mapping to resource elements (k,l) on antenna port p not reserved for other purposes shall be in increasing order of first the index k and then the index l, where $l = 0, \ldots, L-1$ and $L \leq 3$ correspond to the value transmitted on the PCFICH. In the case of the PDCCHs being transmitted using antenna port 0 only, the mapping operation shall assume reference signals corresponding to antenna port 0 and antenna port 1 being present; otherwise, the mapping operation shall assume reference signals being present corresponding to the actual antenna ports used for transmission of the PDCCH.

(v) Dealing with the PBCH
The block of complex-valued symbols $y^{(p)}(0), \ldots, y^{(p)}(M_{symb}-1)$ for each antenna port is transmitted during four consecutive radio frames and shall be mapped in sequence starting with $d(0)$ $y(0)$ to physical resource blocks number $(N_{RB}^{DL}-1)/2 - 3$ to $(N_{RB}^{DL}-1)/2 + 2$ if N_{RB}^{DL} is an odd number and $N_{RB}^{DL}/2 - 3$ to $N_{RB}^{DL}/2 + 2$ if N_{RB}^{DL} is an even number. The mapping to resource elements (k,l) not reserved for transmission of reference signals shall be in increasing order of first the index k, then the index l in subframe 0, then the slot number and finally the radio frame number. For frame structure type 2, only subframe 0 in the first half-frame of a radio frame is used for PBCH transmission. The set of values of the index l to be used in subframe 0 in each of the four radio frames during which the physical broadcast channel is transmitted is detailed in the specification.

(vi) Dealing with the PMCH
The physical multicast channel shall be processed and mapped to resource elements in the same way as the PDSCH, with the following exception:

- For transmission on a single antenna port, layer mapping and precoding shall be done assuming a single antenna port and the transmission shall use antenna port 4, i.e. no transmit diversity is specified.

(vii) Dealing with the PCFICH
The block of vectors $y(i) = [y^{(0)}(i) \cdots y^{(P-1)}(i)]^T$, $i = 0, \ldots, 15$ shall be mapped to physical resource elements in the first OFDM symbol in a downlink subframe.

(viii) Dealing with the PHICH
The block of modulation symbols $d(0), \ldots, d(M_{symb}-1)$ shall be mapped to layers according to one of Sections 6.3.3.1 or 6.3.3.3 and precoded according to one of Sections 6.3.4.1

or 6.3.4.3, resulting in a block of vectors $y(i) = [y^{(0)}(i) \; \cdots \; y^{(P-1)}(i)]^T$, where $y^{(p)}(i)$ represents the signal for antenna port p and where $p = 0, \ldots, P-1$ and the number of antenna ports $P \in \{1,2,4\}$.

(ix) Regarding the Two-Dimensional Reference Signal Sequence $r_{m,n}$

Let us consider the single antenna case first. The reference signals it shall be mapped to resource atoms $a_{k,l}$ according to the following scheme, given in Figure 3.47.

However, as the numbers of downlink transmit antennas equals 1, 2, 3 or 4, the reference signals have to be transmitted according to the following rules:

- Each antenna shall use some dedicated resource elements as reference signals.
- Each antenna element shall transmit nothing on a resource element used as a reference signal on another antenna.

Figure 3.47 provides a mapping example for a type 1 frame structure with normal cyclic prefix. For conciseness, the similar case (type 1 extended prefix, type 2 normal and extended prefix, MBSFN reference signals case) are not shown but can be found in the 36.211 specification.

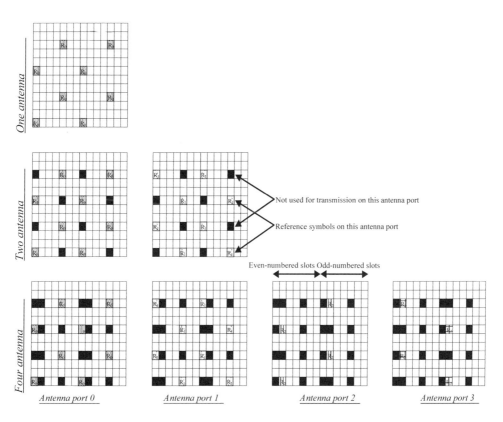

Figure 3.47 Mapping of downlink reference signals.

(x) Regarding the Synchronization Signal

The primary synchronization signal is transmitted on 72 active subcarriers, centered on the DC subcarrier. The primary synchronization signal $d(n)$, whose use is described in Section 3.3.15, shall be mapped to the resource atoms $a_{k,l}$ according to

$$a_{k,l} = \begin{cases} d(k+36) & k = -36, \cdots, -1 \\ d(k+35) & k = 1, \cdots, 36. \end{cases} \tag{3.59}$$

The primary synchronization signal is only transmitted in subframe l.
Other types of reference signals can also be mentioned:

- The MBSFN reference signals: MBSFN reference signals shall only be transmitted in subframes allocated for MBSFN transmissions. MBSFN reference signals are transmitted on antenna port 4.
- The UE-specific reference signals: they are only supported for frame structure type 2. UE-specific reference signals are transmitted on antenna port 5. The UE is informed by higher layers, whether the UE-specific reference signal is present and is a valid phase reference or not.

3.7.11 E-MBMS Concepts

The support of MBMS (Multimedia Broadcast Multicast Services) is an essential requirement for E-UTRAN. The so-called E-MBMS will therefore be an integral part of EPS.

E-MBMS provides a transport feature to send the same content information to a given set of users in a cell to all the users (broadcast) or to a given set of users (multicast) for which a notion of subscription applies in order to restrict the multicast services to a given set of users.

But, as will be described in detail in Chapter 6, the E-MBMS multicast concept should not be confused with the 'IP multicast service' as, in the case of 'IP multicast', there is no sharing of a given radio resource among all interested users, as it is only a pure IP notion (duplication of some IP packets at some routers in the networks) and not a 'radio' one.

MBMS (Multimedia Broadcast Multicast System) broadcast only depends on the users' terminals' ability to receive the broadcast channel (like a TV receiver) and is a very fashionable system, as its main commercial application is that it allows to carry efficiently the TV channels broadcast on the small size screen of the mobile terminals in a given cell. For instance, a medium data rate of 128 kb/s is enough to carry a video signal in an average radio cell to a phone screen with a good perceived quality.

E-MBMS (which is the evolved version of the legacy MBMS system) should be able to make use of some MIMOs' open loop scheme advantages. In E-MBMS, there will be a single (single-cell broadcast) or multiple transmitting Node Bs and multiple receiving UEs.

E-MBMS is a good application to demonstrate what MIMO can bring to the system. Indeed, in the case of broadcast of the same signal on the same frequency band (see SFN below), the transmission power has to be chosen so that the far mobiles should receive the signal with good quality. To reduce the required power, increasing the number of transmit and receive antennas is a good solution. MIMO options, like spatial multiplexing, is possible in the MBMS context.

(i) SFN Networks
In E-UTRAN, MBMS transmissions may be performed as single-cell transmissions or as multi-cell transmissions. In the case of multi-cell transmission, the cells and content are synchronized to enable for the terminal to soft-combine the energy from multiple transmissions. The superimposed signal looks like multipaths to the terminal. This concept is also known as Single Frequency Network (SFN). The E-UTRAN can configure which cells are parts of an SFN for transmission of an MBMS service.

A MBMS Single Frequency Network is called a MBSFN. MBSFN is envisaged for delivering services such as mobile TV using the LTE infrastructure, and is expected to be a competitor to DVB-H-based TV broadcasts.

In MBSFN, the transmission happens from a time-synchronized set of eNodeBs using the same resource block. The Cyclic Prefix (CP) used for MBSFN is slightly longer, and this enables the UE to combine transmissions from different eNodeBs located far away from each other, thus somewhat negating some of the advantages of SFN operation. There will be six symbols in a slot of 0.5 ms for MBSFN operation versus seven symbols in a slot of 0.5 ms for non-SFN operation.

For MBSFN operation, 3GPP is currently defining a SYNC protocol between the E-MBMS gateway and the eNodeBs to ensure that the same content is sent over the air from all the eNodeBs. As shown in Figure 3.48, a broadcast server is the source of the MBMS traffic, and a gateway (E-MBMS gateway) is responsible for distributing the traffic to the different eNodeBs of the MBSFN area.

'IP Multicast' may be used for distributing the traffic from the E-MBMS gateway to the different eNodeBs.

3GPP has defined a control plane entity, known as the MBMS Coordination Entity (MCE), that ensures that the same resource block is allocated for a given service across all the eNodeBs of a given MBSFN area. It is the task of the MCE to ensure that the RLC/MAC layers at the eNodeBs are appropriately configured for MBSFN operation.

3GPP has currently assumed that header compression for MBMS services will be performed by the E-MBMS gateway. In this case, the scrambling should be identical for all cells involved in the MBSFN transmission.

Historically, the notion of SFN has been introduced in 3GPP for 3GPP R7, as, so far, each of the broadcasting cells used to use its own frequency/scrambling code, meaning that even if the same content is broadcasted from multiple cells, their signals interfere due to scrambling. In SFN networks for UTRAN, the same scrambling/frequency is used in order to superimpose signals like multipaths allowing HSDPA terminals to constructively combine them and leading to a capacity gain of the broadcast network of up to three. However, in order for each signal to fall inside the reception windows of the terminal, i.e. not exceeding the CP length to cause inter-symbol interference, precise synchronization mechanisms of base stations need to be achieved, for instance, based on the GPS system.

The MBMS traffic can share the same carrier with the unicast traffic or be sent on a separate carrier.

In the case of subframes carrying MBMS SFN data, specific reference signals are used. MBMS data are carried on the MBMS traffic channel (MTCH) as a logical channel. This

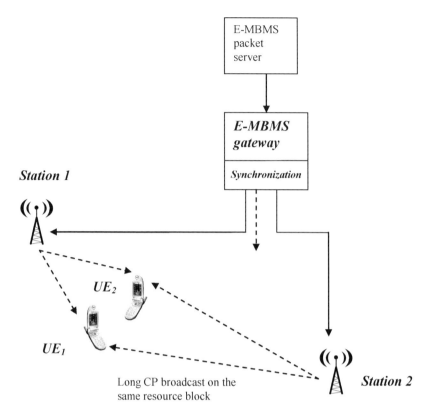

Figure 3.48 MBSFN concept.

logical channel is then mapped either on the MCH transport channel or on the DL-SCH downlink shared channel. In the case of mapping on the MCH channel, the physical channel associated is the PMCH.

When used with several cells, the reference symbols needed for MBMS reception are identical in all cells to be considered for combining and transmission scheduling has to be such that received packets are received in a range substantially less than the cyclic prefix.

MBMS transmissions may share the same carrier with unicast, or 'dedicated' traffic. For instance, for mobile TV, MBMS data can be sent on a separate carrier, not carrying anything other than broadcast/MBMS-related information. Unicast multiplexing schemes such as FDM or TDM are allowed (in the case of TDM, unicast transmissions are not sharing the same subframe).

In case MBMS transmissions are handled using a separate carrier:

- There is only TDM multiplexing between different services.
- Only long CP is considered.

The physical layer coding and modulation chain for MBMS transmissions is the same as unicast transmissions as a baseline.

When used with MBMS, feedback signalling from the UE may not be feasible, including CQI. In the case of multi-codeword spatial multiplexing, dynamic adaptation of modulation and coding, etc. for each codeword is not possible due to the absence of channel quality feedback. However, different codewords can potentially use different modulation and coding and/or power offsets in a semi-static fashion in order to enable efficient interference cancellation at the UE receiver.

Since the baseline UE has only two antennas, the number of broadcast codewords are limited to two. E-MBMS for UE limited to single codeword reception capability should be further considered. E-MBMS signals from NodeB with more than two transmit antennas should be transparent to the UE.

Finally, MBMS should be supported in paired or unpaired spectrum.

3.7.12 Downlink Link Adaptation

According to the position of the mobile in the cell and the interferences which it tries out (signal/interference ratio reports, error rate, etc.), the operator of the network will be able to optimize:

- The rate of the communications in the cell.
- The rate and the latency time of transmissions of the blocks jointly.

The radio resources allocator of the base station can manage the power of the transmitted packets, the coding channel – meaning the number of packets to be transmitted for a given volume of information – or the type of modulation.

It should be possible, for example, to transmit all the packets at maximum available power in the sector of the cell and to adapt the choice of MCS (Modulation and Coding Scheme) according to the radio conditions, resulting in a rate error value on the blocks [BLER (Block Error Rate)] allowing to get either:

- a maximum capacity with a BLER not necessarily too weak (coding scheme/repetition compromise); or
- a lower rate but a shorter latency time of block transmission because of an optimal BLER.

As indicated in Section 3.4.8, the dynamic choice of a MIMO scheme is also part of the global link adaptation process (Figure 3.49).

3.7.13 HARQ

(i) Architectural Considerations

One of the architecture decisions made in 3GPP was to exclude ARQ functionality from the Core Network Serving Gateway and to terminate ARQ in UE and eNodeB. This leads to the question of whether E-UTRAN should use only a single (H)ARQ protocol or a two-layer approach with HARQ and ARQ located at the RLC layer on top. It can be shown that achieving the required reliability with a single HARQ layer can be very costly in terms of resources needed for HARQ feedback. Thus, it has been decided that a two-layered ARQ/HARQ

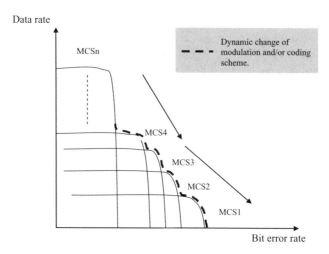

Figure 3.49 Link adaptation principle.

approach was the best way of achieving both high reliability and low resource cost for the ARQ/HARQ feedback.

Considering that (H)ARQ in E-UTRAN is terminated in the eNodeB, using only one HARQ protocol to ensure reliability may be attractive, as a second protocol may add complexity, additional control signalling overhead, and harmful protocol interactions. However, this approach suffers from some drawbacks, i.e. from the robustness constraint on the HARQ feedback mechanism so as to obtain the expected error rate (10^{-6}). Since the end-to-end RTT (Round Trip Time) needs to be low to achieve a high throughput, it is desirable to have fast and frequent HARQ feedback to correct transmission errors as soon as possible. It is natural to consider the same approach used in HSPA, where a synchronous one-bit ACK/NACK signal is sent every transmission attempt and the timing of the feedback message is used to identify the corresponding data transmission, achieving the fastest possible feedback while minimizing the information in the feedback message. However, this binary feedback is susceptible to transmission errors and, in particular, NACK reception errors (i.e. erroneously interpreted as ACKs at the receiver) lead to data losses at the HARQ layer.

Thus, the reliability of the HARQ layer is bounded by the error rate of the feedback and not the error rate of the data transmission and it is costly to achieve a sufficient HARQ feedback reliability for the frequently transmitted feedback message, as the transmitter power requirements are high due to channel fading and the constraint that a single bit cannot be protected with good Forward Error Correction (FEC) codes.

One solution for achieving a higher reliability without excessive expenditure on HARQ feedback is to apply a second layer of ARQ on top of the MAC HARQ layer, e.g. to use an RLC (Radio Link Control) Acknowledged Mode, as done for HSPA. Note that in this case, the second ARQ layer is mainly responsible for correcting error events due to HARQ feedback errors and not transmission errors themselves. One main benefit of the additional ARQ protocol is that it provides a much more reliable feedback mechanism based on asynchronous status reports with explicit sequence numbers that are protected by a cyclic redundancy check (CRC).

This implies that the receiver of the status report can detect any errors in the report through the CRC. The reliable transmission of the feedback information is further enhanced in several ways. First, the status messages are protected by turbo codes. Secondly, HARQ is also applied to status messages. Thirdly, the status messages are accumulative. Even if the transmission of a status report fails, the subsequent status includes the information of the lost one.

(ii) HARQ Process
The N-channel Stop-and-Wait protocol is used for downlink HARQ. Downlink hybrid ARQ (HARQ) is based on Incremental Redundancy (IR). This means that the retransmissions are basically nonidentical (chase or soft combining case). For instance, in the case of turbo encoding of packets, different rate matching can be used between retransmissions. The relative number of parity bits to systematic bits varies between retransmissions. Obviously, this solution requires more memory in the user equipment. Practically, the various encoding between different retransmissions can be done in 'real time' for each transmission or the data can be simultaneously encoded at the same time and stored in a buffer.

HARQ can be also classified as synchronous or asynchronous. In principle, synchronous operation with an arbitrary number of simultaneous active processes at a time instant could be envisioned. Asynchronous operation already supports an arbitrary number of simultaneous active processes at a time instant. Asynchronous HARQ offers the flexibility of scheduling retransmissions based on air interface conditions.

Synchronous HARQ implies that (re)transmissions for a certain HARQ process are restricted to occur at known time instants. No explicit signalling of the HARQ process number is required, as the process number can be derived from, for example, the subframe number. The various forms of HARQ schemes are further classified as adaptive or nonadaptive in terms of transmission attributes, e.g. the Resource Block (RB) allocation, modulation and transport block size, and duration of the retransmission. Control channel requirements are described for each case.

Synchronous HARQ transmission entails operating the system on the basis of a predefined sequence of retransmission packet format and timing but the benefits of synchronous HARQ operation when compared to asynchronous HARQ operation are:

- Reduction of control signalling overhead. (No signalling of the HARQ channel process number.)
- Lower operational complexity if nonadaptive operation is chosen.
- Possibility to soft combine control signalling information across retransmissions for enhanced decoding performance if nonadaptive operation is chosen.

Adaptive implies the transmitter may change some or all of the transmission attributes used in each retransmission as compared to the initial transmissions (e.g. due to changes in the radio conditions). Hence, the associated control information needs to be transmitted with the retransmission. The changes considered are:

- Modulation.
- Resource block allocation.
- Duration of transmission.

With those definitions, the HS-DSCH (shared transport channel in HSDPA) in 3G/UMTS uses an adaptive, asynchronous HARQ scheme, while E-DCH (improved dedicated channel of HSUPA standard) in 3G/UMTS uses a synchronous, nonadaptive HARQ scheme.

To sum up, the considerations provided in this section lead to the following ARQ concept for E-UTRAN downlink:

- HARQ handles transmission errors and uses binary, synchronous feedback.
- The HARQ retransmission unit is a transport block that may contain data from more than one radio bearer (MAC multiplexing).
- ARQ handles residual HARQ errors, i.e. it retransmits data for which the HARQ process failed.
- ARQ retransmission unit is an RLC PDU.
- RLC performs segmentation or concatenation according to scheduler decisions. An RLC PDU contains either a segment of a Service Data Unit (SDU), a complete SDU, or it may contain data of several SDUs (concatenation).
- In the case of no MAC multiplexing, there is a one-to-one mapping between HARQ and ARQ retransmission units.
- RLC performs in-order delivery to higher layers.

In the downlink, the current working assumption is to use an adaptive, asynchronous HARQ scheme based on incremental redundancy (IR) for E-UTRAN. On the basis of the CQI reports from the UEs, the scheduler in NodeB selects the time of the transmission and the transmission attributes for the initial transmission and the retransmissions.

3.7.14 Downlink Packet Scheduling

The Node B scheduler (for unicast transmission) dynamically controls which time/frequency resources are allocated to a certain user at a given time. Downlink control signalling informs UE(s) what resources and respective transmission formats have been allocated. The scheduler can dynamically choose the best multiplexing strategy from the available methods, e.g. localized or distributed allocation. Obviously, scheduling is tightly interacting with link adaptation and HARQ. The decision of which user transmissions to multiplex within a given subframe may, for example, be based on:

- Minimum and maximum data rate.
- Available power to share among mobiles.
- BER target requirements according to the service.
- Latency requirement, depending on the service.
- Quality of Service parameters and measurements,

Physical Layer of E-UTRAN

- Payloads buffered in the Node-B ready for scheduling,
- Pending retransmissions,
- CQI (Channel Quality Indicator) reports from the UEs.
- UE capabilities.
- UE sleep cycles and measurement gaps/periods.
- System parameters such as bandwidth and interference level/patterns, etc.

Methods to reduce the control signalling overhead, e.g. pre-configuring the scheduling instants (persistent scheduling for applications like VoIP, for instance) and grouping for conversational services, should be considered.

Due to signalling constraints, only a given number of mobiles can be scheduled on the same TTI (e.g. eight).

(i) Interactions between HARQ and Link Adaptation for Packet Scheduling
Figure 3.50 illustrates the interactions between the different entities involved in packet scheduling (PS), which are located at the base station (eNodeB) in order to facilitate fast channel-dependent scheduling in shortening the round trip delay. The basic time–frequency resource available for data transmission is the physical resource block (PRB), which consists of a fixed number of adjacent OFDM subcarriers and represents the minimum scheduling resolution in the frequency domain. The Packet Scheduler is the controlling entity in the overall scheduling process. It can consult the Link Adaptation (LA) module to obtain an estimate of the supported data rate for certain users in the cell, for different allocations of PRBs. LA may utilize frequency-selective CQI feedback from the users, as well as Ack/Nacks from past transmissions, to ensure that the estimate of supported data rate corresponds to a certain BLER target for the first transmissions. Further, the offset calculation module in the link-adaptation process may be used to stabilize the BLER performance in the presence of LA uncertainties. It provides a user-based adaptive offset on a subframe interval that is applied to

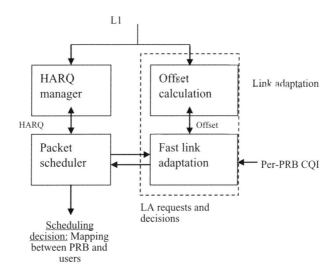

Figure 3.50 A packet-scheduling framework.

the received CQI reports in order to reduce the impact of biased CQI errors on LA performance. The aim of the scheduler is to optimize the cell throughput for the given load condition under the applied scheduling policy in time and frequency. The HARQ manager provides buffer status information as well as transmission format of the pending HARQ retransmissions.

Among the possible scheduling policies, we can mention the:

- **Fair allocation scheme** in which each mobile (in DL or UL) is allocated the same amount of available PRBs. The number of allocated PRBs per UE changes only when the number of UE in the cell changes (handover).
- **Proportional allocation scheme** in which the user bandwidth is adapted to the changing channel conditions while trying to match the signal-to-noise ratio by means of power control.

It can also be noted that the interference system in the frequency domain is somehow highly dependent on the way the spectrum is used in the respective cells of the network. Frequency planning tricks close to the ones well known in FDMA/TDMA systems – like GSM – may be valid, including efficient packet scheduling under fractional re-use of the spectrum, e.g. the whole spectrum is not used in the whole system in case of lack of traffic, to decrease cell edge interference.

3.7.15 Cell Search and Acquisition

Cell search is the procedure by which a UE acquires time and frequency synchronization with a cell and detects the cell identity of that cell. E-UTRAN cell search is assumed to be based on two signals ('channels') transmitted in the downlink: the 'SCH' (Synchronization Channel) and the 'BCH' (Broadcast Channel), which is a transport channel carried by the PBCH.

The SCH (primary and secondary) enable acquisition of the symbol timing and the frequency of the downlink signal.

The BCH carries cell/system-specific information (like for 3G/UTRAN). Indeed, the UE must acquire at least:

- The overall transmission bandwidth of the cell.
- Cell ID.
- Radio frame timing information when this is not directly given by the SCH timing (SCH can be transmitted more than once every radio frame).
- Information regarding the antenna configuration of the cell (number of transmitter antennas).
- Information regarding the BCH bandwidth (multiple transmission bandwidths of the BCH can be defined).
- CP length information regarding the subframe in which the SCH and/or BCH are transmitted.

(i) SCH and BCH Timing Arrangement
The SCH and the BCH bursts are transmitted once or several times in a 10-ms radio time frame and their respective numbers may not be the same (more SCH bursts than BCH ones). The

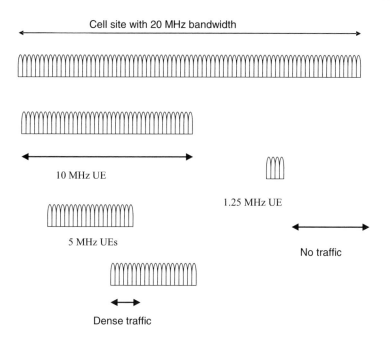

Figure 3.51 The multiple bandwidth problem.

BCH is placed at a well defined time instant after/before the downlink SCH by the time delay/advance of τ. The optimum number of SCH/BCH bursts depends on the cell detection time specification.

SCH symbols are always placed in the same place within a subframe, i.e. the two last symbols of the subframe. The reason is to ease the post-detection process (averaging of multiple periodic detections).

(ii) Frequency Arrangement

The main problem faced is related to the multiplicity of possible supported bandwidths (Figure 3.51). Indeed, E-UTRAN offers system flexibility by supporting systems and UEs of multiple bandwidths.

This introduces a challenge in synchronization and bandwidth detection.

The chosen solution is that the SCH and the BCH symbols are placed at the center of the transmission bandwidth. The simplified process is then the following one as shown in Figure 3.52:

- The UE detects the central part of the spectrum, regardless of the transmission bandwidth capability of the UE and that of the cell site.
- The UE moves to the transmission bandwidth according to the UE capability of the actual communication.

The bandwidth used to transmit the SCH is set to 1.25 MHz (72 subcarriers), whereas, for diversity reasons, the BCH may use a bandwidth of 1.25 or 5 MHz.

Example: 10 MHz UE in 2- MHz cell site, SCH bandwidth = 1.25 MHz and BCH bandwidth = 1.25MHz

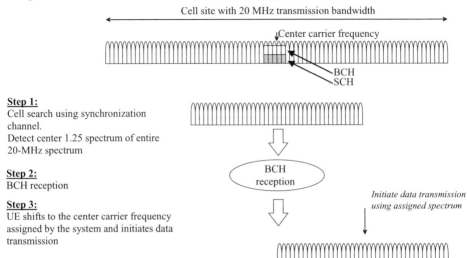

Step 1:
Cell search using synchronization channel.
Detect center 1.25 spectrum of entire 20-MHz spectrum

Step 2:
BCH reception

Step 3:
UE shifts to the center carrier frequency assigned by the system and initiates data transmission

Figure 3.52 The cell search algorithm.

(iii) Transmit Diversity Arrangement
As the SCH is the first physical channel for a UE to acquire, it must be received without a priori knowledge of the number of transmitter antennas of the cell. Thus, transmit diversity methods that do not require knowledge of the number of transmit antennas can only be considered [e.g. time-switched transmit diversity (TSTD), frequency switched transmit diversity (FSTD), and delay diversity including cyclic delay diversity (CDD)].

These diversity schemes can also be considered for the BCH in order to improve the packet error rate (PER). Moreover, should the configuration of the transmitter antennas of the cell be provided using the SCH or reference symbols, block code-based transmit diversity can be considered for BCH.

(iv) SCH Signal Structure and Cell Search Procedure
Two principal structures of the SCH have been discussed by the 3GPP, depending on whether the synchronization acquisition and the cell ID detection are obtained from the different SCH signals (hierarchical SCH, like in UTRA), or from the same SCH signal (nonhierarchical SCH). Performances criteria in terms of time, complexity and overhead will drive the need for such and such SCH. The last decision of 3GPP tends to prefer a hierarchical structure of SCH, like the one described below. Independently of the SCH structure, the basic cell search procedure is provided in Figure 3.53.

- SCH symbol timing detection can be performed by several correlation methods, depending on the type (hierarchical or not) of the received SCH.
- Radio frame timing detection can be done on the SCH or can be obtained by decoding the BCH.

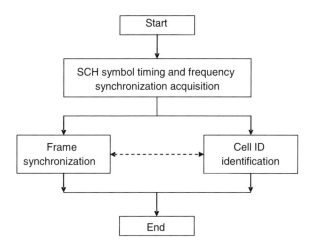

Figure 3.53 The cell search procedure.

- The cell ID identification can be indicated directly by the SCH sequence, or the SCH sequence can indicate a group of cell IDs, on which an exhaustive search can be done, as in 3G/UTRAN.

Figure 3.54 provides the downlink physical channels for synchronization of the mobile and system parameters acquisition.

From a hierarchical structure point of view, the SCH may consist of two signals: the primary SCH (P-SCH), used to obtain subframe level synchronization, and the secondary synchronization signal (S-SCH), used to obtain frame level synchronization.

The initial cell energy detection (closest cell) and subframe synchronization is done through correlation (matched filter) with a sequence selected among three different ones on the primary SCH which is transmitted over the centre 72 subcarriers on the last slots 0 and 10 in the first and sixth subframe of each frame.

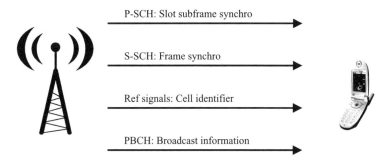

Figure 3.54 The physical channel roles for sysnchronization and system parameters acquisition.

The secondary SCH is transmitted over the centre of 72 subcarriers on the second to last symbol of slots 0 and 10 in the first and sixth subframe of each frame.

This signal in every cell carries the one among the 170 unique cell group identifiers, which are basically pseudo-random sequences. Once the UE tunes into one of these 170 sequences, the UE is subframe and frame synchronized.

Then, the mobile has still to decode the broadcast channel BCH (PBCH) information to find the cell identifiers that are sent by the reference signal. The secondary SCH are arranged so that each sequence maps to a cell group identifier of three cell-specific identifiers; therefore, once the frame synchronization is obtained, there are only three cell-specific possible identifiers (0,1 or 2 are the indexes of the cell identity group).

The primary synchronization signal is generated from a frequency-domain Zadoff-Chu sequence. The primary synchronization signal is transmitted on 72 active subcarriers, centered on the DC subcarrier according to

$$d_u(n) = \begin{cases} e^{-j\frac{\pi u n(n+1)}{63}} & n = 0, 1, ..., 30 \\ e^{-j\frac{\pi u(n+1)(n+2)}{63}} & n = 31, 32, ..., 61 \end{cases}, \quad (3.60)$$

where the Zadoff-Chu root sequence index u is 25, 29 or 34 if the cell identity group is 0, 1 or 2, respectively.

The mapping of the primary synchronization signal depends on the frame structure.

For frame structure type 1, the primary synchronization signal $d(n)$ is only transmitted in slots 0 and 10 and shall be mapped to the resource elements according to

$$a_{k,l} = d(n), \quad k = n - 31 + \left\lfloor \frac{N_{RB}^{DL} N_{SC}^{RB}}{2} \right\rfloor, \quad l = N_{symb}^{DL} - 1, \quad n = 0, \ldots, 61. \quad (3.61)$$

Resource elements (k, l) in slots 0 and 10 where

$$k = n - 31 + \left\lfloor \frac{N_{RB}^{DL} N_{SC}^{RB}}{2} \right\rfloor, \quad l = N_{symb}^{DL} - 1, \quad n = -5, -4, \ldots, -1, 62, 63, \ldots, 66 \quad (3.62)$$

are reserved and not used for transmission of the primary synchronization signal.

For frame structure type 2, the primary synchronization signal is transmitted in the DwPTS field. The second synchronization signal $d(n)$ is a binary sequence. The sequence used for the second synchronization signal is an interleaved concatenation of two length-31 binary sequences obtained as cyclic shifts of a single length-31 M sequence generated by $x^5 + x^2 + 1$. The concatenated sequence is scrambled with a scrambling sequence given by the primary synchronization signal.

The secondary synchronization signal is transmitted on 72 active subcarriers, centered on the DC subcarrier. The mapping of the secondary synchronization signal depends on the frame structure.

For frame structure type 1, the secondary synchronization signal $d(n)$ shall be mapped to the resource elements according to

$$a_{k,l} = d(n), \quad k = n - 31 + \left\lfloor \frac{N_{RB}^{DL} N_{SC}^{RB}}{2} \right\rfloor, \quad l = N_{symb}^{DL} - 2, \quad n = 0, \cdots, 61. \quad (3.63)$$

Resource elements (k,l) in slots 0 and 10, where

$$k = n-31 + \left\lfloor \frac{N_{RB}^{DL}N_{sc}^{RB}}{2} \right\rfloor, \quad l = N_{symb}^{DL}-2, \quad n = -5, -4, \ldots, -1, 62, 63, \ldots, 66, \quad (3.64)$$

are reserved and not used for transmission of the secondary synchronization signal, where $l = N_{symb}^{DL}-2$. The secondary synchronization signal is transmitted in and only in slots where the primary synchronization signal is transmitted.

For frame structure type 2, the secondary synchronization signal is transmitted in the last OFDM symbol of subframe 0.

The BCH then carries the system information: the physical layer parameters (e.g. bandwidth), system frame number (SFN), scheduling information of the most frequently repeated scheduling unit (a group of system information which has the same periodicity): network identities, tracking area code, cell identity, cell baring status, scheduling information of other scheduling unit, etc.

3.7.16 Methods of Limiting the Inter-Cell Interference

To make the best use of the whole available spectrum and limit the complexity of frequency planning, it is planned usually to use the whole spectrum in any cell, i.e. the re-use factor is set to 1. However, granted that in that case, the cell edge users may suffer from interference of the neighbouring cells, some approaches to mitigate these interferences may be required. Three approaches to inter-cell interference mitigation, not necessarily mutually exclusive, are being considered:

- Inter-cell-interference randomization. Methods considered for inter-cell-interference randomizations include: **cell-specific scrambling** [or applying (pseudo) random scrambling after channel coding/interleaving], **cell-specific interleaving** [also known as Interleaved Division Multiple Access (IDMA)] and a third method, which consists of applying different kinds of frequency hopping.
- Inter-cell-interference cancellation. Fundamentally, inter-cell-interference cancellation aims at interference suppression at the UE beyond what can be achieved by just exploiting the processing gain. For instance, spatial suppression by means of multiple antennas at the UE can be considered. interference cancellation based on detection/subtraction of the inter-cell interference also. One example is the application of cell-specific interleaving (IDMA) to enable inter-cell-interference cancellation.
- Inter-cell-interference coordination/avoidance. Based on measurements performed by the UE and communicated to the NodeB (CQI, path loss, average interference, etc.) and on measurements performed by different network nodes and exchanged between them (which requires inter-eNodeB synchronization) a better downlink allocation can be done to mitigate interferences. For instance, soft frequency re-use (Figure 3.55) can be achieved. This consists of having, for the cell edge users, a primary band with a frequency re-use pattern of 1/3, for example, served by high-power transmission with a good SNR, and a secondary band for the cell center users with the remaining spectrum and power.

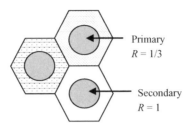

Figure 3.55 Soft frequency re-use.

(i) The IDMA Scheme
This scheme is a new scheme proposed for E-UTRAN: Interleave Division Multiple Access (IDMA) as a way to mitigate the inter-cell interference (ICI) in a downlink EUTRA system. The principle of IDMA is to employ distinct interleaving patterns in the neighbouring cells so that the UE can distinguish the cells by means of cell-specific interleavers. IDMA has a similar characteristic, with scrambling in whitening ICI when the traditional 'single-user (NodeB)' receiver is used.

Figure 3.56 illustrates the use of IDMA in the downlink case, in which UE1 and UE2 are respectively served by NodeB1 and NodeB2 but allocated the same time–frequency resource (chunk). Suppose NodeB1 interleaves the signal for UE1 with interleaving pattern1, while NodeB2 interleaves the signal for UE2 with interleaving pattern2 (different from pattern1), then UE1 (UE2) may distinguish the signals from the two NodeBs by means of different interleavers.

(ii) IDMA with Iterative Receiver
Let us assume that the UEs can perform an iterative decoding of information coming both from Stations 1 and 2.

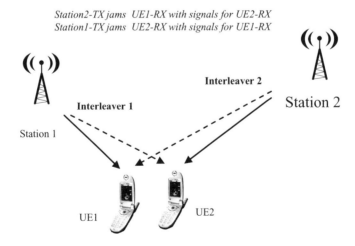

Figure 3.56 Using IDMA to suppress inter-cell interference.

In the case of using the single-cell receiver, the interference from the other NodeB will be whitened to noise. When iterative multi-cell receiver is used, the interference could be effectively cancelled. Note that IDMA can be employed not only between neighbouring NodeBs, but also potentially between neighbouring sectors.

The iterative multi-cell receiver is based on the interference cancellation and iterative decoding. Let us briefly consider a two-cell case. In the first iteration, the single-user decoding is performed for Cell 1. Assuming, after the decoding, a given information bit in the frame is relatively unreliable [log-likelihood ratio (LLR) is small]. Then, the information bits are re-encoded. Thus, the unreliable information bit is converted into a given number of N unreliable code bits. The N unreliable code bits are scrambled to distributed positions after re-interleaving for Cell 1. Then, the Cell 2 signal is obtained by subtracting the Cell 1 from the received signal. After the interference subtraction, the N unreliable bits in the Cell 1 frame are affecting the corresponding bits in the Cell 2 frame, but then the Cell 2 signal is fed to the de-interleaver of Cell 2. If the two cells used the same interleaver pattern, the N unreliable bits would be reassembled together. However, if IDMA is used, Cell 2 is employing a different interleaver pattern from Cell 1. Hence, the N unreliable bits in the frame will be scrambled to another series of distributed positions, providing at the second iteration a good estimation of the N bits that were previously doubtful.

3.7.17 Downlink Physical Layer Measurements

This paragraph deals with the important subject of measurements performed by the UE on the downlink. These measurements are useful for scheduling purpose and mobility management.

(i) Scheduling

The UE has to report to the Node B the channel quality of one resource block or a group of resource blocks in the form of a CQI. This CQI is measured on a multiple of 25 or 50 subcarriers' bandwidth and is a key parameter to tune the following:

- Time/frequency-selective scheduling.
- Link adaptation.
- Interference management.
- Power control of downlink physical channel.

(ii) Mobility

The classical functions of mobility are based on measurements so as to be able to perform the following:

- The PLMN selection according to their quality.
- The cell selection and reselection. (Detection of the most suitable cells.)
- The handover decision, including intra-frequency handover (the measurement are made on a same carrier frequency band or the UE at least receives the common channel of the target cell) or inter-frequency handover (the measurements are made on a different carrier

frequency band or the UE at least does not receive the common channel of the target cell) or even inter-RAT handover. Classically, the UE has to perform enough uplink/downlink idle periods to efficiently manage the measurements on the serving cell and its neighbourhood.

3.8 Uplink Scheme: SC-FDMA (FDD/TDD)

3.8.1 Uplink Physical Channel and Signals

An uplink physical channel corresponds to a set of resource atoms carrying information originating from higher layers. The following uplink physical channels are defined (Figure 3.57):

- Physical Uplink Shared Channel, PUSCH, used for uplink shared data transmission.
- Physical Uplink Control Channel, PUCCH. The PUCCH shall be transmitted on a reserved frequency region in the uplink. It is used to carry ACK/NACK, CQI for downlink transmission and scheduling request for uplink transmission.
- Physical Random Access Channel, PRACH.

An uplink signal is used by the physical layer but does not carry information originating from higher layers. The following uplink physical signals are defined:

- Reference signal.

3.8.2 SC-FDMA

The basic uplink transmission scheme is single-carrier transmission (SC-FDMA) with cyclic prefix to achieve uplink inter-user orthogonality and to enable efficient frequency–domain equalization at the receiver side. Frequency–domain generation of the signal, sometimes known as DFT-SOFDM (Discrete Fourier Transform Spread Orthogonal Frequency Division Multiplex), is assumed and illustrated in Figure 3.58. This allows a relatively high degree of commonality with the downlink OFDM scheme and the same parameters, e.g. clock frequency, can be re-used.

The subcarrier mapping determines which part of the spectrum is used for transmission by inserting a suitable number of zeros at the upper and/or lower end in Figure 3.59. Between each DFT output, sample L-1 zeros are inserted. A mapping with $L=1$ corresponds to localized transmissions, i.e. transmissions where the DFT outputs are mapped to consecutive subcarriers. With $L>1$, distributed transmissions happens, which is considered as a

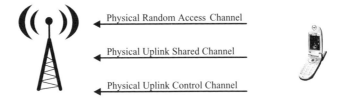

Figure 3.57 The uplink physical channels.

Physical Layer of E-UTRAN

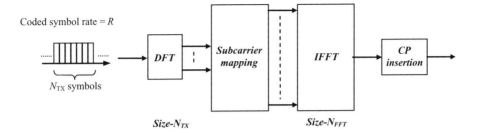

Figure 3.58 Transmitter structure for SC-FDMA.

complement to localized transmissions for additional frequency diversity. However, although distributed mapping was planned originally for uplink, recent standardization decisions will only allow a localized mapping and the frequency diversity can be obtained via intra or inter-TTI frequency hopping.

The physical mapping to the N available subcarriers per **one** DFT-SOFDM symbol in the RF spectrum shall be performed as illustrated in Figure 3.59, where f_c is the carrier frequency and where the transmission BW is 1.25/2.5/5/10/15/20 MHz, and N is 75/150/300/600/900/1200, and N_n is 38/75/150/300/450/600, respectively.

3.8.3 Uplink Subframe Structure

As for the downlink, there are also two types of frame structure: type 1 and type 2, designed for compatibility purposes with the LCR UTRA TDD.

The transmitted signal in each slot is described by the contents of N_{symb}^{UL} SC-FDMA symbols, numbered from 0 to $N_{symb}^{UL}-1$. Each SC-FDMA symbol carries multiple

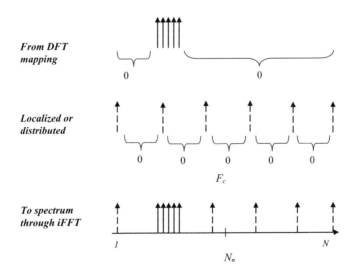

Figure 3.59 Localized mapping (left) and distributed mapping (right).

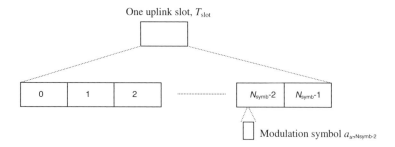

Figure 3.60 Uplink slot format (type 1 frame structure).

complex-valued modulation symbols $a_{u,l}$ representing the contents of resource element (u,l), where u is the time index within the SC-FDMA symbol l.

For the type 1 frame structure, all SC-FDMA symbols are of the same size. The uplink slot structure for the type 1 frame structure is illustrated in Figure 3.60.

For the type 2 frame structure, SC-FDMA symbol 1 and $N_{symb}^{UL}-2$ are denoted by short SC-FDMA symbols, used to carry the uplink demodulation reference signals. The uplink slot structure for the type 2 frame structure is illustrated in Figure 3.61.

The number of SC-FDMA symbols in a slot depends on the cyclic prefix length configured by higher layers and is given in Table 3.9.

In the case of type 2 frame structures, the long symbols are used for control and/or data transmission, while the short ones are used for reference signals (pilot symbol for coherent demodulation and control and/or data transmission).

The basic TTI duration is twice the slot duration, i.e. 1 ms, containing 12 long OFDM symbols; however, as for the downlink, several subframes can be concatenated to offer longer TTI to potentially reduce higher layer protocol overhead (IP packet segmentation, RLC-MAC header, etc.). This TTI duration can be dynamically adjusted through higher layer signalling in a semi-static way or controlled by the eNodeB in a more dynamic way in order to improve HARQ process, for instance.

Dealing with the size in terms of number of occupied subcarriers or samples for both types of blocks, a long block is consuming twice the number of subcarriers or samples as a short one. Each block is separated by a cyclic prefix, whose duration depends on the total available bandwidth. A longer cyclic prefix is called an 'extended cyclic prefix'.

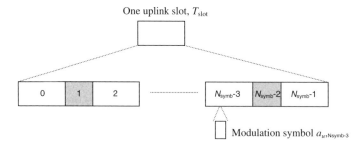

Figure 3.61 Uplink slot format (type 2 frame structure).

Table 3.9 Resource block parameters.

Configuration	N_{sc}^{RB}	N_{symb}^{UL} Frame structure type 1	N_{symb}^{UL} Frame structure type 2
Normal cyclic prefix	12	7	9
Extended cyclic prefix	12	6	8

3.8.4 Resource Grid

The transmitted signal in each slot is described by a resource grid of $N_{RB}^{UL} N_{sc}^{RB}$ subcarriers and N_{symb}^{UL} SC-FDMA symbols. The quantity N_{RB}^{UL} depends on the uplink transmission bandwidth configured in the cell and shall fulfill $6 \leq N_{RB}^{UL} \leq 110$. Only the resources grid of type 1 is provided in Figure 3.62.

Similarly to the downlink, each element in the resource grid is called a resource element and is uniquely defined by the index pair (k,l) in a slot where k and l are the indices in the frequency

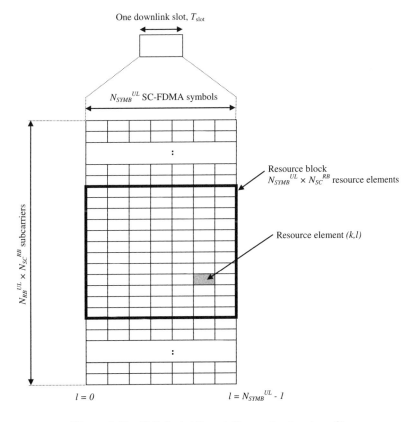

Figure 3.62 Uplink slot format (frame structure type 1).

and time domains, respectively. Resource element (k,l) corresponds to one complex-valued modulation symbol $a_{k,l}$ where $k = 0, \cdots, N_{RB}^{UL} N_{sc}^{RB} - 1$ and $l = 0, \cdots, N_{symb}^{UL} - 1$. Modulation symbols $a_{k,l}$ corresponding to resource elements not used for transmission of a physical channel or a physical signal in a slot shall be set to zero.

A resource block is defined as N_{symb}^{UL} consecutive SC-FDMA symbols in the time domain and N_{sc}^{RB} consecutive subcarriers in the frequency domain, where N_{symb}^{UL} and N_{sc}^{RB} are given in Table 3.9. A resource block in the uplink thus consists of $N_{symb}^{UL} \times N_{sc}^{RB}$ resource elements, corresponding to one slot in the time domain and 180 kHz in the frequency domain.

Assuming the TTI is 1 ms, the basic uplink resource units are:

- Frequency resources: 12 subcarriers = 180 kHz.
- Symbols: 1 ms.180 kHz = 14 OFDM symbols*12 subcarriers = 168 modulation symbols.

At the opposite of the downlink, no unused subcarrier is defined as the DFT based precoding reduces the PAPR impact of it, spreading it over the M modulation symbols.

3.8.5 PUSCH Physical Characteristics

The following general steps can be identified for transmission of the physical uplink shared channel (Figure 3.63):

- Scrambling.
- Modulation of scrambled bits to generate complex-valued symbols.
- DFT-precoding to generate complex-valued modulation symbols.
- Mapping of complex-valued modulation symbols to resource elements.
- Generation of complex-valued time–domain SC-FDMA signal for each antenna port.

(i) Scrambling

If scrambling is configured, the block of bits $b(0), \ldots, b(M_{bit} - 1)$, where M_{bit} is the number of bits to be transmitted on the physical uplink shared channel, shall be scrambled with a UE-specific scrambling sequence prior to modulation, resulting in a block of scrambled bits $c(0), \ldots, c(M_{bit} - 1)$.

(ii) DFT-Precoding

The block of complex-valued symbols $d(0), \ldots, d(M_{symb} - 1)$ is divided into M_{symb}/M_{sc} sets, each corresponding to one SC-FDMA symbol. DFT-precoding shall be applied according to

$$z(l \cdot M_{sc} + k) = \sum_{i=0}^{M_{sc}-1} d(l \cdot M_{sc} + i) e^{-j\frac{2\pi i k}{M_{sc}}}, \quad (3.65)$$

$$k = 0, \ldots, M_{sc} - 1$$
$$l = 0, \ldots, M_{symb}/M_{sc} - 1$$

→ Scrambling → Modulation mapper → DFT precoder → Resource element mapper → SC-FDMA signal gen. →

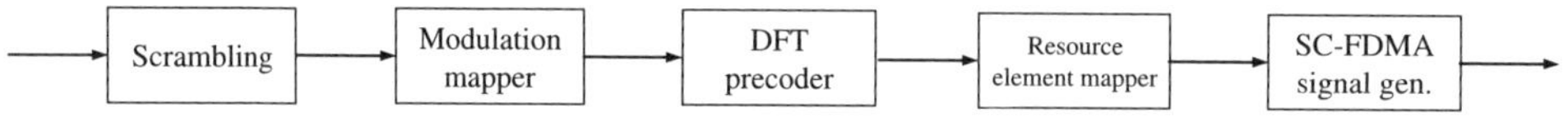

Figure 3.63 Overview of uplink physical channel processing.

resulting in a block of complex-valued modulation symbols $z(0), \ldots, z(M_{\text{symb}} - 1)$. The variable M_{sc} represents the number of scheduled subcarriers used for PUSCH transmission in an SC-FDMA symbol and shall fulfill

$$M_{\text{sc}} = N_{\text{sc}}^{\text{RB}} \cdot 2^{\alpha_2} \cdot 3^{\alpha_3} \cdot 5^{\alpha_5} \leq N_{\text{sc}}^{\text{RB}} \cdot N_{\text{RB}}^{\text{UL}}, \quad (3.66)$$

where $\alpha_2, \alpha_3, \alpha_5$ is a set of non-negative integers.

(iii) Modulation Scheme
The supported uplink modulation schemes are QPSK, 16 QAM and 64 QAM. The PUSCH can be QPSK, 16-QAM or 64-QAM modulated.

(iv) Mapping
The block of complex-valued symbols $z(0), \ldots, z(M_{\text{symb}} - 1)$ shall be multiplied with an amplitude scaling factor named β_{PUSCH} and mapped in sequence starting with $z(0)$ to resource blocks assigned for transmission of PUSCH. The mapping to resource elements (k,l) not used for transmission of reference signals shall be in increasing order of first the index k and then the index l, starting with the first slot in the subframe. The index k is given by

$$k = k_0 + f_{\text{hop}}(\cdot), \ldots, k_0 + f_{\text{hop}}(\cdot) + M_{\text{sc}} - 1, \quad (3.67)$$

where $f_{\text{hop}}(\cdot)$ denotes the frequency-hopping pattern and k_0 is given by the scheduling decision.

Frequency hopping provides additional frequency diversity, assuming that the hops are in the same order or larger than the channel coherency bandwidths.

3.8.6 PUCCH Physical Characteristics

(i) Scrambling
This process is identical to the PUSCH case.

(ii) Modulation
The physical uplink control channel, PUCCH, carries uplink control information. The PUCCH is never transmitted simultaneously with the PUSCH.

The physical uplink control channel supports multiple formats, as shown in Table 3.10, but only BPSK and QPSK modulations are supported, depending on the PUCCH format.

The scrambled signal $c(0), \ldots, c(M_{\text{bit}} - 1)$ is modulated according to Table 3.10, depending on its format, to provide $d(0), \ldots, d(M_{\text{symb}} - 1)$; the result is multiplied bit per

Table 3.10 Slot formats supported by the PUCCH.

Format	Number of reference symbols per slot	Modulation scheme	Number of bits per subframe, M_{bit}	
			Normal cyclic prefix	Extended cyclic prefix
0	3	OOK	1	1
1	3	BPSK	2	2
2	3	QPSK	20	20

bit with two sequences, respectively, for PUCCH formats 0 and 1 and one sequence only for PUCCH format 2. For the former case, there are a first cyclically shifted length $N_{ZC} = 12$ Zadoff-Chu sequence Zc(i) and a second orthogonal sequence $w(i)$, depending on the mobile. This signal aims to provide a high orthogonality between the respective signalling channels of all the mobiles in the cells. For frame structure type 1, the w sequence is four length-4 Hadamard sequences.

The result $d(i).Z_c(i).w(i)$ is mapped to resource elements assigned for transmission of PUCCH. For PUCCH format 2, the modulated signal is only modulated by the cyclically shifted base sequence.

(iii) Mapping

The block of complex-valued symbols $z(i)$ shall be mapped in sequence starting with $z(0)$ to resource elements assigned for transmission of PUCCH. The mapping to resource elements (k,l) not used for transmission of reference signals shall start with the first slot in the subframe. The set of values for index k shall be different in the first and second slots of the subframe, resulting in frequency hopping at the slot boundary.

3.8.7 Uplink Multiplexing Including Reference Signals

As already discussed, the channel-coded, interleaved and data modulation information can be arranged on a given number of resources units (RU), each being localized (LRU) or distributed (DRU). A localized LRU consists of M consecutive subcarriers during N long blocks. A distributed LRU consists of M equally spaced nonconsecutive subcarriers during N long blocks.

With $M = 25$ (other choices are possible), we have Table 3.11.

Several RUs can be assigned by the NodeB to the UE. In the case of LRU, the allocation should be contiguous in the frequency domain.

Table 3.11 Number of resource units, dependent on bandwidth.

Bandwidth (MHz)	1.25	2.5	5.0	10.0	15.0	20.0
Bandwidth (kHz) occupied by a resource unit	375	375	375	375	375	375
Number of available resource units	3	6	12	24	36	48

3.8.8 Reference Signals

As for the downlink, the reference signals are basically used by the eNodeB for channel estimation purposes and quality estimation for packet scheduling. Reference signals are transmitted within the two short symbols or 'blocks' of the slot, which are time-multiplexed with long 'blocks' and with an instantaneous bandwidth equal to the bandwidth of the data transmission. Multi-antenna mobiles may require multiple orthogonal reference signals for MIMO purposes, but can also be allocated to different UEs controlled by the same eNodeB.

Therefore, two types of uplink reference signals are supported:

- Demodulation reference signal, associated with transmission of uplink data and/or control signalling.
- Sounding reference signal, not associated with uplink data transmission with a much larger bandwidth than the former.

The same set of base sequences is used for demodulation and sounding reference signals.

The reference signals can be transmitted in a distributed or localized way. In the nominal case, the orthogonality between uplink reference signals can be achieved in the Frequency Domain (FDM), but it could be also done in the code domain, i.e. several reference signals could be Code Division Multiplexed (CDM) on a contiguous subcarriers set. The uplink reference signals are based on well known CAZAC sequences, leading to minimum intercorrelation products. Various phase shift of a single CAZAC sequence may be also used as a way of multiplexing the reference signals in the code domain. Within neighbours cells, the uplink reference signals should be based on different ZC sequences.

A combination of the two above-mentioned methods may also be used.

The base sequence $r(0), \ldots, r(M_{sc}^{RS}-1)$ of length M_{sc}^{RS} is defined by a time–domain cyclic shift in the frequency–domain truncated or extended Zadoff-Chu sequence (see Section 3.4.12). Note that different cyclic shifts can be used in different slots of a subframe. The cyclic shift to use in the first slot of the subframe is given by the uplink scheduling grant in the case of multiple shifts within the cell.

Demodulation reference signals for PUSCH and PUCCH and sounding reference signals are constructed from the base sequence.

Such or such method of multiplexing the reference signals will be selected, depending on the scenario of the mobile spectrum allocation (same band or not), eNodeB selection, and also antenna configuration (MIMO).

3.8.9 Multiplexing of L1/L2 Control Signalling

The uplink channel carries three types of information:

- The data.
- The data-associated signalling mandatory for uplink data demodulation. This includes the transport format or the HARQ information (retransmission sequence number for synchronous HARQ or hybrid ARQ process number, redundancy version, etc. in the case of asynchronous ARQ).
- The signalling nonassociated to data, like, for instance, the information associated with the downlink transmission, such as the downlink CQI (Channel Quality Indicator), ACK/NACK due to downlink transmissions and scheduling request for uplink transmission assigned by the eNodeB, Synchronous or asynchronous random access and MIMO-related feedback information if required.

All of this information is time-multiplexed within the subframe and then mapped in the time–frequency plan.

From a frequency point of view, information of various types can be time-multiplexed on a given subcarrier part of a resource unit. For instance, the frequency resources can successively carry the pilot, some control information for multiple UEs, some data for a given set of UEs, etc.

From a time perspective, information of various types can be frequency-multiplexed on subframe duration. For instance, some control information for multiple UEs, some data for a given set of UEs and multiple pilots can be transmitted at the same time on several resource units.

3.8.10 Channel Coding and Physical Channel Mapping

Channel coding is based on UTRAN release 6 turbo-coding schemes, i.e. turbo code with $R = 1/3$. Other FEC (Forward Error Correction) schemes are also envisaged to cope with additional E-UTRAN requirements, like codes polynomial for lower rates or repetition coding for higher processing gain. This is similar to the downlink.

3.8.11 SC-FDMA Signal Generation

The SC-FDMA symbols in a slot shall be transmitted in increasing order of l. The time-continuous signal $s_l(t)$ in SC-FDMA symbol l in an uplink slot is defined by

$$s_l(t) = \sum_{k=-\lfloor N_{RB}^{UL} N_{sc}^{RB}/2 \rfloor}^{\lceil N_{RB}^{UL} N_{sc}^{RB}/2 \rceil - 1} a_{k^{(-)},l} \cdot e^{j2\pi(k+1/2)\Delta f(t - N_{CP,l} T_s)} \quad (3.68)$$

for $0 \leq t < (N_{CP,l} + N) \times T_s$, where $k^{(-)} = k + \lfloor N_{RB}^{UL} N_{sc}^{RB}/2 \rfloor$. The variable $N = 2048$ and $\Delta f = 15$ kHz.

Table 3.12 lists the values of $N_{CP,l}$ that shall be used for the two frame structures. Note that different SC-FDMA symbols within a slot may have different cyclic prefix lengths. For frame structure type 2, note that the SC-FDMA symbols do not fill all uplink subframes completely, as the last part is used for the guard interval.

3.8.12 The Random Access Channel

As mentioned, the process can be synchronized or non-synchronized. In the non-synchronized case, the random access burst is transmitted with no prior synchronization of the uplink with respect to eNodeB. The minimum used bandwidth is 1.25 MHz. In the frequency domain, the random access burst occupies a bandwidth corresponding to $N_{BW}^{RA} = 72$ subcarriers for both frame structures. Higher layers configure the location in frequency of the random access burst.

Table 3.12 SC-FDMA parameters.

	Cyclic prefix length $N_{CP,l}$		
Configuration	Frame structure type 1	Frame structure type 2	Guard interval
Normal cyclic prefix	160 for $l=0$ 144 for $l=1, 2, \ldots, 6$	224 for $l=0, l, \ldots, 8$	288
Extended cyclic prefix	512 for $l=0, 1, \ldots, 5$	512 for $l=0, 1, \ldots, 7$	256

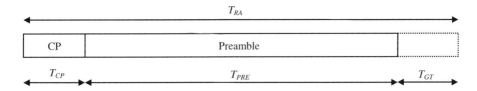

Figure 3.64 Random access preamble format (frame structure type 1).

The preamble sequence occupies $T_{PRE} = 0.8$ ms and the cyclic prefix occupies $T_{CP} = 0.1$ ms, resulting in a guard period of $T_{GT} = 0.1$ ms in case one subframe pair is reserved for random access. Higher layer signalling controls in which subframe pairs random access preamble transmission is allowed.

The physical layer random access burst, illustrated in Figure 3.64, consists of a cyclic prefix of length T_{CP}, a preamble of length T_{PRE}, and a guard time T_{GT}, during which nothing is transmitted. The parameter values are listed in Table 3.13 and depend on the frame structure and the random access configuration. Higher layers control the preamble format. For instance, up to four formats are allowed for random access burst parameters for type 1 frame structure.

For frame structure type 2, the start of the random access burst depends on the burst length configured. For the normal burst length, the burst shall start T_{RA} before the end of the UpPTS at the UE. For the extended burst length, the start of the random access burst shall be aligned with the start of uplink subframe 1.

In the frequency domain, the random access burst occupies a bandwidth corresponding to $N_{BW}^{RA} = 72$ subcarriers for both frame structures. Higher layers configure the location in the frequency of the random access burst.

Open loop power control is used to determine the initial transmit power level. It is possible to vary the random access burst transmit power between successive bursts using power ramping with configurable step size including zero step size for both the FDD and TDD cases. Classically, the access burst comprises mainly a preamble used for signature detection and time alignment and a few bits payload message.

The random access preambles are generated from Zadoff-Chu sequences with zero correlation zones, generated from one or several root Zadoff-Chu sequences. The network configures the set of preamble sequences the UE is allowed to use.

Table 3.13 Random access burst parameters.

Frame structure	Burst format	T_{CP}	T_{PRE}
Type 1	0	$3152 \times T_s$	$24576 \times T_s$
	1	$21012 \times T_s$	$24576 \times T_s$
	2	$6224 \times T_s$	$2 \times 24576 \times T_s$
	3	$21012 \times T_s$	$2 \times 24576 \times T_s$
Type 2	0	$0 \times T_s$	$4096 \times T_s$
	1	$0 \times T_s$	$16384 \times T_s$
	2		

Table 3.14 Random access preamble sequence parameters.

Frame structure	Burst format	N_{ZC}	N_{CS}	Number of preambles	Preamble sequences per cell
Type 1	0–3	839			64
Type 2	0	139		552	16
	1	557			

The uth root Zadoff-Chu sequence is defined by

$$x_u(n) = e^{-j\frac{\pi u n(n+1)}{N_{ZC}}}, \quad 0 \leq n \leq N_{ZC}-1, \tag{3.69}$$

where the length N_{ZC} of the Zadoff-Chu sequence is given by Table 3.13.

From the uth root Zadoff-Chu sequence, random access preambles with zero correlation zones are defined by cyclic shifts of multiples of N_{CS} according to $x_{u,v}(n) = x_u(n + vN_{CS})$ mod N_{ZC}), where N_{CS} is given by Table 3.14.

About the Zadoff-Chu sequence:

In 1972, Chu and, almost at the same time, Zadoff described a method to construct a polyphase perfect sequence without constraint on the sequence length N. They observed that original sequences always have the best peak-to-side peak ratios over all possible decimations and shift values. In addition, the ratios grow linearly with the square root of the length N of the sequences.

All out-of-phase periodic autocorrelation values are 0.

The PAPR and dynamic range of Zadoff-Chu sequences are far better than the Walsh codes.

The so-called **polyphase Zadoff-Chu** sequences are among the most important class of CAZAC (Constant Amplitude Zero Auto-Correlation) sequences, whose elements are roots of unity. It has been shown that such sequences exist for all periods L. If L is odd, an L-phase sequence can be constructed; if L is even, two L-phases are needed.

CAZAC sequences are useful for channel estimation and fast start-up equalization. They are shown to be optimal under certain assumptions. The most important members of this family are the polyphase sequences of lengths that are a power of 2 because of their suitability for Fast Fourier Transform processing.

(i) Base Band Signals Generation for Random Access

The time-continuous random access signal $s(t)$ is defined by

$$s(t) = \beta_{PRACH} \sum_{k=0}^{N_{ZC}-1} \sum_{n=0}^{N_{ZC}-1} x_{u,v}(n) \cdot e^{-j\frac{2\pi nk}{N_{ZC}}} \cdot e^{j2\pi(k+\varphi+K(k_0+1/2))\Delta f_{RA}(t-T_{CP})}, \tag{3.70}$$

where β_{PRACH} is an amplitude scaling factor, $0 < T_{PRE} + T_{CP}$ and $k_0 = k_{RA}N_{SC}^{RB} - N_{RB}^{UL}N_{SC}^{RB}/2$. The location in the frequency domain is controlled by the parameter k_{RA}, expressed as a resource block number configured by higher layers and fulfilling $0 \leq k_{RA} < N_{RB}^{UL}-6$. The factor $K = T_{PRE}/(2048 \cdot T_s)$ accounts for the difference in subcarrier spacing between the random access preamble and uplink data transmission and the variable $\varphi = 12$ is a fixed offset determining the frequency–domain location of the random access preamble within the resource blocks.

(ii) The Access Procedure
Figure 3.65 provides one possible example of a random access procedure: the eNodeB responds to the nonsynchronized random access attempt preamble with timing information and resource allocation for transmission of scheduling request (and possibly any additional control signalling or data).

(iii) Timing Information
On receiving a random access burst from a UE, the network determines if the UE needs a timing advance (TA) adjustment and, if so, signals to the UE a TA indicator which is a multiple of 0.52 µs and is applied as a one-step adjustment relative to the random access channel preamble transmit timing.

A TA indicator received by the UE will be a multiple of 0.52 µs and is applied as a one-step adjustment relative to current uplink timing.

UE then sends the scheduling request at the assigned time–frequency resource using the shared data channel or physical random access channel (for coexisting LCR-TDD-based frame structure). The eNodeB adjusts the resource allocation according to the scheduling request from the UE.

In the case of the synchronized procedure, the UE uplink is synchronized with the eNodeB so that the latency time of the access procedure can be reduced. The minimum bandwidth is equal to the bandwidth of the uplink RU, i.e. 375 kHz, but can be wider.

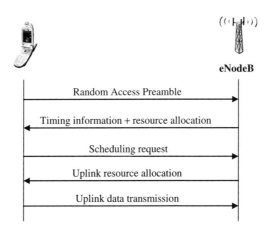

Figure 3.65 Example of nonsynchronized RACH procedure.

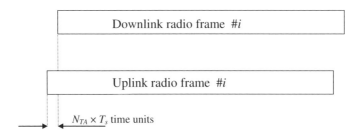

Figure 3.66 Uplink–downlink timing relation.

3.8.13 Uplink–Downlink Frame Timing (Figure 3.66)

Transmission of the uplink radio frame number i from the UE shall start $N_{TA} \times T_s$ seconds before the start of the corresponding downlink radio frame at the UE. Note that not all slots in a radio frame may be transmitted. One example hereof is TDD, where only a subset of the slots in a radio frame is transmitted.

3.8.14 Scheduling

Downlink control signalling informs UE(s) about resources and respective transmission formats to be allocated. The decision of which user transmissions to multiplex within a given subframe may, for example, be based on:

- Type of required services (BER, min and max data rate, latency, etc.).
- Quality of Service parameters and measurements.
- Pending retransmissions.
- Uplink channel quality measurements.
- UE capabilities.
- UE sleep cycles and measurement gaps/periods.
- System parameters such as bandwidth and interference level/patterns.
- etc.

The states of buffers inside the mobiles are unknown to the eNodeB; therefore, the scheduling cannot be based also on this type of information, as for the downlink.

However, some time–frequency resources can be allocated for contention-based access. Within these time–frequency resources, UEs can transmit without first being scheduled. As a minimum, contention-based access should be used for random-access and for request-to-be-scheduled signalling.

In unpaired spectrums, system capacity may be improved through the use of localized FDMA contention-based access channels. The UE may select the access channel based upon knowledge of the channel state information measured on a recent downlink subframe.

3.8.15 Link Adaptation

In a large sense, the uplink link adaptation process deals, according to the radio channel conditions, with the following:

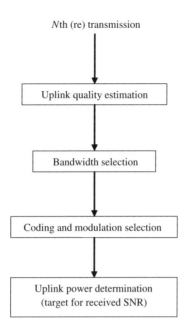

Figure 3.67 The uplink link adaptation process.

- Adaptive transmission bandwidth.
- Transmission power control.
- Adaptive modulation and channel coding. The same coding and modulation is applied to all resource units assigned to which the same L2 PDU is mapped on the shared data channel scheduled for a user within a TTI.

Therefore, the global scheme for uplink link adaptation can be seen in Figure 3.67.

3.8.16 Uplink HARQ

As for the downlink case, there are two levels of retransmissions for providing reliability, namely the Hybrid Automatic Repeat reQuest (HARQ) at the MAC layer and an outer ARQ at the RLC layer. The outer ARQ is required to handle residual errors that are not corrected by HARQ that is kept simple by the use of a single bit error-feedback mechanism. An N-process stop-and-wait HARQ is employed that has synchronous retransmissions in the UL.

Thus, in the current standard for the uplink, a synchronous, nonadaptive HARQ is used for SC-FDMA on the uplink of E-UTRAN. The main advantages of synchronous nonadaptive HARQ are:

- Reduced control signaling.
- Lower complexity for HARQ operation.
- The possibility of soft-combining control information.

In the case of synchronous HARQ, the uplink attributes of each of the retransmissions may remain the same as for the first transmission.

Reference documents about E-UTRAN physical layer:

3GPP technical specifications:

- 36.201, 'LTE Physical Layer: General Description'
- 36.211, 'Physical Channels and Modulation'
- 36.212, 'Multiplexing and Channel Coding'
- 36.213, 'Physical Layer Procedures'
- 36.214, 'Physical Layer Measurements'

Other documents:

- Moisseev (2006), 'System load model for the OFDMA network', *IEEE Communication Letters*, 10(8).
- Meyer, Saford, Cheng (2006), 'ARQ concepts for UMTS long-term-evolution', IEEE, Vehicular Technology Conference.
- Hicheri, Terre, Fino, 'CDMA,OFDM, MC-CDMA quel choix pour une liaison descendante', Conservatoire National des Arts et Metiers.
- Milewski (1983), 'Periodic sequences with optimal properties for channel estimation and fast start-up equalization', *IBM J. Res. Develop.*, 27(5).
- Common Public Radio Interface, available online at www.CPRI.info.
- *Improved Feedback for MIMO Precoding*, IEEE C802.16e-04/527r4, Intel Corporation.

4

Evolved UMTS Architecture

The aim of this chapter is to enter into the details of Evolved UMTS architecture, and complete the general overview provided in Chapter 2. For that purpose, this chapter will describe in more detail the functional entities of the network as well as the interfaces and protocols. A special focus will be given on the radio interface, but the IMS protocol stack, built on top of Evolved UMTS architecture, will also be described.

4.1 Overall Architecture

Figure 4.1 describes a simplified view of the EPS architecture. Not all the network nodes and interfaces are represented here (as this was already done in Chapter 2). This picture rather focuses on E-UTRAN/EPC interactions and user signalling and data connectivity and architectural aspects which will be developed in this chapter.

Starting from the E-UTRAN part, the X2 interface shall be seen as a meshed interface rather than a point-to-point between two specific E-UTRAN nodes. This optional interface has been defined for the main purpose of forwarding packets between eNodeB so as to limit packet loss for intra E-UTRAN user mobility.

The S1 interface shall also not be seen as a simple interface between one eNodeB and one MME/Serving gateway (represented above as a single box for simplicity), since an eNodeB can possibly be connected to more than one MME. This flexibility is known as S1-flex (an equivalent of the Iu-flex 3G/UMTS option) described later in this chapter.

When MME and Serving GW are deployed as two separate physical boxes, the S1 interface is split into two parts:

- The S1-U (for User plane) – which carries user data between eNodeB and Serving GW.
- The S1-C (for Control plane) – which is a signalling-only interface between the eNodeB and the MME.

On the S5 side, a serving GW may also be linked to different PDN GW, corresponding to the fact that EPC may provide user connectivity to several different and separated IP networks.

Evolved Packet System (EPS) P. Lescuyer and T. Lucidarme
Copyright © 2008 John Wiley & Sons, Ltd.

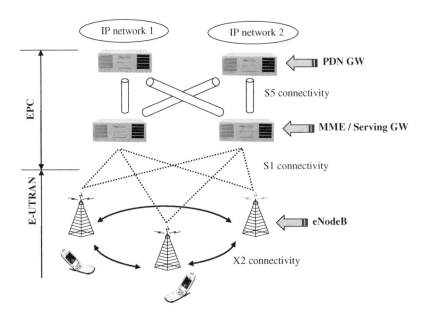

Figure 4.1 EPS architecture – user and control plane connectivity.

The following parts of this section provide a more detailed description of the EPC and E-UTRAN nodes which were introduced in Chapter 2. In a second step, the specific aspects of the S1 and X2 interface are presented.

4.1.1 Evolved UMTS Node Features

(i) In E-UTRAN
As presented in Chapter 2, the eNodeB is the only type of logical node present in the E-UTRAN part of the network. The main features an eNodeB supports are the following:

- Radio Bearer management – this includes Radio Bearer setup and release and also involves radio resource management features for initial admission control and bearer allocation. This set of functions is under the control of the MME through the S1 interface during session setup, release and modification phases.
- Radio interface transmission and reception – this includes radio channel modulation/demodulation as well as channel coding and decoding.
- Uplink and Downlink Dynamic radio resource management and data packet scheduling – this is probably the most critical function which requires the eNodeB to cope with many different constraints (like radio-link quality, user priority and requested Quality of Service) so as to be able to multiplex different data flows over the radio interface and make use of available resources in the most efficient way.
- Radio Mobility management – this function relates to terminal mobility handling while the terminal is in an active state. This function implies radio measurement configuration and processing as well as the handover algorithms for mobility decision and target cell

determination. Radio Mobility has to be distinguished from Mobility Management in Idle, which is a feature handled by the Packet Core.
- User data IP header compression and encryption – this item is key to radio interface data transmission. It answers to the requirements to maintain privacy over the radio interface and transmit IP packets in the most efficient way.
- Network signalling security – because of the sensitivity of signalling messages exchanged between the eNodeB itself and the terminal, or between the MME and the terminal, all this set of information is protected against eavesdropping and alteration.

In addition, the eNodeB also supports some additional functions, which are less obvious but still mandatory to make the overall system work:

- Scheduling and transmission of broadcast information – this function is present in most, if not all, of the cellular systems. It refers to system information broadcasting so that idle terminals can learn network characteristics and be able to access and register to it.
- Scheduling and transmission of paging messages – this function is essential for the network to be able to set up mobile terminated sessions. In addition, paging is also used for nonidle but inactive terminals which the network needs to join.
- Selection of MME at terminal attachment – this nonessential feature may be used to increase network resilience to EPC node failure, and also helps to cope with network load management. It is part of the 'S1 flexibility' option described hereafter.

As described further in this chapter, the eNodeB is defined by the 3GPP using the traditional OSI-like layered model. From that perspective, all the functions listed above are handled by different layers, like the physical layer, the RLC/MAC data link layer, or the Radio Resource Control signalling layer.

(ii) In EPC
Putting aside the HSS (Home Subscriber Server) already described in Chapter 2, the EPC comprises three logical nodes:

- The MME (Mobility Management Entity) associated with the Control plane, or terminal to network signalling handling.
- Two packet data gateways associated with the User plane: the Serving GW and the PDN GW.

The main features supported by the MME are the following:

- NAS signalling support – NAS (Non Access Stratum) signalling refers to the signalling layer being used between the Packet Core and the terminal supporting functions such as network attachment and data session setup.
- Active session mobility support – this refers to user context transfer in the case of active session mobility, either between 2G and 3G systems (which involves user context transfer over the S3 interface) or between MME nodes (which involves S10 support).
- Idle mode terminal Mobility Management – this function is also known as terminal location tracking. It allows the EPC to know where to page terminals in case of user-terminated sessions.
- Authentication and Key Agreement (AKA) – this refers to user and network-mutual authentication and session key agreement between terminal and EPC.

- Determination of Serving and PDN GW at bearer establishment – this function is the EPC equivalent of the GGSN selection function which is performed in 2G/GPRS and 3G/UMTS networks by the SGSN.

The main features supported by the Serving GW are:

- Packet routing between E-UTRAN and EPC.
- Mobility anchoring – the Serving GW is actually the User plane anchor point in case of active session mobility between 2G and 3G systems (which involves the S4 interface) or between eNodeB in E-UTRAN. This is described in more detail in the mobility part of Chapter 5.

In addition, the Serving GW shall support some buffering capabilities in case of network-initiated service requests. This typically occurs when the network receives a downlink packet for a terminal while no Radio Bearer is available and the terminal is inactive. In such a case, MME needs to page the terminal and the Serving GW shall buffer downlink packets until the Radio Bearer is established.

The main PDN GW features are:

- Packet routing between the EPC and external PDN (Packet Data Network). In this context, PDN is a very generic term which covers any kind of IP network as well as IMS domain.
- Policy enforcement – based on the rules provided by the PCRF.
- Charging support – as being the EPC edge router, the PDN GW is in charge of applying specific data-flow charging rules.
- IP address allocation for terminals – the IP address allocation is performed when the initial bearer is set up during the network attachment procedure, as described in Chapter 5.

(iii) Moved Functions

The architecture of E-UTRAN has introduced some modifications about the location of the main functions within the network. As a result, there are some differences in the way packets are handled. These changes are presented in Figure 4.2, which focuses on the downlink User plane.

In UMTS, taking high-speed HSDPA as an example, data packets used to be buffered twice, as a result of the separation of the two data retransmission ARQ (Automatic Repeat Request) loops. The outer loop supported by the RLC is located in the RNC, whereas the MAC Hybrid ARQ loop is located in the NodeB, as HARQ is based on link adaptation mechanisms close to the radio interface. When RLC is configured in AM (Acknowledged Mode) or UM (Unacknowledged Mode) – which is the case for IP-based services transported over HSDPA – the RLC layer is also in charge of data encryption.

The consequence of this physical separation of RNC and NodeB and related functions is that UMTS data packet handling requires two separated packet buffers, one for each of the repetition loops. This implies the implementation of flow control mechanism over the Iub interface in order to avoid two potential issues:

- NodeB buffer overflow – if the NodeB receives more packets from the RNC than it can actually buffer.

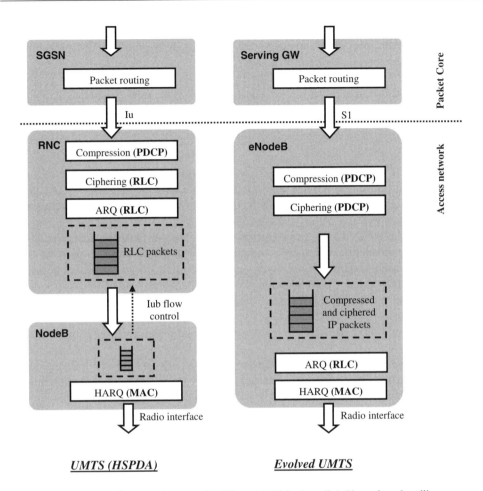

Figure 4.2 Differences between UMTS and EPS in downlink User plane handling.

- NodeB data shortage – if the packet arrival rate at the NodeB is too small when compared to the available radio resources. This means that the radio bandwidth is not used in an optimum way.

In Evolved UMTS, Compression and Ciphering are both supported by the PDCP layer, located in the eNodeB. As all retransmission mechanisms are also located in the eNodeB, data packet processing requires only one buffer of header compressed and ciphered IP packets.

In addition, it is interesting to note that, in the UMTS/HSDPA case, the whole NodeB data buffer will be lost when the terminal changes NodeB during its HSDPA session. Of course, the lost data will be retransmitted, thanks to the overall RLC ARQ loop (when it is configured in Acknowledged Mode) or even higher at the TCP application level (if RLC is configured in Unacknowledged Mode and if TCP is used). The price to pay is an increased data recovery time, as the reaction time at RLC and TCP levels is much higher than in the NodeB MAC. This is the reason why the amount of buffered packets at the NodeB level shall be minimized by the RNC.

In the E-UTRAN case, thanks to the X2 inter-eNodeB interface, the data buffer may be forwarded between source and target eNodeB, which helps to minimize the probability of packet loss at the lowest radio protocol level.

The last point to note is that there is no difference in the Packet Core gateway node as regards to packet handling. From a User plane perspective, the role of the 2/3G SGSN or the Serving SAE GW is limited to packet data routing, as the compression and ciphering features remain in the Access network.

4.1.2 E-UTRAN Network Interfaces

The aim of this section is to provide some general information about the E-UTRAN S1 and X2 network interfaces. Those two interfaces follow the same model, which is described in Figure 4.3.

Similarly to the 3G/UTRAN network interface model, the E-UTRAN model is composed of two main parts: the radio network layer – which encompasses the top-level protocols of the interface – and the transport network layer – which only refers to as the way radio network layer data are transported. This separation allows independence between the two layers, so that, for example, the application part can evolve without impacting the transport layer, or the other way around.

In addition to this OSI-like vertical separation, each interface is split between a User plane and a Control plane.

The User plane transports all information considered as user data from the interface point of view. This consists of pure user data (such as voice or video packets) as well as application-level signalling (such as SIP, SDP or RTCP packets). Before transmission over the interface,

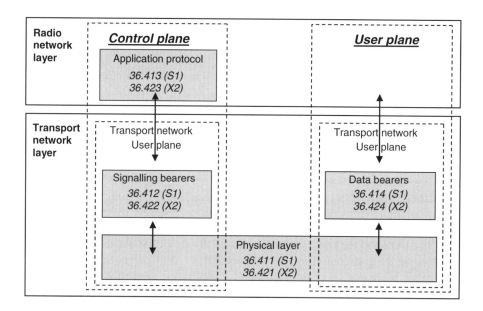

Figure 4.3 The E-UTRAN network interface model.

the different packets are simply submitted to the transport layer, without any kind of processing. This is the reason why the radio network box of the User plane is void.

The Control plane relates to all messages and procedures strictly related to the interface-supported features. This includes, for example, control messages for the handover management or bearer management.

The physical layer, as part the transport layer, is common to both User and Control planes. Aside from that, User and Control planes use specific protocol boxes which allow defining a different and independent transport stack and bearers for each of the planes. As described below, Control plane information is more constraining in terms of security, reliability and data loss, whereas the User plane information can rely on simpler and less secured routing protocols.

As in 3G/UTRAN, the E-UTRAN interfaces are fully 'open', meaning that S1 and X2 are completely defined by the 3GPP in all details and any manufacturers implementing an E-UTRAN node shall conform to the way the interfaces are specified. This allows – in principle – eNodeB from different manufacturers to be deployed in a single network and be inter-connected over the X2 interface or with MME or Serving GW nodes over the S1 interface.

Although being shared between the Access and Core Networks, the S1 interface is under the responsibility of the 3GPP Radio Access Network (RAN) group. This is due to the fact that most of the functions supported by this interface are tightly related to E-UTRAN features, such as the user mobility or Radio Bearer management.

Reference documents about the S1 interface

3GPP Technical specifications:

- 36.410, 'S1 General Aspects and Principles'
- 36.411, 'S1 Layer 1'
- 36.412, 'S1 Signalling Transport'
- 36.413, 'S1 Protocol Specification'
- 36.414, 'S1 Data Transport'

Reference documents about the X2 interface

3GPP Technical specifications:

- 36.420, 'X2 General Aspects and Principles'
- 36.421, 'X2 Layer 1'
- 36.422, 'X2 Signalling Transport'
- 36.423, 'X2 Protocol Specification'
- 36.424, 'X2 Data Transport'

4.1.3 S1 Interface

(i) S1 User Plane Interface

The S1-U (or S1 User plane interface) role is to transport user data packet between the eNodeB and the Serving GW. This interface makes use of a very simple 'GTP over UDP/IP' transport protocol stack which only provides user data encapsulation. There is no flow control or error control, or any mechanism to guarantee data delivery over the S1-U interface.

The GTP (GPRS Tunnelling Protocol) is actually inherited from 2G/GPRS and 3G/UMTS networks. In 2G/GPRS networks, GTP is used between GPRS nodes (the SGSN and the GGSN). In 3G/UMTS networks, GTP is also used over the Iu-PS interface (between the RNC and the SGSN).

(ii) S1 Control Plane Interface
The S1-C (or S1 Control plane interface) is a signalling interface which supports a set of functions and procedures between the eNodeB and the MME. All the S1-C signalling procedures belong to four main groups:

- Bearer-level procedures – this set corresponds to all procedures related to bearer setup, modification and release. On the scope of the S1 interface, a bearer corresponds to the S1 segment of a session plus the radio interface path. These procedures are typically used during the establishment or the release of a communication session.
- Handover procedures – which encompasses all the S1 functions related to user mobility between eNodeB or with 2G or 3G 3GPP technologies.
- NAS signalling transport – this corresponds to the transport of terminal–MME signalling over the S1 interface. The terminal–MME signalling is also called NAS (Non Access Stratum signalling), as it is transparent to the eNodeB. Due to the importance of these messages, they are transported over S1-C using specific procedures, rather than the nonguaranteed S1-U GTP.
- Paging procedure – which is used in case of user terminated session. Through the paging procedure, the MME request the eNodeB to page to terminal in a given set of cells.

The S1-C interface shall provide a high level of reliability in order to avoid message retransmission and unnecessary delay in control plane procedure execution.

Depending on transport network deployment, there may be some cases in which the UDP/IP transport is not reliable enough. Besides, in case the transport network is not owned by the mobile radio network operator, it may happen that the transport network Quality of Service cannot be guaranteed all the time. This is the reason why the S1-C interface makes use of a reliable transport network Layer, which is set up end-to-end (between the eNodeB and the MME nodes).

In the EPS architecture, this service is ensured by SCTP.

(iii) About SCTP
STCP (Stream Control Transmission Protocol) is a reliable connection-oriented transport protocol which is very similar to the well known and widely used TCP. As TCP, STCP implements congestion and flow control, detection of data corruption, loss or duplication of data and supports a selective retransmission mechanism.

As TCP, STCP works in connected mode, so that an 'association' (the actual STCP term for 'connection') needs to be set up between peers before data transmission can occur. In SCTP, an association is defined by a (Source IP, Source Port, Destination IP, Destination Port) group.

When comparing TCP and SCTP from a functional perspective, SCTP provides two key features which TCP does not support:

- The multi-streaming.
- The multi-homing.

In the SCTP domain, a stream is a unidirectional sequence of user messages to be delivered to upper layers. As a consequence, bi-directional communication between two entities involves at least a pair of streams, one for each direction. The multi-streaming is the feature from which the STCP name is actually derived. It allows setting up several independent streams between two peers. In such a case, when a transmission error occurs on one of the stream, it does not affect data transmission on the other streams.

In contrast, TCP only provides one stream for a given connection between IP peers, which may cause additional data transmission delay when a packet or group of packets is lost. When a transmission loss occurs on a TCP connection, packet delivery is suspended until the missing parts are restored, as in-sequence data delivery (or data sequence preservation) is a key TCP feature.

This important characteristic of TCP is not necessary in all cases. For example, in the case of a multimedia document such as a Web page, multiple parallel streams may be opened in parallel to retrieve the whole page content. Content delivery is more critical than content order for such a kind of application. The same applies to independent signalling flows which are transferred between two network nodes, such as the MME and the eNodeB. The delivery order of each signalling flow (e.g. corresponding to one mobile-network connection) needs to be preserved; however, all the flows can be delivered independently.

The other core added value of SCTP is multi-homing. This allows a SCTP endpoint to be reached through multiple network addresses. The interest of multi-homing is about redundancy, as it improves the resilience when network failures occur. In case of transmission errors, retransmitted packets may be sent to alternate addresses in order to increase the probability of successful transmission.

Of course, there are other differences between TCP and SCTP. The two which are worth mentioning here are:

- SCTP framing: SCTP works at the message level whereas TCP is an octet stream protocol. This was one of the main reasons for 3GPP adoption in E-UTRAN signalling transport.
- SCTP built-in cookie-based protection against denial of service attacks, which is described hereafter.

(iv) Why SCTP?

There have been some debates in 3GPP working groups on which transport protocol would be the most suitable for the E-UTRAN Control plane, to support signalling message exchange between network nodes. Among the three most obvious candidates, UDP was quickly ruled out as not being reliable enough.

From a high-level perspective, SCTP and TCP are quite close to each other, as they both support reliable and ordered data delivery, as well as congestion control to regulate network data flow. What really made the difference in favour of STCP was the following:

- The multi-streaming feature.
- The fact that SCTP is message-oriented and supports framing of individual messages as opposed to TCP, which is octet stream-oriented and does not preserve transmitted data structure. In SCTP, messages are transmitted as a whole set of bytes (provided the maximum length is not reached) which helps to improve transmission efficiency.

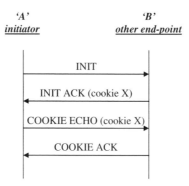

Figure 4.4 The steps of SCTP association setup.

- The resilience of SCTP against some types of denial of service attack TCP is vulnerable to, like the 'SYN flood'.

The 'SYN flood' attack is special kind of attack, causing a TCP endpoint to receive a connection request to reserve a resource context and memory for some incoming connections that will never be fully set up by the initiator. At some point, this process can exhaust all memory or processing resources in the receiver. To counter this, SCTP makes use of a 'cookie' mechanism. The actual seizure of association resource is only performed once the initiator successfully answers with the correct cookie.

For illustration, Figure 4.4 describes the four steps of a SCTP association establishment.

On reception of the INIT message, the receiver builds a cookie and sends it to the initiator using the INIT ACK message. To enable the association, the initiator must answer a COOKIE ECHO containing the same cookie as received in the INIT ACK. Resource reservation related to the association is only performed by the 'B' side on reception of a COOKIE ECHO. At the end, the COOKIE ACK is sent back to the initiator to acknowledge the association setup.

Resource attack is prevented by building the 'cookie' in a special way. In principle, the receiver of the INIT message is using a secret key and a hash mechanism to create it, so that on reception of the COOKIE ECHO, it can then validate that the cookie was actually previously generated by the receiver.

This protection is based on the fact that the receiving entity (the 'B' part in the diagram) does not reserve resources or keep context pending during the INIT phase. Resource activation is only performed when a valid 'COOKIE ECHO' message is received. Of course, this assumes the rogue initiator does not process the answers, which is generally the case for denial of service attacks.

The cookie structure is not fully specified by the SCTP recommendation, but it may possibly contain a Timestamp corresponding to its creation time.

(v) SCTP in E-UTRAN Transport Network
In the S1 interface (and the same applies to the X2 interface described below), SCTP is used over the usual IP network layer. There is only one association per instance of S1 interface (or 'eNodeB to MME' relation). Over this association, one SCTP stream is used for all common procedures – such as the paging procedure – between two pieces of equipment.

Regarding all dedicated procedures – which include all procedures which apply to a specific communication context – they all are supported over a limited number of SCTP streams.

(vi) IP in E-UTRAN Transport Network
The transport network of the S1 and X2 interfaces makes use of the legacy IP network layer for both User and Control planes. In addition to the basic services provided by this protocol, IP in E-UTRAN shall also support the following:

- NDS/IP (Network Domain Security for IP) – which refers to a set of IP-level security features (in the sense of confidentiality and integrity) defined by 3GPP for data exchange between network entities. NDS/IP is further described in the security section of Chapter 5.
- Diffserv (Differentiated Services) – which refers to an enhancement of the IP protocol providing service discrimination.

DiffServ re-uses an existing field present in the IP header [the IPv4 TOS field (Type of Service) or IPv6 Traffic Class field] so as to define a new field known as the DS field (for (DiffServ field). This DS field, assigned by IP edge routers, allows intermediate routers to classify the packets and apply specific (or differentiated) processing known as PHB (Per Hop Behaviour). Depending on the value of the DS field, an intermediate router is then able to, for example, assign each packet to a given waiting queue and provide a better precedence to any IP flow which would be marked as 'high priority' by an edge router.

> *Reference documents*
>
> SCTP IETF documents:
>
> - RFC2960, 'Stream Control Transmission Protocol'
> - RFC3286, 'An Introduction to the Stream Control Transmission Protocol'
>
> Diffserv IETF document:
>
> - RFC2474, 'Definition of the Differentiated Services Field in the IPv4 and IPv6 Headers'
> - RFC2475, 'An Architecture for Differentiated Services'

4.1.4 S1 Flexibility

In traditional 2G and 3G cellular networks, the connectivity between the Core Network and Access Network part was defined as a one-to-multi hierarchical relationship: a Core Network node (either the MSC on the Circuit domain or the SGSN in the Packet domain) serves a set of radio Controllers (the 2G BSC or the 3G RNC), and a given controller is only assigned to one Core Network node within a domain. In other words, each Core Network node is connected to its own set of radio Controllers, having no intersections with other sets.

In Release 5 of the 3G/UMTS standard, a new feature was introduced, allowing more flexibility in the inter-connection between Access and Core nodes, breaking the usual network hierarchy. This feature has been introduced from the beginning in the EPS standard and is known as 'S1-flex'.

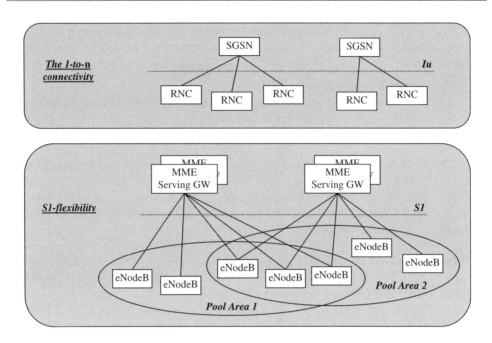

Figure 4.5 Traditional Access–Core connectivity and S1-flex.

As represented in Figure 4.5, S1-flex allows an eNodeB to be connected to more than one MME or Serving GW node. In this picture, MME and Serving GW are combined in one node for simplicity, but the S1 flexibility applies to both MME and Serving GW, independently. This picture also introduces the notion of pool area.

Some definitions

- A pool area is an area which a terminal may move into without a need to change its serving Core Network node.
- A pool area is composed of a predefined set of eNodeB, corresponding to one or several TA (Tracking Areas).
- Pool areas can overlap.
- A pool area can be served by one or several MME and/or Serving GW.
- A given MME or Serving GW node may serve one or several pool areas.

Although an eNodeB can be connected to several MME, a terminal is only associated to one MME at a time, due to the fact that the sessions of a subscriber are always under the control of a single Core Network MME node.

S1 flexibility has many advantages:

- By extending the usual service area seen by a Core Network node, the S1 flexibility allows the reduction of the number of inter-Core Network node handover procedures (in Connected

mode) or Tracking Area updates (in Idle mode). This extends the possibility for a MME to maintain the connectivity with a moving terminal, as long as the terminal remains in the same pool area. As a consequence, the S1 flexibility helps to reduce the HSS load generated by the change in MME.
- The S1 flexibility also helps to define network architectures shared by different operators. As an example, part of the E-UTRAN network – represented by a set of eNodeB in a given geographical – may be simultaneously operated by two different business entities. In such a case, when a terminal attempts to register, the eNodeB can forward the initial registration message to the MME, which corresponds to the network operator of the subscriber.
- S1 flexibility allows the network to become more robust to Core Node failure, as the loss of one Core Network node will be compensated for by other nodes associated to the same pool areas. This increased service availability is, however, not dynamic, meaning that in case of failure, on-going communication sessions are not automatically transferred to a new node.

And, at last, S1 flexibility has some advantages as regards to capacity upgrade and network load management. Opening the possibility for an eNodeB to be connected to more than one MME allows balancing and possibly redistributing the load by directing incoming terminal connection requests to less loaded Core Network nodes.

The S1 flexibility relies on a new field of information which is actually a sub-part of the temporary subscriber identity (S-TMSI). This new field uniquely identifies a MME in an area served by multiple MMEs and is then used by the eNodeB to direct an initial connection request towards the right MME – or set of MME – in case of network sharing, or to send the terminal initial message to the MME it was registered to.

4.1.5 X2 Interface

(i) X2 User Plane Interface
The X2-U (or X2 User plane interface) role is to transport user data packets between eNodeBs. This interface is only used for limited periods of time, when the terminal moves from one eNodeB to another, and provides buffered packet data forwarding. X2-U makes use of the same GTP tunnelling protocol already used over the S1-U interface.

(ii) X2 Control Plane Interface
The X2-C (or X2 Control plane interface) is a signalling interface which supports a set of functions and procedures between eNodeBs. The X2-C procedures are very limited in number and are all related to user mobility between eNodeBs, so as to exchange user context information between nodes (including allocated bearers, security material, etc.).

In addition, the X2-C interface proposes the *Load Indicator* procedure whose purpose is to allow an eNodeB to signal its load condition to neighbouring eNodeBs. The detailed use of this function is not further detailed by the standard (as it relates to algorithms under the control of the equipment manufacturer). The aim of this procedure is to help the support of load-balancing management, or to optimize handover thresholds and handover decisions.

The need for a reliable transport of signalling between nodes is the same as over the S1-C interface. This is the reason why X2-C also uses an 'SCTP over IP' transport layer.

4.2 User and Control Planes

This section describes the overall end-to-end protocol structure of Evolved UMTS for the User and Control planes, which correspond respectively to user data transmission and signalling transmission.

4.2.1 User Plane Architecture

From the wireless network perspective – including both Access and Core parts – the User plane not only includes user data such as voice packets or Web content, but also the signalling associated to the application services such as the SIP or RTCP, which are described further in this chapter. Although being considered as control information by the application layers, the high-level signalling is transmitted via the User plane.

The end-to-end User plane is described in Figure 4.6, from the terminal up to the application server. In this picture, the application layer, only present in the terminal and application server, is based on an IP transport. The application-level packets are routed through Packet Core Gateways before reaching the destination. In this example, the application layer may comprise a very large set of protocol-like end-to-end transport protocols (e.g. TCP or UDP) and RTP (Real Time Protocol) for user data transport, as well as application-level signalling protocols mentioned above (SIP, SDP, RTCP, etc.). Further, in this chapter, a section dedicated to IMS protocol stack describes the set of 'application-level' protocols which can be used to support IMS-based services.

In this picture, L1 and L2 refer respectively to physical and data link interfaces of S1, S5 and SGi-fixed network interfaces. For those layers, the EPS standard is quite flexible and proposes many possible options suitable to IP networks.

(i) About GTP Data Tunnelling

In the telecommunication world, a tunnel is a generic term which designates a two-way point-to-point communication path established between two entities. In 3GPP networks, the main purpose of data tunnelling is to work out packet-routing issues consecutive to moving terminals.

Considering packets arriving from the external network to the PDN GW, the actual route such a packet has to follow may change during the session lifetime. For example, a change of

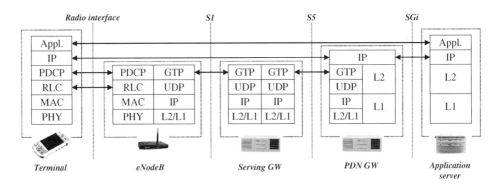

Figure 4.6 The User plane protocol stack.

Evolved UMTS Architecture

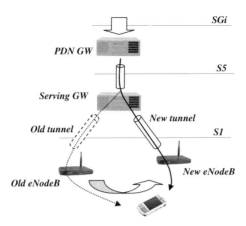

Figure 4.7 USE of GTP tunnels in case of terminal mobility.

serving eNodeB or, to a lesser extent, a change of Serving GW may occur during a streaming or Web-browsing session, so that the packet data path needs to be modified in order to maintain service continuity.

In IP networks, there exist some solutions to cope with terminal dynamic changing locations, the most popular being MIP (Mobile IP) defined by the IETF in RFC3220, 'IP Mobility Support for IPv4'. In MIP, special-purpose routers known as HA (Home Agent) and FA (Foreign Agent) get updated about terminal location information, so that user data can be forwarded – or tunnelled – to the terminal in its current position within the overall IP network.

The 3GPP Packet Core answer to terminal mobility is not based on MIP, although being very close to it from a functional perspective. Like in MIP, the relevant Serving or PDN Gateways are updated as the terminal is changing serving node. Tunnels towards new serving nodes are set up as appropriate, in order to maintain the data path.

Figure 4.7 illustrates a case of terminal mobility requiring tunnel adaptation. In this example, the terminal moves to a new serving eNodeB. The tunnel over S5 interface is not changed; however, a new tunnel needs to be established between the serving GW and the new eNodeB. A similar process would apply between the PDN GW and Serving GW if the Serving GW were to be changed during the session.

In the 3GPP definition, user data tunnelling between network nodes is ensured by the GTP layer (GPRS Tunnelling Protocol), inherited from the 2G/GPRS standard. This protocol is actually composed of two parts:

- The User plane part (or GTP-U), which provides user data encapsulation and transmission between two nodes.
- The Control plane part (or GTP-C), which is used on the EPC part of the network and provides all the procedures and messages for tunnel management (to set up, modify and release tunnels) and location management (to exchange moving user information between nodes).

For illustration, Figure 4.8 shows how the GTP encapsulation process works. The data packet itself – including the header and payload – is preserved and kept unchanged. The

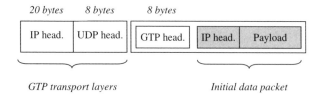

Figure 4.8 The effect of GTP tunnelling (using IPv4 for GTP transport).

packet is just added a GTP header (containing mainly Tunnel endpoint identifiers and optional GTP PDU sequence number) used by the receiving end to identify which tunnel the packet is associated to. This GTP encapsulated packet is transported between the two tunnel endpoints using a traditional UDP/IP stack.

For information, Table 4.1 describes the header of a GTP packet. The first byte contains some usual fields such as 'Version' and 'PT' (Payload Type – which indicates if the packet is pure user data or a GTP control message). The Tunnel Endpoint Identifier uniquely identifies the receiving protocol entity. Each Tunnel Identifier corresponds to a Packet Data Protocol (PDP) context (in the case of 2G/GPRS or 3G/UMTS packet core) or EPS bearer setup (in the case of an EPC packet core) between the terminal and the network for the support of one or several packet services.

Taking the example of Voice over IP data tunnelling, the 'initial data packet' is about 72 bytes long (including a 20-IP header, an 8-byte UDP header, a 12-byte RTP header, plus 32 bytes of AMR 12.2 Kb/s of encoded speech). The cost of GTP tunnelling over transport interfaces is 36 bytes. This represents 33% of the overall packet size, which is quite significant.

Hopefully, on the radio interface segment, the picture is completely different. There is no GTP tunnelling anymore, and the protocol headers of the 'initial data packet' are dramatically reduced, thanks to the PDCP radio interface layer described hereafter.

The tunnel setup is part of the data session establishment. For a given terminal, the network builds as many sets of tunnels as separate EPS bearers or PDP contexts. The GTP tunnelling

Table 4.1 The GTP-PDU format.

Octets	Bits							
	8	7	6	5	4	3	2	1
1	Version			PT	0	E	S	PN
2	Message Type							
3	Length (1st Octet)							
4	Length (2nd Octet)							
5	Tunnel Endpoint Identifier (1st Octet)							
6	Tunnel Endpoint Identifier (2nd Octet)							
7	Tunnel Endpoint Identifier (3rd Octet)							
8	Tunnel Endpoint Identifier (4th Octet)							
...	...							

process is completely transparent to the terminal and application server, as its only purpose is to cope with intermediate route updates between EPC and E-UTRAN network nodes.

> *Reference documents about mobility and data tunnelling*
>
> 3GPP Technical specifications:
>
> - 29.060, 'GPRS Tunnelling Protocol (GTP) across the Gn and Gp Interface'
>   IETF documents about MIP (Mobile IP):
> - RFC3344, 'IP Mobility Support for IPv4'

(ii) About the Radio Interface
Radio interface distinguishes from wired transmission on the cost and scarcity of the medium, as well as a much higher transmission error rate. Because of these characteristics, the radio interface protocol stack is very specific. It is composed of the following layers:

- PHY (Physical Layer).
- MAC (Medium Access Control) – in charge of packet scheduling and fast repetition.
- RLC (Radio Link Control) – responsible for reliable data transmission.
- PDCP (Packet Data Convergence Protocol) – provides protocol header compression and implements data encryption.

As in the OSI model, the E-UTRAN radio interface is composed of the traditional Layer 1 (or 'Physical Layer') implemented by the PHY part, and Layer 2 (or 'Data Link Layer') supported by both RLC and MAC parts.

As in 2G/GPRS and 3G/UMTS radio interface definitions, Layer 2 is not supported by a single box. The MAC and RLC differentiation comes from the need to design a flexible model, generic enough to accommodate changes or evolutions in the physical layer. This probably shows the limit of strict application of layered modelling. Some Layer 2 functions are actually much more efficient when being designed according to the specific characteristics and constraints of the physical layer they are supposed to work with.

This is the reason why pure Layer 2 functions independent to the physical aspects of the interface (like reliable data transmission and packet in-sequence delivery) are grouped into a separate RLC layer, whereas other medium-dependent functions (like data scheduling, of fast-repetition HARQ mechanisms) are part of the MAC layer.

In 3G/UMTS radio interface, several types of MAC layers (or sub-layers) have been defined in order to better cope with the specific aspects of transmission over the possible physical channels, including the DCH (Dedicated transport channel), the FACH shared channel and HSDPA (high-speed downlink shared channel), or any solution which may be designed in the future.

On the other hand, the RLC layer is common to all MAC layers and was not modified in the Standard, as new transport channel options were introduced.

To maintain the same degree of flexibility, the radio interface of E-UTRAN is following the same modelling principle. It is worth noting that this principle is actually inherited from the beginning of IEEE Local Area Network standardization work. As described in Figure 4.9, the IEEE 802 model for LAN and WLAN networking is based on LLC (Link Layer Control,

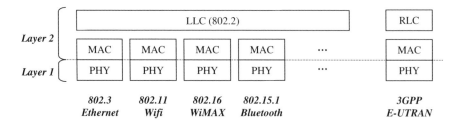

Figure 4.9 Radio interface modelling principle.

the equivalent of RLC) and multiple MAC layers, each of them specific to a type of physical access technology, like Wifi, WiMAX or wired Ethernet.

4.2.2 Control Plane Architecture

The Control plane corresponds to the information flows actually considered as signalling by E-UTRAN and EPC. For example this includes all the RRC (Radio Resource Control) E-UTRAN signalling (supporting functions such as Radio Bearer management, radio mobility, user paging) and NAS (Non Access Stratum) signalling, which refers to functions and services being independent from the access technology. This later includes the GMM (GPRS Mobility Management) and SM (Session Management) layers in charge of all the signalling procedures between the user terminal and the Packet Core network MME for session and EPS bearer management, security control and authentication.

Figure 4.10 describes the Control plane protocol stack. The stack stops at the MME level because the top-level protocols terminate in the MME. On the radio interface, the Control plane uses the same PDCP, RLC, MAC and PHY stack to transport both RRC (Radio Resource Control) and Core Network NAS signalling. The RLC, MAC and PHY layers support the same functions for both the User and Control planes.

However, this does not mean the User and Control plane information is transmitted the same way. Several Radio Bearers can be established between the terminal and the network, each of them corresponding to a specific transmission scheme, radio protection and priority handling. This is the purpose of the radio channels, presented in the next section.

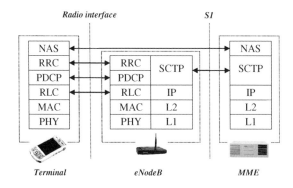

Figure 4.10 The Control plane protocol stack.

4.3 Radio Interface Protocols

4.3.1 The E-UTRAN Radio Layered Architecture

Figure 4.11, which introduces some new vocabulary, is an overview of radio protocol structure, which is further described in the next section. It briefly describes the main purpose of the different layers and how they interact with each other. This picture only describes the protocol layering on the eNodeB side, but there, of course, exist similar – or dual – functions and layers on the terminal side.

Starting from the top of the picture, the RRC layer (Radio Resource Control) supports all the signalling procedures between the terminal and the eNodeB. This includes mobility procedures as well as terminal connection management. The signalling from the EPC Control plane (e.g. for terminal registration or authentication) is transferred to the terminal through the RRC protocol, hence the link between the RRC and upper layers.

The **PDCP** layer (whose main role consists of header compression and implementation of security such as encryption and integrity) is offered to Radio Bearers by E-UTRAN lower layers. Each of these bearers corresponds to a specific information flow such as User plane data (e.g. voice frames, streaming data, IMS signalling) or Control plane signalling (such as RRC or NAS signalling issued by the EPC). Due to their specific purpose and handling, information flows generated by 'System Information Broadcast' and 'Paging' functions are transparent to the PDCP layer.

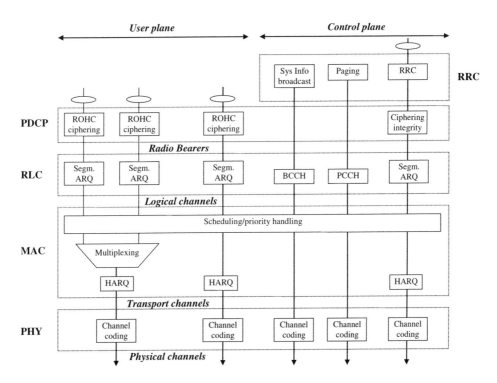

Figure 4.11 Protocol layered structure in eNodeB for downlink channels.

The **RLC** layer provides to the PDCP layer basic OSI-like Layer 2 services such as packet data segmentation and ARQ (Automatic Repeat Request) as an error-correction mechanism. There is one-to-one mapping between each RLC input flow and Logical channels provided by RLC to the MAC layer.

The **MAC** layer's main task is to map and multiplex the logical channels onto the transport channels after having performed priority handling on the data flows received from the RLC layer. The flows being multiplexed on a single transport channel may be originated by a single user (e.g. one or more instances of DCCH and DTCH) or multiple users (e.g. several DTCH from different users). The MAC also supports HARQ (Hybrid ARQ), which is a fast repetition process.

Finally, the MAC delivers the transport flows to the **PHY** layer, which will apply the channel coding and modulation before transmission over the radio interface.

4.3.2 The Radio Channels

(i) About Channel Types

As for most radio communication systems, the radio interface of E-UTRAN faces many challenges. In terms of requirements, the E-UTRAN shall be able to transmit high-rate and low-latency information in the most efficient way. However, not all the information flows require the same protection against transmission errors or Quality of Service handling.

In general, it is critical, especially in the case of radio mobility, that E-UTRAN signalling messages are transmitted as fast as possible, using the best error-protection scheme. On the other hand, voice or data streaming applications can accept a reasonable frame loss due to radio transmission. Interactive connection-oriented applications (such as Web browsing) are also different, as the end-to-end retransmission can help to recover from radio propagation issues.

In order to be flexible and allow different schemes for data transmission, the E-UTRAN specifications introduce several types of channels:

- The **logical** channels – what is transmitted.
- The **transport** channels – how it is transmitted.
- The **physical** channels.

The logical channels correspond to data-transfer services offered by the radio interface protocols to upper layers. Basically, there are only two types of logical channels: the control channels (for the transfer of Control plane information) and the traffic channels (for the transfer of User plane information). Each of the channels of these two categories corresponds to a certain type of information flow.

The E-UTRAN **logical control channels** are:

- The BCCH (Broadcast Control Channel): this channel is a downlink common channel, used by the network to broadcast E-UTRAN system information to the terminals presents in the radio cell. This information is used by the terminal, e.g. to know serving cell network operator, to get information about the configuration of the cell common channels, how to access to the network, etc.
- The PCCH (Paging Control Channel): the PCCH is a downlink common channel which transfers paging information to terminals presents in the cell, e.g. in case of mobile-terminated communication session.

- The CCCH (Common Control Channel): the CCCH is a special kind of transport channel, used for communication between the terminal and E-UTRAN when no RRC connection is available. Typically, this channel is used in the very early phase of a communication establishment.
- The MCCH (Multicast Control Channel): this channel is used for the transmission of MBMS (Multimedia Broadcast and Multicast Service) information from the network to one or several terminals.
- The DCCH (Dedicated Control Channel): the DCCH is a point-to-point bi-directional channel supporting control information between a given terminal and the network. In the DCCH context, the control information only includes the RRC and the NAS signalling. The spplication-level signalling (such as SIP of RTCP) is not handled by the DCCH.

The E-UTRAN **logical traffic channels** are:

- The DTCH (Dedicated Traffic Channel): the DCCH, the DTCH is a point-to-point bi-directional channel, used between a given terminal and the network. It can support the transmission of user data, which include the data themselves as well as application-level signalling associated to the data flow.
- The MTCH (Multicast Traffic Channel): a point-to-multipoint data channel for the transmission of traffic data from the network to one or several terminals. As for the MCCH, this channel is associated to the MBMS service (Multimedia Broadcast and Multicast Service), described in more detail in Chapter 6.

The **transport channels** describe how and with what characteristics data are transferred over the radio interface. For example, the transport channels describe how the data are protected against transmission errors, the type of channel coding, CRC protection or interleaving which is being used, the size of data packets sent over the radio interface, etc. All this set of information is known as the 'Transport Format'.

As in the specification, the transport channels are classified into two categories: the downlink transport channels (from the network to the terminal) and the uplink transport channels (from the terminal to the network).

The E-UTRAN **downlink transport channels** are:

- The BCH (Broadcast Channel), associated to the BCCH logical channel. The BCH has a fixed and predefined Transport Format, and shall cover the whole cell area.
- The PCH (Paging Channel), associated to the BCCH.
- The DL-SCH (Downlink Shared Channel), which is used to transport user control or traffic data.
- The MCH (Multicast Channel), which is associated to MBMS user of control information transport.

The E-UTRAN **uplink transport channels** are:

- The UL-SCH (Uplink Shared Channel), which is the uplink equivalent of the DL-SCH.
- The RACH (Random Access Channel), which is a specific transport channel supporting limited control information, e.g. during the early phases of communication establishment or in case of RRC state change.

The **physical** channels are the actual implementation of the transport channel over the radio interface. They are only known to the physical layer of E-UTRAN and their structure is tightly dependent on physical interface OFDM characteristics, described in detail in Chapter 3.

The physical channels defined in the **downlink** are the:

- Physical Downlink Shared Channel (PDSCH) – which carries user data and higher-layer signalling.
- Physical Downlink Control Channel (PDCCH) – this channel carries scheduling assignments for the uplink.
- Physical Multicast Channel (PMCH) – which carries Multicast/Broadcast information.
- Physical Broadcast Channel (PBCH) – which carries System Information.
- Physical Control Format Indicator Channel (PCFICH) – which informs the UE about the number of OFDM symbols used for the PDCCH.
- Physical Hybrid ARQ Indicator Channel (PHICH) – which carries ACK and NACK eNodeB responses to uplink transmission, relative to the HARQ mechanism.

The physical channels defined in the **uplink** are the:

- Physical Uplink Shared Channel (PUSCH) – which carries user data and higher-layer signalling.
- Physical Uplink Control Channel (PUCCH) – this channel carries uplink control information, including ACK and NACK responses from the terminal to downlink transmission, relative to the HARQ mechanism.
- Physical Random Access Channel (PRACH) – which carries the random access preamble sent by terminals to access to the network.

In addition to physical channels, the physical layer makes use of **physical signals**, which include:

- Reference signals – one signal being transmitted per downlink antenna port.
- Synchronization signals – split into a primary and a secondary synchronization signals.

(ii) Mapping between Channels

Figure 4.12 represents the mapping between logical, transport and physical channels presented above. For obvious reasons, not all combinations are allowed, as some of the logical channels have specific constraints.

PCCH and BCCH logical channels have particular transport and physical characteristics so that the transport and physical channel mapping is specific to them. The mapping of the BCCH on the BCH and DL-SCH transport channels is not an option. This comes from the fact that the System Information is actually composed of two parts:

- Critical system information which has a fixed format and requires frequent update – this one is mapped on the PBCH.
- Dynamic and less critical information which is mapped on a transport channel offering more flexibility in terms of bandwidth and repetition period – the DL-SCH.

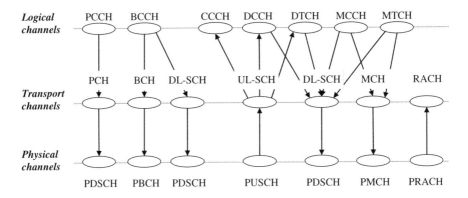

Figure 4.12 E-UTRAN mapping between channel types (as in the network).

On the other hand, some logical channels can benefit from different possible options as regards to mapping to the transport channel. Typically, this is the case for the MCCH and MTCH Multicast channels, which are mapped on a specific MCH transport channel in case of multi-cell MBMS service provision. When an MBMS service is provided in a single cell, MCCH and MTCH channels are mapped over conventional DL-SCH channels.

The other physical channels (such as PUCCH, PDCCH, PCFICH and PHICH) do not carry information from upper layers (such as RRC signalling or user data). They are only intended for the purpose of the physical layer, as they carry information related to the coding of physical blocks, or HARQ-related information. This is the reason why those channels are not mapped to any of the transport channels of the radio interface.

The RACH is a specific case of transport channel, having no logical channel equivalent. This comes from the fact that the RACH only carries RACH preamble (which is basically the very first set of bits the terminal sends to the network to request access). Once access is granted by the network and physical uplink resources are allocated to the terminal, the RACH is no longer used by the terminal.

(iii) Some Remarks about E-UTRAN and UTRAN Channels
Not surprisingly, the E-UTRAN channel model has been inherited from the UTRAN channel model. The concept of separation between logical, transport and physical channels was already present in the initial UTRAN model.

Figure 4.13 presents the logical-to-transport channel mapping performed in the MAC layer of UTRAN RNC (Radio Network Controller) node for the FDD mode only. The first obvious point is that the UTRAN and E-UTRAN models share almost the same logical channel structure, showing that radio layers from both systems will actually provide the same types of services to upper layers, i.e.:

- Broadcast and Paging services (associated to BCCH and PCCH), which are the basis of all cellular systems.
- Dedicated – or point-to-point – information transfer (supported by DCCH and DTCH).
- Multicast – or point-to-multipoint – information transfer (supported by MCCH and MTCH).

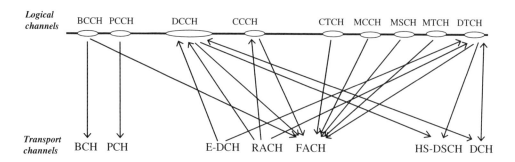

Figure 4.13 The UTRAN/FDD logical-to-transport channel mapping in the network (from 3GPP 25.301).

However, when looking at the transport channel level, the two models are completely different. The DCH (Dedicated transport Channel) present in the UTRAN model has disappeared from the E-UTRAN model, which only supports shared transport channels. This channel was designed for constant bit rate and real-time constraining services, such as voice or streaming applications.

In the E-UTRAN model, all point-to-point data services are packetized, and supported by only one kind of transport channel: the SCH (Shared Channel) for both uplink and downlink transmission. This is an interesting evolution, as the radio interface concepts are following the same 'all-IP' direction as the Packet Core and service evolution. The newly introduced SCH can actually be seen as an evolution of both HS-DSCH – the support of the well known HSDPA (High Speed Downlink Packet Access) – and E-DCH – the high-speed evolution for the uplink.

At the end, the channel model of E-UTRAN looks much simpler, as the number of transport channels and cross-mapping between channel types has been greatly simplified and reduced.

4.3.3 PHY

The role of the PHY layer is to provide data transport services on physical channels to the upper RLC and MAC layers. The PHY layer of E-UTRAN as well as the principles being used for transmission over the radio interface is extensively described in Chapter 3. This section is therefore limited to the description of the physical layer from a functional perspective and also in terms of interactions with other radio interface layers.

Figure 4.14 describes the eNodeB physical layer model in the example of a downlink SCH transport channel as being the most generic scheme. Of course, similar models exist for the uplink (in the terminal), and for all the other transport channels listed in this chapter.

At each TTI (Transmission Time Interval), the physical layer receives a certain number of Transport Blocks for transmission. To each Transport Block is added a CRC (Cyclic Redundancy Check) or set of bits used by the receiving end to detect transmission errors.

The Blocks are then protected by a robust channel-encoding scheme (like convolutional or turbo coding) and size-adapted to make sure the encoded packet matches the physical channel

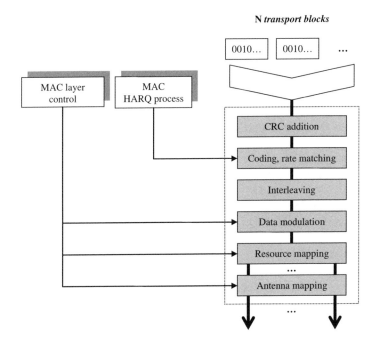

Figure 4.14 The Downlink Shared Channel PHY layer model.

size. This phase is under the control of the MAC HARQ (Hybrid ARQ) process which may adapt the channel coding rate (meaning the robustness to transmission errors) based on the information reported by the receiving entity.

Interleaving is a process to improve robustness to radio transmission errors. When an error occurs on an encoded packet transmitted over the radio interface, it will affect multiple consecutive bits or symbols. On the receiving side, the action of de-interleaving will have the effect of spreading the erroneous bits on the whole transmitted sequence on different Transport Blocks. This will make it easier for the channel decoder to recover the exact bits transmitted initially, as a single block will only be affected by a smaller part.

In the data-modulation process, the actual modulation is under the control of the MAC scheduler. Resource mapping relates to the segmentation of transmitted data into resource blocks. Antenna mapping relates to the mapping of resource blocks (as above) on available antenna ports (MIMO).

CRC and interleaving processes are not controlled by higher layers. For those two operations, the PHY layer uses static parameters and algorithms specified by the E-UTRAN standard.

As mentioned above, there exist similar models for other transport channels. However, there are more or fewer subsets of the shared channel model presented above. Transmission over other transport channels such as the PCCH for paging or BCCH for system information broadcasting is not flexible in terms of channel coding or modulation. For this type of transport channel, the E-UTRAN standard does not propose any options or alternatives.

4.3.4 MAC

The Medium Access Control (MAC) radio protocol layer of E-UTRAN's main purpose is to provide an efficient coupling between the RLC Layer 2 services and the physical layer. From that perspective, the MAC supports four main functions:

- Mapping between logical channels and transport channels – when the standard offers different options for the transport of data for a given logical channel, it is up to the MAC layer to choose the transport channel according to the configuration defined by the operator.
- Transport format selection – this refers to, for example, the choice of Transport Block size and modulation scheme made by the MAC layer and provided as input parameters to the physical layer.
- Priority handling between logical channels of one terminal as well as between terminals.
- Error correction through HARQ (Hybrid ARQ) mechanism.

(i) About Priority Handling

Priority handling is one of the main functions supported by the MAC layer. Priority handling refers to the process which selects the packets from the different waiting queues to be submitted to the underlying physical layer for transmission on the radio interface. This process is complex, as it takes into account the different flows of information to be transmitted – including pure user data (the DTCH logical channel) as well as signalling initiated by the E-UTRAN or the EPC (the DCCH logical channel) – with their relative priority, as well packet repetition in case an already transmitted packet has not been correctly received by the other end. For that reason, the 'priority handling' part of the MAC layer is tightly coupled with the 'Hybrid ARQ' part.

In addition, the MAC layer on the network side is also responsible for uplink priority handling, as it arbitrates between all the uplink scheduling requests from all the terminals which share the same UL-SCH transport channel (Figure 4.15).

On the terminal side, the MAC layer only mixes flows from the terminal for uplink transmission and has to arbitrate between its own information flows for uplink scheduling requests and transmission. In contrast, for the Downlink Shared Channel, the eNodeB has to consider all the flows (or logical channels) sent to all the users in the cell.

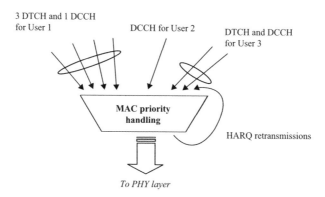

Figure 4.15 An illustration of priority handling in the eNodeB MAC layer.

(ii) About HARQ

The principle of hybrid ARQ is to buffer blocks that were not received correctly and consequently combine the buffered data with retransmissions. The actual method of doing soft combining depends on the HARQ combining scheme selected. In the Chase combining scheme, initial transmission and retransmission are identical. The receiver always combines the full retransmission of the failed block. In the IR (Incremental Redundancy) schemes, new parity bits are transmitted together with the failed block. The receiver receives coded symbols, which introduce new information to the first transmitted block.

In E-UTRAN, HARQ is composed of several parallel parts, so that transmission can continue on other processes while one of them is stuck with retransmissions.

In the downlink, HARQ is based on asynchronous retransmissions with adaptive transmission parameters. In the uplink, HARQ is based on synchronous retransmissions.

In the synchronous scheme, retransmission can only occur at certain subframe numbers following the first transmission. In the asynchronous scheme, packet retransmissions are not constrained in terms of frame time.

Synchronous retransmission is preferred in uplink, because there is less protocol overhead. Synchronous retransmission does not require to explicitly signal the HARQ process number, since it can be deduced from the subframe number.

The HARQ in E-UTRAN is similar to those in 3G HSDPA (for the downlink transmission) and E-DCH/HSUPA (for the uplink transmission).

4.3.5 RLC

The main purpose of the Radio Link Control (RLC) E-UTRAN protocol layer is to receive/deliver a data packet from/to its peer RLC entity. For that purpose, the RLC proposes three modes of transmission TM (Transparent Mode), UM (Unacknowledged Mode) and AM (Acknowledged Mode).

The **TM mode** is the simplest one, as it does not change or alter the upper layer data. This mode is typically used for BCCH or PCCH logical channel transmissions which require no specific treatment from the RLC layer. The RLC Transparent Mode Entity receives data from the upper layers and simply passes it to the underlying MAC layer. There is no RLC header addition, data segmentation or concatenation.

The added value of the **UM mode** is to allow the detection of packet loss (the receiving entity can detect that a RLC packet has not been received correctly) and provides packet re-ordering and re-assembly. These operations can be performed thanks to the presence of a Sequence Number in the RLC packet header. The UM mode can apply to any Dedicated or Multicast logical channel, depending on the types of application and expected Quality of Service.

Packet re-ordering refers to the re-sequencing of packets in case they have not been received in order (which may happen in the case of HARQ repetition). Packet re-assembly is performed when an upper-layer packet has been segmented by the sending RLC entity before transmission.

Finally, the **AM mode** is the most complex one. In addition to UM mode-supported features, an AM RLC entity is able to ask its peer for packet retransmission in case a loss is detected. This mechanism, specific to the AM mode, is known as ARQ (Automatic Repeat Request). For that reason, the AM mode only applies to DCCH or DTCH logical channels.

> **Some words bout the ARQ process of RLC**
>
> - The ARQ process is only available in RLC AM mode
> - ARQ retransmissions are triggered either by RLC status report exchanged between peers (the receiving entity has detected missing packets), or on the sending side following a failure of a HARQ process to deliver a Transport Block.

(i) The RLC Block Structure

Figure 4.16 describes the structure of a RLC PDU. RLC receives as input data the blocs provided by the above PDCP layer (those blocks are referred to as RLC SDU for Service Data Unit). In order to fill the payload part of the PDU, RLC makes use of two mechanisms used in most of the data link layer protocols and known as **segmentation** and **concatenation**.

When a RLC-SDU cannot be added to a given payload because the remaining size of the PDU is too short, the SDU is then segmented (as for SDU 'n' and '$n+3$' in the picture) and therefore transmitted using two different PDUs. Inversely, if the SDU size is smaller than the PDU, the RLC layer will concatenate as many as possible in order to fill the payload.

As in the figure, the segmentation and concatenation are always done in sequence, so that the RLC receiving entity can deliver the SDU to its PDCP upper layer in the correct order.

4.3.6 RRC

The RRC (Radio Resource Control) layer is a key signalling protocol which supports many functions between the terminal and the eNodeB. The log list of procedures proposed by the RRC layer can be classified into the following:

- RRC connection management – which includes the establishment and the release of the RRC connection between the terminal and the eNodeB.
- Establishment and release of radio resources – which relates to the allocation of resources for the transport of signalling messages or user data between the terminal and eNodeB.
- Broadcast of system information – this is performed through the BCCH logical control channel. The information broadcast from the RRC layer is either related to the Access Network (such as radio-related parameters) or to the Packet Core (for information such as the cell corresponding geographical area or network identity).

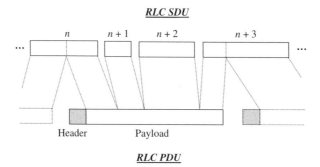

Figure 4.16 RLC AM or UM PDU structure.

- Paging – this is performed through the PCCH logical control channel.
- Transmission of signalling messages to and from the EPC – these messages (known as NAS for Non Access Stratum) are transferred to and from the terminal via the RRC; they are, however, treated by RRC as transparent messages.

The RRC also supports a set of functions related to end-user mobility for terminals in RRC Connected state. This includes:

- Measurement control – which refers to the configuration of measurements to be performed by the terminal as well as the method to report them to the eNodeB.
- Support of inter-cell mobility procedures – which are also known as handover. More information on handover procedures from an overall system perspective is provided in Chapter 5.
- User context transfer between eNodeB at handover.

(i) The RRC States

The main function of the RRC protocol is to manage the connection between the terminal and the E-UTRAN access network. To achieve this, RRC protocol states have been defined (Figure 4.17). Each of them actually corresponds to the states of the connection, and describes how the network and the terminal shall handle special functions like terminal mobility, paging message processing and network system information broadcasting.

In E-UTRAN, the RRC state machine is very simple and limited to two states only:

- RRC_IDLE, and
- RRC_CONNECTED.

In the RRC_IDLE state, there is no connection between the terminal and the eNodeB, meaning that the terminal is actually not known by the E-UTRAN Access Network.

The terminal user is inactive from an application-level perspective, which does not mean at all that nothing happens at the radio interface level. Nevertheless, the terminal behaviour is specified in order to save as much battery power as possible and is actually limited to three main items:

- Periodic decoding of System Information Broadcast by E-UTRAN – this process is required in case the information is dynamically updated by the network.

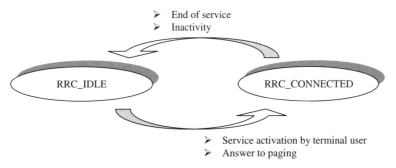

Figure 4.17 The RRC states.

- Decoding of paging messages – so that the terminal can further connect to the network in case of an incoming session.
- Cell reselection – the terminal periodically evaluates the best cell it should camp on through its own radio measurements and based on network System Information parameters. When the condition is reached, the terminal autonomously performs a selection of a new serving cell.

Following cell reselection, it may happen that the terminal changes geographical area (or Tracking Area). Should this occur, the terminal is required to update the network in order to be still able to receive paging messages. For that purpose, the terminal has to leave temporarily the RRC_IDLE state so as to be able to exchange the necessary signalling information with the network. When the update procedure is over and if no service has been activated or answered by the user in the meantime, the terminal returns to the RRC_IDLE state.

In the RRC_CONNECTED state, there is an active connection between the terminal and the eNodeB, which implies a communication context being stored within the eNodeB for this terminal. Both sides can exchange user data and or signalling messages over logical channels.

Unlike the RRC_IDLE state, the terminal location is known at the cell level. Terminal mobility is under the control of the network using the handover procedure, which decision is based on many possible criteria including measurement reported by the terminal of by the physical layer of the eNodeB itself.

How the RRC states machine relates to the other state machine handled at the EPC level is described in Chapter 5.

4.3.7 PDCP

The main purpose of the Packet Data Convergence Protocol (PDCP) E-UTRAN protocol layer is to receive/deliver a data packet from/to its peer PDCP entity. In principle, this function is ensured by the RLC layer. From that perspective, the PDCP layer provides some additional features, which are – in principle – out of the scope of a generic OSI-compliant Layer 2 protocol.

The PDCP added value relies in four main functions:

- **Layer 2-related features,** such as re-ordering of RLC packets in case of inter-eNodeB mobility, or duplicate detection of RLC packets. As in the RLC layer, this function is ensured thanks to a Sequence Number contained in the PDCP header.
- **IP packet header compression and decompression.** As opposed to its 3G/UMTS equivalent, PDCP supports only one compression scheme, known as ROHC (Robust Header Compression).
- **Ciphering of data and signalling.** 'Data' relates to user data as well as application-level signalling such as SIP or RTCP. 'Signalling' relates to both RRC signalling messages sent by the eNodeB and NAS (Non Access Stratum) signalling messages issued by the Evolved Packet Core. The NAS part is further described in this chapter.
- **Integrity protection for signalling.** In the scope of PDCP, 'signalling' includes both NAS and RRC messages. Integrity is a mechanism which helps to prevent 'man in the middle' types of attacks, by providing the receiving side with a mean to determine whether or not the signalling message has been altered during the transmission.

Evolved UMTS Architecture

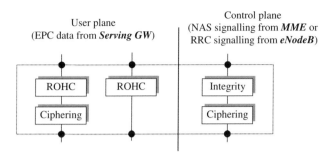

Figure 4.18 PDCP layer model.

Figure 4.18 is an overview of the functional block model and the different combinations offered by the PDCP layer. Ciphering and integrity are mandatory for the Control plane. However, ciphering may optionally be used for user data. ROHC header compression is not an optional process; however, ROHC supports a transparent 'uncompressed' mode, which does not provide any sort of header change or compression.

(i) About Ciphering and Integrity
Chapter 5 provides more details regarding security mechanisms, such as keys, algorithms and the overall EPS security framework. However, in the scope of E-UTRAN radio protocol, it is interesting to consider the impact of such mechanisms on the initial packet shape and format.

Ciphering protection makes use of block or stream-shared secret key ciphering algorithms which have no impact on data length. The principle of such algorithms is that both sending and receiving entities generate a set of bits – or keystream – using the secret key. This keystream is then added (using the 'exclusive OR' logical operation) to the plain data for ciphering at the sending side, and added the same way on the receiving side for deciphering.

In contrast, integrity requires additional information to be sent, for the receiving entity to be able to check the data are actually coming from a reliable source. This additional information is known as the MAC (Message Authentication Code).

At the end, the input packet submitted by higher layers on the Control plane (which is either the EPC Core Network or the RRC layer) is modified by the PDCP layer, as in Figure 4.19. A header containing the PDCP Sequence Number (for re-ordering in inter-eNodeB mobility case) is added to the data as well as the MAC code for integrity check.

The User plane is equivalent, except that, in this case, the integrity and MAC code do not apply. The other difference is that ciphering applies to the output of the ROHC compressor engine of the PDCP layer.

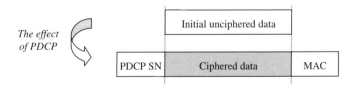

Figure 4.19 Overview of a PDCP packet (Control plane).

(ii) ROHC Based Header Compression

All of the services that will be proposed by Evolved UMTS will actually be IP-based. Even in the case of the called party still belonging to the circuit-switched telephony world, all the UTRAN end Evolved Packet Core part of the service will be supported by IP and IETF protocols, the signalling and bearer conversion being supported by wireless network edge nodes when necessary.

From a radio perspective, the key point of this evolution resides in the fact that IETF protocols suffer from large overhead. Those protocols are quite verbose and the messages have significant header sizes at the network level (IP), transport level (TCP, UDP) as well as the application level (RTP, SIP, etc.).

Besides, those network and transport protocols contain lots of information which becomes redundant when the transmission occurs over a point-to-point medium such as the E-UTRAN radio interface. In this case, lots of information fields are repeated from packet to packet (such as the source and destination IP address) without any changes at all.

When considering wired high bit rate Internet, this is not much an issue, as reliable high-transmission capacity networks are quite transparent to this extra bandwidth requirement. However, when looking at the radio interface, the picture is not the same. Although the radio physical layer is providing powerful error-protection schemes, the radio interface is still characterized by a comparatively high error rate. Besides, radio resource is scarce and expensive. Hence the need to apply an efficient compression scheme for the transmission over the radio interface which could help avoiding sending redundant information while being resilient to transmission errors.

There exist quite a lot of compression schemes proposed by the IETF world. However, they are either limited to only one protocol level (such as the initial IP header compression scheme described in RFC2507, which was the compression scheme defined for the first release of UMTS) or have significant bandwidth impact in case of erroneous or lost packets; this requires lots of information transfer to resynchronize compression and decompression machines. This latter is a critical point in the domain of radio transmission, characterized by a much higher error rate than fixed transmissions.

ROHC (Robust Header Compression), defined by the IETF, is an answer which tries to address all those concerns.

ROHC has been defined within IETF as a general and extendable framework. The first release of ROHC was documented in RFC3095, 'RObust Header Compression (ROHC): Framework and Four Profiles: RTP, UDP, ESP, and Uncompressed'.

At some time, it was decided to improve this initial specification by clearly separating the ROHC framework defined in RFC3095 (such as the concepts, definitions, packet types and formats, etc. which really constitute the basis of ROHC) from the profile definition which defines the compressor and decompressor behaviour for the most typical applications. Eventually, ROHC version 2 (or ROHCv2) describes all the frameworks of ROHC (inherited and kept unchanged from version one) in a separate document, and introduces new profiles, providing some substantial improvements in terms of overall complexity and implementation simplicity, as well as robustness to out-of-order packet delivery.

ROHC basically works on the notion of 'profile'. Each profile describes the behaviour of the compressor and the decompressor for specific protocol use cases, as well as the machine state transitions, and the behaviour when transmission starts and errors occur.

Table 4.2 Profiles supported by ROHC versions 1 and 2.

Profile	RFC (version)	Use case
Uncompressed	RFC3095 (v1)	Used for packets for which compression has not been defined (e.g. the RTCP packets) or when compression is not possible, e.g. due to resource constraints
RTP	RFC3095 (v1)	Applicable to the RTP/UDP/IP stack, mainly used for voice and audio/video streaming applications
UDP	RFC3095 (v1)	Applicable to UDP/IP stack, mainly used for conversational or streaming applications not being RTP-based
ESP	RFC3095 (v1)	Used when ESP (Encryption Security Payload) is applied
IP	RFC3843 (v1)	The IP only profile, very similar to the UDP profile, the UDP part being put aside
RTP	Draft RFC (v2)	ROHCv2 equivalent of the version 1 profile
UDP	Draft RFC (v2)	ROHCv2 equivalent of the version 1 profile
ESP	Draft RFC (v2)	ROHCv2 equivalent of the version 1 profile
IP	Draft RFC (v2)	ROHCv2 equivalent of the version 1 profile
TCP	RFC4996 (v2)	Applicable to TCP/IP flows

Notes: At the time of writing, the reference for the ROHCv2 RFC for RTP, UDP, ESP and IP profiles is not known and the document is still in the form of an IETF draft, known as 'draft-ietf-rohc-rfc3095bis-rohcv2-profiles'. The 'IP' profile was actually not supported by 3G/UMTS, which only supported ROHCv1 profiles.

Table 4.2 describes all the profiles supported by ROHC in its first and second versions.

(iii) Some More Details on How ROHC Works
Figure 4.20 illustrates how ROHC works. Although the picture only shows one-way compressed transmission, the compression can be applied to both. However, each direction is handled independently from the other, having its own context, compression and decompression state machine, and parameters.

The ROHCv2 decompressor is based on a three-level state machine: NC (No Context), IC (Initial Context) and FC (Full Context). When starting, the decompressor enters the NC state and stays in this state until it receives a packet with a correct ROHC header. In the FC states, the decompressor successfully receives optimized compressed packets. As decompression failures occur, the state machine transits back to IC or even NC states, and corresponding negative feedback indications are sent to the compressor.

The compressor starts in unidirectional mode until it receives feedback from the compressor. As feedback is received, the compressor assumes the decompressor context is valid, and can increase the compression efficiency of transmitted packets.

As opposed to the v1, ROHCv2 does not specify the compressor state machine in a deep level of detail. The compressor is, however, still required to ensure the decompressor gets the proper information for packet decompression and works in the most efficient state, based or not on the information it receives about the feedback channel.

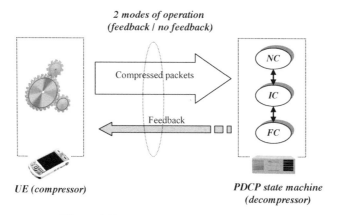

Figure 4.20 ROHCv2 decompressor states.

When the feedback channel is used, the decompressor can provide three kinds of information:

- **ACK,** which acknowledges the reception of compressed packets up to the transmission of the ACK.
- **NACK,** which indicates that some of the dynamic fields received by the decompressor are no longer valid.
- **STATIC-NACK,** which indicates that the static part of the decompressor context is no longer valid, which is actually an explicit request for a complete context update.

On the compressor side, two categories of packets are defined:

- **IR** (for Initialization and Refresh), which contains the static and dynamic fields of the packet header. This kind of packet is used when the transmission starts, or if negative feedback is sent by the decompressor.
- **CO** (for Compressed), which is a standard packet with compressed header followed by user data information. Depending on feedback received and algorithm in the compressor, not all the fields may be compressed. A CO packet may contain uncompressed information which the compressor chooses to send as is.

In any case, IR and CO packets include a CRC in order to protect packet header integrity against transmission errors. When received, the CRC is checked by the decompressor, which further decides if the packet is valid or not and possibly provides feedback information to the compressor.

(iv) About ROHC Compression Methods and Efficiency
ROHC does not propose any magic compression encoding method. It rather reuses a set of well known techniques to apply to each header field, depending on its type. These methods are based on the fact that most of the IP, UDP or RTP header fields are either static (meaning that they don't vary once a transport session is set up), inferred (meaning their value can be deduced for other values, such as the 'frame size' field) or changing within a limited value set of ranges or in a predictable manner.

As an illustration, ROHC makes use of the two following methods for field compression for changing fields:

- The LSB (Least Significant bit) method.
- The scaled RTP Timestamp encoding.

The LSB method is suitable to large bit fields varying by a small value at each occurrence (such as the 16-bit RTP packet sequence number incremented by one at each frame). In such a case, the compressor builds and sends a reduced range field using the least significant bits rather than sending the whole set of bits.

The scaled RTP Timestamp method is a kind of variation of the LSB method. As explained later in this chapter, in case of voice application, the 32-bit long RTP Timestamp may be incremented by large and constant values (e.g. 160) every packet. Rather than sending the whole set of bits, the compressor builds a downscaled value. The original value can be reconstructed by the decompressor by interpolation with a linear function, such as '$y = a.x + b$', which the 'a' and 'b' parameters have been provided by the compressor.

As described above, ROHC is based on the notion of a profile which specifies, for each of the header fields, what shall be the initialization value and the compression method to be used.

As an example, Table 4.3, extracted from the ROHC profile document, describes the field classification for an IPv4 header. It basically defines the way the compressor (and the decompressor) shall process all of the individual header fields for each profile. This table also illustrates the potential gain that a sensible method for header compression may provide. When looking into the details of the 160 bits of an IPv4 header, it appears that 115 of them are either static or can be inferred (or deduced) from other information. The remaining fields of information are changing randomly, so that they cannot be compressed using the ROHC methods.

Finally, Figure 4.21 illustrates the performances of ROHC in the example of a RTP/UDP/IP voice application. This example shows a typical 32-byte payload audio packet. Before compression, the header part is even larger than the payload. Thanks to the ROHC

Table 4.3 IPv4 header field classification.

Field name	Size (bits)	Class
Version	4	Static
Header length	4	Static
Type of service	8	Changing
Packet length	16	Inferred
Identification	16	Changing
Flags	3	Static
Time to live	8	Changing
Protocol	8	Static
Header checksum	16	Inferred
Source address	32	Static
Destination address	32	Static

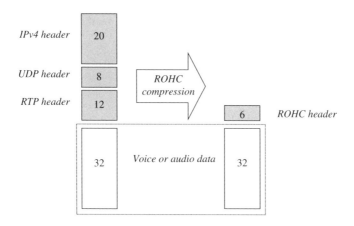

Figure 4.21 An example of ROHC performances on a VoIP flow.

compression process, the 40 bytes of RTP/UDP/IPv4 headers are reduced to a mean value of 6 bytes.

> *Reference documents about ROHC*
>
> ROHC IETF documents:
>
> - RFC3095, 'ROHC: Framework and Four Profiles: RTP, UDP, ESP, and Uncompressed'
> - RFC3843, 'ROHC: A Compression Profile for IP'
> - RFC4995, 'The RObust Header Compression (ROHC) Framework'
> - RFC4996, 'RObust Header Compression (ROHC): A Profile for TCP/IP (ROHC-TCP)'
> - Draft RFC, 'RObust Header Compression Version 2 (ROHCv2): Profiles for RTP, UDP, IP, ESP, and UDP Lite'

4.3.8 NAS Protocols

(i) About the AS/NAS Model

In order to better define the scope of NAS protocols, Figure 4.22 presents a model which was introduced at the beginning of 2G/GSM and is still valid for the evolution of UMTS. This model presents two areas – the AS (Access Stratum) and NAS (Non-Access Stratum) – which span over several entities (Terminal, Access Network and Packet Core) and aims at classifying the main functions supported by the whole set.

The Access Stratum corresponds to features linked to the radio interface. It is, however, not limited to the access network and the radio part of the terminal, as it also mandates for a specific support from the Packet Core. As an illustration, the main features supported by AS are:

- Radio Bearer management – which includes Radio Bearer allocation, establishment, modification and release.

Evolved UMTS Architecture

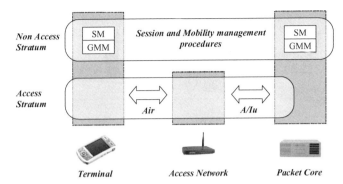

Figure 4.22 Access and Non-Access Stratum model.

- Radio channel processing – including channel coding and modulation.
- Ciphering – this only refers to the ciphering process itself. The initiation of ciphering and the choice of security algorithm are under the responsibility of the Non-Access Stratum. In addition, and depending on applications, end-to-end ciphering main also be used, as in VPN (Virtual Private Network).
- Radio mobility – also known as handover.

In 2G/UMTS, the AS features are actually supported by the PHY, MAC, RLC, RRC and PDCP radio protocol layers, as well as all procedures defined within the Iu interface, between the Packet Core and UTRAN Access Network.

In contrast, the Non-Access Stratum corresponds to functions and procedures which are completely independent from the access technology. This includes features such as:

- Session management – which includes session establishment, modification and release as well as Quality of Service negotiation.
- Subscriber management – which corresponds to user data management, attach and detach features initiated at mobile switch-on or off.
- Security management – this includes mutual user-network authentication as well as ciphering initiation.
- Charging.

In 2G/GSM, the NAS features are supported by two sets of protocols, being the GMM (GPRS Mobility Management) and SM (Session Management) layers, defined by 3GPP specifications (Figure 4.23). Thanks to this independence from the access technology, the NAS layer has evolved in a backward-compatible way. When UTRAN specifications were defined in 1999, the corresponding NAS layers were actually inherited from the existing GSM ones. Some additions were introduced, corresponding to UMTS enhancements (like, for example, the Quality of Service handling) but most of the rest was actually duplicated from the legacy GSM NAS layers.

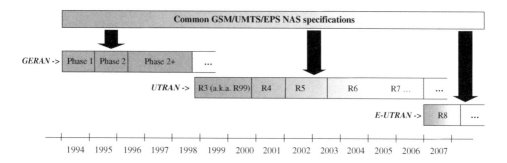

Figure 4.23 The relationship between NAS and access protocol versions in 3GPP.

(ii) About GMM and SM Protocols

The **GMM** layer is in charge of supporting the mobility of user terminals. In this context, the term 'mobility' does not refer to radio mobility, such as handover between cells, as radio mobility is already supported by E-UTRAN. GMM rather refers to 'terminal location management', which is required, for example, to join the terminal in case of mobile-terminated session initiation. For that purpose, the Mobility Management layer supports the following set of procedures:

- User attachment, which is part of the registration process initiated when the terminal is switched on. A similar procedure is also used when the mobile is switched off.
- The terminal location updating procedures, performed as the terminal is moving or periodically to refresh and keep updated the information stored in the Packet Core.

In addition to Mobility Management, the GMM layer also supports security functions, such as:

- Mutual subscriber and network authentication.
- Activation of ciphering or integrity protection.
- Management of terminal states (Detached, Idle and Active).

The **SM** layer is built on top of the GMM and uses GMM services for the management of sessions. The main function of SM is to support user terminal PDP (Packet Data Protocol) context management and bearer management between the terminal and the SGSN. This includes procedures for the activation, modification or deactivation of session context and associated bearer.

Briefly, a PDP context is a context for packet transmission determined by a terminal address (being an IP one) and a set of 'Quality of Service' attributes such as maximum bit rate, guaranteed bit rate, transmission delay, etc. which are negotiated between network and terminal when the PDP context is set up. This concept, as well as examples of Session Management procedures, is further described in Chapter 5.

(iii) About NAS Protocols in EPS

Not surprisingly, most of the NAS functions and procedures which are used in EPS networks are based on the set of concepts and procedures inherited from GSM and UMTS networks.

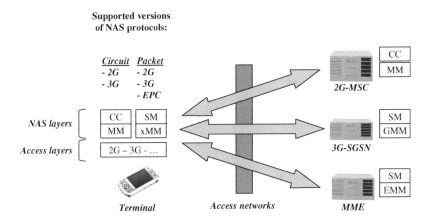

Figure 4.24 Combination of NAS protocols in a multi-access terminal.

The only exception is circuit-switched-based services and the corresponding MM (Mobility Management) and CC (Call Control) layers (which are the circuit-switched equivalents of GMM and SM). As Evolved UMTS is 'Packet Only', all the CS Core domain-related procedures of the NAS layer are no longer applicable to Evolved UMTS. However, a multi-access terminal able to access to 2G/GSM, 3G/UMTS and EPS networks will have to support all of the procedures, as illustrated in Figure 4.24.

The Mobility Management layer of the EPS network is named EMM (for EPS Mobility Management) and supports the same functions as its 2G and 3G GMM equivalents. Most of these functions (like terminal location management and terminal state management) are further detailed in Chapter 5.

As regards Session Management, the EPS SM layer supports the basic function set for EPS Bearer establishment, modification and release, including bearer Quality of Service negotiation. More examples about Session Management procedures are described in Chapter 5.

In 2G/GSM and 3G/UMTS, Mobility and Session Management-level procedures were defined in a quite independent way, allowing procedures to be handled or performed separately. Although re-using the same concepts, the way EPS NAS protocol is specified in a fairly different way so as to limit as much as possible the amount of signalling exchanged. For that reason, Mobility or Session Management-level information is often transferred over the interface as part of other procedures such as S1 Bearer Management or RRC Connection Management messages. Some examples to illustrate this are provided in Chapter 5.

4.4 IMS Protocols

The aim of this section is to briefly describe the main application protocols used by IMS (IP Multimedia Subsystem). All of these protocols have been defined by the IETF (Internet Engineering Task Force), which is the international community dedicated to the evolution of the Internet. Because they were designed in the scope of wired Internet communication session setup, 3GPP had to adapt them to specific IMS architectures and requirements.

However, re-using worldwide standards for packet-based services within IMS has major advantages:

- The development cycle for terminals and network nodes is reduced as IMS based applications development only require adaptations to protocol stacks that already exist
- The inter-working between other packet networks is easier, as IETF application protocols are now widely used in public and private networks.

The rest of this section describes the following:

- SIP (Session Initiation Protocol), which is the protocol for packet session control (establishment, modification and termination).
- SDP (Session Description Protocol), which is the standard for describing and negotiating media components.
- RTP (Real Time Protocol), being the transport stack used by real-time sensitive packet-based services.

The intention here is not to provide an extensive view of IETF protocols, but rather to give an overview of the main characteristics (the main concepts and principles) and describe how they fit with the other elements of evolved UMTS architecture.

For further information, the reader is invited to refer to standard documents, whose references are provided in each section.

4.4.1 The IMS Protocol Stack

As an introduction to the IETF and IMS protocol description, Figure 4.25 describes an overview of the protocol stack architecture for IMS services. The figure represents two end-user terminals as well as an intermediate IP router. The name on the left corresponds to the well known OSI model terminology.

Starting from the bottom, the Physical and Data Link layers are parts of Local Area Network technologies, ensuring reliable data frame transmission between two points. From an implementation perspective, these two layers may be supported by a lot of technologies, from wired or wireless Ethernet to E-UTRAN access network. The Network layer is in charge of packet routing through the network. This is generally implemented by the IP, which proposes a solution enabling packet routing, intermediate nodes and endpoint addressing.

Above the Network layer, the Transport layer is an end-to-end service ensuring reliable data transmission to application-layer protocols like SIP, RTP or RTCP. The Transport layer may provide a reliable connected mode data-transmission service – as with TCP (Transmission Control Protocol) – or an unreliable connectionless transmission service – as with UDP (User Datagram Protocol). In most of the cases, transactional and interactive services like email transfer or Web browsing make use of TCP, whereas UDP is more often associated to real-time flows like streaming of Voice over IP.

4.4.2 SIP

This section briefly describes the SIP (Session Initiation Protocol), one of the most popular application protocols for establishing and terminating communication sessions over the

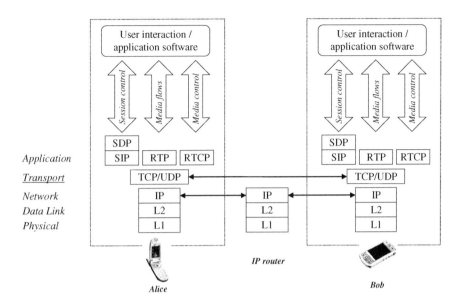

Figure 4.25 IMS protocol architecture.

Internet. Like other session-control protocols, SIP supports the following two main features:

- An addressing scheme which allows endpoints (could be a server or an end-user) to discover each other.
- Session state management, from establishment to termination, including on-line session modifications (such as the addition or removal of a media component).

SIP's worldwide success and acceptance can be explained by many factors. The main ones which are worth mentioning in the scope of this section are: its relative simplicity, the protocol flexibility and its openness to future extensions.

- Simplicity: SIP protocol states are limited in number and the interaction with other sub-protocols (like SDP) was made simple. In terms of protocol format, SIP messages are text-based, meaning that debugging can be performed using standard interface dump tools. For those two reasons, SIP is considered as comparatively easier to implement than other session control protocols like H.323, which is intrinsically more complex.
- Flexibility: SIP can handle a very large set of real-time or nonreal-time session types, like voice, video telephony, multimedia communications involving one or more participants and also text services like 'presence' or 'event notification'. Besides, SIP was designed as being independent from the underlying transport layers, and can be deployed over UDP (the most common case) or TCP.
- Extensibility: SIP can easily be customized and extended, as it was made in order to deal with specific IMS requirements. New message headers are defined in a backward-compatible way (all unknown information is simply ignored).

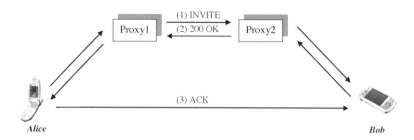

Figure 4.26 A simplified example of SIP session setup (the so-called SIP trapezoid).

SIP was defined as a generic protocol and is therefore not self-sufficient to establish a session. Typically, SIP does not provide any means for media format description or Quality of Service negotiation. For this reason, it is generally used in combination with SDP (the protocol for media session description) and RTP/RTCP (the protocol for transport and control of real-time application packet bearers) which are further described in this chapter.

(i) SIP Architecture Elements
Figure 4.26 describes an example of SIP session establishment. Basically, the whole process looks like a three-way handshake:

- The initial INVITE request sent by Alice contains the address of Bob in the form of an URI (Uniform Resource Identifier), but Alice does not know the exact location of Bob. This is the reason why the INVITE request is sent to a local Proxy1, and forwarded to a Proxy2 serving Bob's domain.
- As Bob accepts the session and decides to answer Alice, a '200 OK' message is sent back to Alice using the same signalling route as the initial INVITE message.
- The final acknowledgement 'ACK' sent by Alice may follow a different route, as each party knows each other's address.

As shown in this example, SIP not only describes the protocol, but also introduces architecture elements, network logical nodes and associated features. As usual, this definition does not mandate any specific network architecture, and a given physical network node may aggregate several logical functions:

- User agent: this is a generic name for SIP clients (the entity which generates the request, which is Alice's terminal in the example above) and servers (the entity that generates the response and could be either a machine like an application server or an end-user terminal like Bob's phone).
- SIP proxy: the role of SIP proxies is to make sure SIP messages (including requests and answers) are routed appropriately. Some proxies (called stateful proxies) may maintain client or server connection states based on the transactions they receive. Other proxies (stateless proxies) just forward the requests and responses they receive.
- Registrar: the registrar is a particular SIP server whose role is to accept and process SIP registration requests from users, in order to maintain end-user location information.

The SIP addressing scheme is based on URI (Uniform Resource Identifier) and is defined by the following generic format:

$$sip:user:password@host:port;uri-parameters?headers$$

This format is very similar to email addresses (like, for example, *mailto:johan@domain.com*). In addition, the SIP URI format may optionally include parameters used, for example, to specify the transport parameters to be used when sending SIP messages to the corresponding client, and headers to be added to the SIP request.

The following example illustrates this flexibility. This address specifies that all SIP messages shall be sent to Alice using a TCP transport connection, and that an 'urgent' priority field shall be included in all requests.

$$sip:alice@atlanta.com;transport=tcp?priority=urgent$$

(ii) The SIP Methods

As in SIP terminology, a 'method' is a primary function being invoked by a request initiated from a client. SIP specification defines a list of methods, which are briefly described below:

- **REGISTER:** The registration is an important process in SIP, as it allows binding a SIP URI (which is the SIP identity which does not give any indication about the actual user location) to a contact address (which is generally an IP address). When a user registers to a SIP Registrar, it provides both its SIP URI and its contact address. Further on, when a user-terminated request is received by the user domain proxy, the contact address is retrieved by the proxy so that the request can be transmitted to its destination.
- **INVITE:** This method is sent by user willing to initiate a session towards a SIP server, which could be an application server or another end-user. Once the dialog is established between the peer entities, the INVITE method can also be used to modify the existing session, for example to add or remove a media stream, or change the addresses and ports being used for the signalling or user packet transmission.
- **BYE:** This method is used to terminate a session initiated with a SIP INVITE.
- **CANCEL:** This one is used to cancel a pending request. This may happen in case an initial INVITE message has been acknowledged by a '200 OK' answer but the final ACK has not been yet received.
- **OPTIONS:** This method allows a User Agent to query another User Agent, or a SIP proxy about its capabilities. This may be used, for example, to know the SIP methods supported by the peer entity, or to check if a specific extension is supported.

As described above, the basic SIP methods specified in RFC3261 have been extended in order to define additional services supported by the SIP framework. Among the multiple extensions designed by IETF, this section will describe two of them, as being part of the IMS domain:

- The Event Notification extension.
- The Instant Messaging extension.

The SIP Event Notification extension
This SIP extension is specified in document RFC 3265, 'Session Initiation Protocol (SIP)-Specific Event Notification', and allows SIP to manage events. This is a useful add-on to support applications such as 'Presence indication'.

In RFC3265, two additional SIP methods are defined, which the IMS terminal shall support. However, in the case of a generic SIP terminal, the support of the RFC3265 extension can be checked using the OPTIONS SIP method:

- **SUBSCRIBE:** This method is used to request current state and state updates from a remote node. Many different event packages have been defined within IETF – such as 'presence events' or 'registration events' – so that the SUBSCRIBE method includes which one the subscription is related to.
- **NOTIFY:** Such messages are sent to inform subscribers of changes in the state to which the subscriber has a subscription.

In the scope of IMS, the SIP Event Notification extension is used in association with the registration process. Once registered, the terminal subscribes to the 'registration event package'. This further allows the network to notify the subscriber that it has been de-registered, or that the terminal needs to re-authenticate.

The SIP Instant Messaging Extension
For the support of Instant Messaging, a new SIP method has been defined in RFC 3428, 'Session Initiation Protocol (SIP) Extension for Instant Messaging':

- **MESSAGE:** This new method allows the transfer of Instant Messages. Correct reception (not meaning the message has been read) is acknowledged by the receiver through a '200 OK' response.

(iii) SIP Messages

SIP Protocol Format
SIP messages are written in ASCII-coded plain text, like many IETF application protocols: HTTP (Hyper Text Transfer Protocol) for Web browsing, POP3 (Post Office Protocol) for email retrieval, SMTP (Simple Mail Transfer Protocol) for email posting, NNTP (Network News Transfer Protocol) for news posting and reading, etc.

The SIP protocol format is actually very similar to the HTTP command/response transaction model. Each request message contains an SIP method which is invoked, and a list of parameters associated with the request. Figure 4.27 is an example of an INVITE message sent by a client.

In this example, the user **bob@proxy2.com** is invited to a SIP session by **alice@proxy1.com** for a session whose subject is '**Project Update Meeting**'. The INVITE message is sent over the **UDP** transport protocol and contains an offer for session expressed by a SDP description part.

In principle, the request is answered using a response code covering the normal completion cases and the possible failures which may occur (the request is not correctly formatted or rejected, etc.).

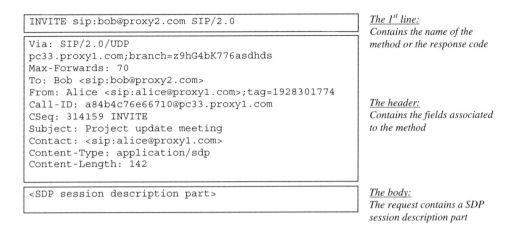

Figure 4.27 An example of an SIP INVITE message.

Using plain text message definition makes the debugging much easier than other coding schemes like ASN.1, which is used by H.323. Messages can easily be captured and displayed by using very basic interface dumping tools like, for example, TCPDUMP. But, on the other hand, this has the drawback of generating quite large messages (size may vary from hundreds of bytes to a few thousands of bytes). This may not be so critical in a broadband wireline IP context, considering the fact that SIP signalling is only exchanged at call setup. However, in a wireless environment characterized by resource scarcity and high bit error rate, a more optimized format is preferable in order to limit bandwidth requirement and keep signalling transmission time as low as possible.

This is the reason why a specific SIP signalling compression has been defined by IETF.

SIP Message Compression
The SIP compression scheme, also known as SigComp, is defined in the following IETF document: RFC3320, 'Signalling Compression (SigComp)'.

Figure 4.28 illustrates the fact that SigComp is not a substitute for the compression process performed at the PDCP level in Evolved in this chapter.

When being sent at the IP level, a SIP message looks like Figure 4.28: the SIP/SDP part contains the SIP message itself, plus a SDP container, when applicable. The transport header (a 20-byte TCP or 8-byte UDP) is added on top of it, as well as the IP header (the iPv4 header is at least 20 bytes long, and the IPv6 is at least 40 bytes).

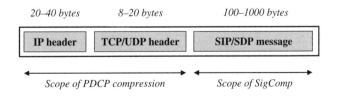

Figure 4.28 Handling of a SIP message.

SigComp only applies to the SIP/SDP part of the IP packet and is independent for each direction (meaning that SIP request from a client may be compressed whereas the responses sent by the server are not).

The main principle of SigComp is that no compression algorithm is specified. This was defined that way to allow each SigComp implementation to find its own compromise between compression efficiency and memory requirements.

For that purpose, SigComp is based on a Universal Decompressor Virtual Machine (UDVM) which makes use of an instruction set, or programming language, specifically designed for the purpose of writing decompression algorithms. This language contains instructions such as basic calculation, memory manipulation, etc.

The compression algorithm is chosen by the message sender, and coded using the UDVM instruction set defined in RFC3320. The corresponding byte code is then included to the first compressed message sent, and used by the receiver each time a compressed message arrives.

In addition, SigComp is based on a static dictionary, defined in RFC3485, 'SIP and SDP Static Dictionary for SigComp'. This document references the most common SIP and SDP well known words which appear in most of the SIP and SDP messages, so that each of these elements can be referred to using a common set of references.

For illustration purposes, Figure 4.29 describes the generic format of a SigComp message. The first five bits of a SigComp message are set to '1'. Because this sequence never occurs in ASCII coding, SigComp message can be sent along noncompressed ASCII-coded messages.

SigComp support is indicated by each entity by using an additional 'comp = sigcomp' parameter in SIP requests. This new information indicates that the requesting entity supports SigComp and is willing to receive compressed messages.

As in IMS subsystems, SigComp is only applied to the signalling path from the user terminal to the P-CSCF call server. Although SigComp support is mandatory in 3GPP specifications, its use is optional and left to terminal and P-CSCF decisions.

(iv) IMS Differences to Standard IETF SIP

As mentioned in the introduction to this section, SIP implementation in the IMS context slightly differs from the principles described in the IETF documents. The most significant additions and differences, as well as the underlying reasons, are described hereafter.

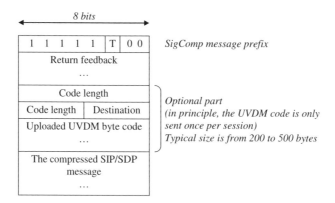

Figure 4.29 An example of a SigComp message.

Need to Register and Authenticate at the SIP Level
As defined by the IETF, SIP does not require an end-user terminal to register before requesting a SIP service through, for example, an INVITE or a SUBSCRIBE method.

In order to increase IMS security, 3GPP specifications mandate a mobile terminal to register to its S-CSCF server before activating any of the SIP-based IMS services.

This IMS registration process is performed through the REGISTER method, and also provides end-user IMS authentication. This authentication is actually a challenge/response mechanism based on credentials stored on the user SIM card and HSS network node.

SIP Signalling Routing
In IMS networks, the SIP message routing is not free and the SIP user agent cannot choose the signalling path, as in the SIP trapezoid example described above.

Because of user service authorization and billing requirements, IMS requires the SIP signalling is routed towards the P, I and S-CSCF call servers. At some point, when the S-SCSF has been reached once by the user SIP request, the I-CSCF server can be removed from the signalling path. The only exception is when a topology hiding feature is being used, as described further in this section.

IMS SIP Node Supported Features
Depending on the session procedure (or SIP method) which is invoked, each of these nodes has a specific role as regards to charging, end-user authentication and SIP signalling handling (like SigComp support in the P-CSCF, or specific treatment to be applied to SIP header fields).

CSCF nodes are not simple or pure SIP proxies and routers, as in the RFC3261 spirit; they are also requested to support specific features in a specific manner.

IMS Topology Hiding Specific Additions
This topology hiding feature is related to two items above, as it requires some specific rules for SIP signalling processing and routing.

For security reasons, an operator may be willing to hide the topology and configuration of its network to other external networks or entities. Practically speaking, this means that such an operator would like to hide the URI of its CSCF nodes to the external world, preventing the outside to know, for example, the addresses of the servers, to evaluate the number of such servers or to detect any reconfiguration action the operator may apply. To achieve this, the I-CSCF, which is basically the network entry point for SIP services, need to support specific features, as illustrated in Figure 4.30.

When an initial incoming SIP request comes from the external world, the I-CSCF supporting THIG adds its address (icscf@net.com in this example) to the SIP 'Path' header, in order to make sure it is kept in the signalling path for further responses.

Further on, when a SIP request or response is to be routed out of the network, the I-CSCF translates all SIP node addresses (like the URI of the S-CSCF for a response to a REGISTER), so that any external receiving entity can retrieve it. Similarly, when a SIP request or response with a translated address arrives at the I-CSCF, the reverse translation is performed.

This process is potentially applied the following SIP headers:

- Via – which describes the path taken by the SIP message so far.

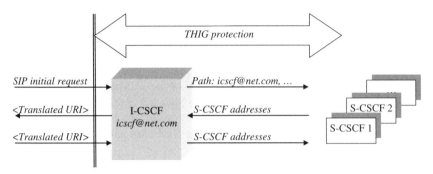

Figure 4.30 Illustration of SIP-based topology hiding.

- Record Route – this header is inserted by proxies willing to force SIP signalling routing (such as the P-CSCF or the I-CSCF implementing THIG).
- Service Route – which contains a list of SIP proxies adding service to the message (like the S-CSCF).

User Identification in IMS

SIP end-user identification is based on URI. As described in the section about SIP methods, the REGISTER method allows binding the user URI to a contact address – being an IP address in most cases. A user may have more than one URI, in which case the user is known by multiple public identities and each of the identities needs to be registered independently if the user is willing to be contacted through each of them.

In IMS, each subscriber is assigned one private identity (also known as IMPI for IP Multimedia Private Identity) and one or more public identities (also known as IMPU for IP Multimedia Public Identity), as illustrated by Figure 4.31.

The IMPI is used during the IMS Authentication and Key Agreement (AKA) process, which aims at mutual user and network authentication as well as security key exchange between network and terminal so as to protect the SIP signalling exchanged between the different entities. The IMS AKA process is described in more detail in Chapter 5.

Briefly, during the IMS registration process, the subscriber has the possibility to register all its IMS public identities using one single REGISTER SIP message.

On reception of the REGISTER message from the terminal, the S-CSCF retrieves the list of public identities associated to it. The S-CSCF stores each of the public user identities and

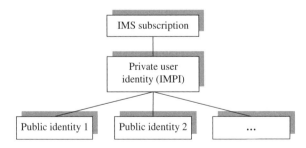

Figure 4.31 IMS and user identities.

returns the full list in the SIP '200 OK' response sent to the terminal. This process is known as 'implicit registration'.

The IMS P-Header Extensions

Specific SIP header extensions (known as Private SIP headers) have been specified for the purpose of SIP IMS implementation. This set of new headers carry IMS-related informational parameters used by the terminal or network CSCF nodes.

The complete list is described in the following IETF documents:

- RFC3455, 'Private Header (P-Header) Extensions to the Session Initiation Protocol (SIP) for the 3rd-Generation Partnership Project (3GPP)'.
- RFC3325, 'Private Extensions to the Session Initiation Protocol (SIP) for Asserted Identity within Trusted Networks'.

For illustration purposes, here is an extract of this list:

- **P-Associated-URI:** is sent from the S-CSCF to the terminal and contains the list of implicitly registered public identities.
- **P-Access Network-Info:** is sent from the terminal to the S-CSCF and contains the current access network and cell identification information.
- **P-Asserted-Identity:** this field represents the IMS subscriber identity once it has been authenticated. In IMS, this field is used to convey user URI from the P-CSCF to the S-CSCF.

Reference documents about SIP

SIP IETF documents:

- RFC3261, 'SIP: Session Initiation Protocol'
- RFC3265, 'Session Initiation Protocol (SIP)-Specific Event Notification'
- RFC3428, 'Session Initiation Protocol (SIP) Extension for Instant Messaging'
- RFC3665, 'Session Initiation Protocol (SIP) Basic Call Flow Examples'

SIP IETF 3GPP specific documents:

- RFC 3455, 'Private Header (P-Header) Extensions to the Session Initiation Protocol (SIP) for the 3rd-Generation Partnership Project (3GPP)'

SigComp IETF documents:

- RFC3320, 'Signaling Compression (SigComp)'
- RFC3485, 'SIP and SDP Static Dictionary for SigComp'
- RFC3486, 'Compressing the Session Initiation Protocol (SIP)'

3GPP technical specifications about SIP usage in IMS:

- 24.229, 'IP Multimedia Call Control Protocol based on SIP and SDP'

4.4.3 SDP

(i) Overview

SDP (Session Description Protocol) is probably not a protocol as such. It is rather a format specification for multimedia session description. SDP is able to handle lots of different information, in order to be able to define a multimedia session in the most accurate manner. The session information items managed by SDP can be classified in the following groups:

- Session name and information: this can be very general and provided for information only. It helps to quickly understand the session purpose.
- Contact information: this describes a contact name, document or Web page that the receiver can read in order to get more information about the session.
- Session starting and stopping times.
- Media types: this describes in detail all the media being used in the session (audio, video, codec, sampling rate, etc.).
- Media connection information: this relates to, for example, the IP address and port to be used for the media transport.

(ii) SDP Message Format

Like SIP, SDP is a plain-text ASCII protocol. This makes SDP very easy to read when captured on the fly, unless it is compressed with SigComp. The SDP format is also very simple, as it consists of a number of lines of text of the form <**type**> = <**value**> describing session parameters. The RFC4566 describing the SDP format specifies the basic set of parameters. This initial list has been further completed and extended in many ways to allow, for example, grouping of media or endpoint capability information exchange.

For illustration purposes, here is the list of the main session parameters supported by SDP:

- **v:** the version of SDP being used.
- **o:** the owner/creator of the session. The (username, session ID, address) T-uple form a unique session ID.
- **s:** the session name.
- **i:** the session information.
- **u:** a URI pointing to session information.
- **e:** email of a session contact person.
- **c:** session connection information.
- **t:** specifies the session start/end time. The coding allows time-limited or endless sessions (e.g. for voice or conversational services).
- **a:** describes a session or media attribute.
- **m:** describes a media.

As illustrated in Figure 4.32, coming from the RFC4566, there are two types of SDP parameters:

Session level parameters	`v=0`
	`o=mhandley 2890844526 2890842807 IN IP4 126.16.64.4`
	`s=SDP Seminar`
	`i=A Seminar on the session description protocol`
	`u=http://www.cs.ucl.ac.uk/staff/M.Handley/sdp.03.ps`
	`e=mjh@isi.edu (Mark Handley)`
	`c=IN IP4 224.2.17.12/127`
	`t=2873397496 2873404696`
	`a=recvonly`

1st media	`m=audio 49170 RTP/AVP 0`
2nd media	`m=video 51372 RTP/AVP 31`
3rd media	`m=application 32416 udp wb` `a=orient:portrait`

Figure 4.32 An example of SDP session description.

- The parameters at the session level.
- The parameters at the media level. In case a SDP session contains more than one media type (for example, voice and video), each media is described using its own set of parameters and attributes

This example describes a time-limited multimedia broadcast session (session attribute is 'recvonly', meaning that the participants are only allowed to receive information).

The session will support three media flows, which are described right after the session level parameters:

- An audio flow on port 49170. The rest of the line specifies that the flow of data is encoded by the Audio and Video Profile (AVP) 0 – which corresponds to the G.711 ITU codec – and supported by RTP (Real Time Protocol).
- A video flow on port 51372. The codec is based on profile 31 – which corresponds to the H.261 ITU coding scheme.
- An application data flow over UDP, which is not further specified in the SDP container, as it is handled by a specific application ('wb' for whiteboard, which is a remote conferencing tool for shared drawing).

(iii) SDP and SIP Interaction

In some cases of multimedia applications, it may happen that terminals are not able to support the audio or video flows in the format proposed by the session initiator. This may happen, for example, in the case of a newly deployed application of an information-coding scheme, or if the called party has limited capabilities for any reason, such as energy-saving concerns, requiring the display and graphical processor to be in power-save mode.

To cope with such a situation, a specific mechanism called 'the Offer/Answer model' has been specified and documented in RFC3264, allowing end-to-end session parameter negotiation

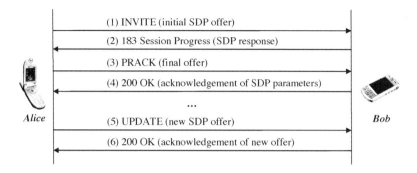

Figure 4.33 The SIP/SDP offer/answer model in IMS.

for SIP/SDP sessions. This mechanism makes use of the three-way handshake in SIP session establishment, and is applied to IMS session setup.

The principle, as it is applied in IMS, is illustrated in Figure 4.33:

- The initial INVITE SIP message contains the list of media types (audio, video, etc.) and a list of codec types the initiator is able to propose, the preferred one being on the top of the list.
- In the '183 Progress' provisional answer, the called party answers with the media codecs it actually supports, chosen from the list received in the INVITE.
- On reception of the '183' response, the initiator chooses the codec which will eventually be used for the session and informs the called party using a PRACK (Provisional ACK) SIP method.
- The last '200 OK' message is the final acknowledgement sent by the called party on session parameters.

Further on, the session characteristics may change because of media removal or addition, or because of a change of codec. In such a case, a SIP UPDATE message is sent, containing new SDP parameters. The UPDATE method is an extension to the SIP protocol, and is defined in:

- RFC3311, 'The Session Initiation Protocol (SIP) UPDATE Method'

(iv) SDP and IMS
SDP implementation in IMS is quite straightforward. In addition to the SDP basis defined in RFC4566, IMS makes use of some extensions, also subjects of IETF documents. This section lists the main ones.

Grouping of Media Lines
This extension is described by RFC3388, 'Grouping of Media Lines in the Session Description Protocol'. It allows indicating a relationship between some of the media of a SDP session. From a practical point of view, this extension makes it possible to group together different media streams (the 'm =' lines in SDP session descriptions) for the purpose of, for example:

- Lip synchronization between an audio and a video stream.
- Flow grouping: in some cases of voice codec – such as 3GPP AMR – several classes of bits are defined, each of them requiring specific handling and error correction. The flow grouping is a way to indicate that each bit flow is actually part of the same voice session.

Handling of Preconditions

This mechanism is described in RFC3312, 'Integration of Resource Management and Session Initiation Protocol'. This extension allows a participant of a multimedia session to specify conditions within a SDP offer. Because the conditions are generally set at the media level, they have been defined by the IETF at the SDP level rather than at the SIP level.

Reference documents about SDP

SDP IETF documents:

- RFC4566, 'SDP: Session Description Protocol'
- RFC3264, 'An Offer/Answer Model with the Session Description Protocol'

4.4.4 RTP

RTP (Real-time Transport Protocol) is an end-to-end transport protocol used for the transport of real-time data over packet networks. This protocol can support a large range of real-time services like Voice over IP, content streaming applications, point-to-point or point-to-multipoint broadcast multimedia sessions.

RTP is described by RFC3550, 'RTP: A Transport Protocol for Real-Time Applications'. As explained below, this document not only specifies the data transport protocol, but also a set of RTP control messages, also known as RTCP (Real-time Control Protocol), which are used to support session setup and operation.

As mentioned in the section about SDP, there might exist simultaneously several RTP flows from one sender to one or several receivers. This leaves the opportunity for terminals with limited capability or connectivity to only receive some of the available flows.

What is the Need for a Real-Time Protocol?

The basis of packet data transmission (as opposed to circuit-switched transmission) is that neither the transmission delay nor the user bandwidth is guaranteed. In principle, this has no impact on nonreal-time services like Web browsing or email transfer, as variations in terms of bandwidth and delay are not noticed by the end-user in most of the cases. However, in the case of real-time-constraining applications like voice or streaming, the end-user experience may suffer from too long transmission delays or variations in packet inter-arrival time.

Figure 4.34 illustrates the effect of real-time packet data transmission over an IP network. On the transmission side, packets are sent in the correct sequence with a period of T. On the reception side, due to network congestions, the packets experience different transmission delays and may even arrive out of sequence. This effect is known as 'jitter'.

The role of RTP is to compensate the jitter so that the application on the receiving side can play the received flows as if the packet inter-arrival time was not altered during the transmission. This

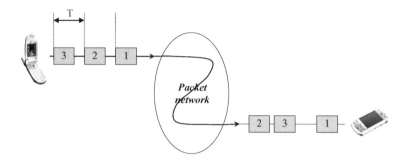

Figure 4.34 The effect of packet network transmission.

can be achieved thanks to the Timestamp – a time indication field contained by the RTP header of each packet, allowing the receiver to restore the correct packet sequencing.

RTP Generic Packet Format

Figure 4.35 describes the format of a RTP data packet. RTCP control packets are approximately of the same format. They are identified by specific values in the PT (Payload Type) field.

The RTP header is at least 12 bytes long and contains the following information:

- **V** (Version) represents the version of RTP protocol being used. The version corresponding to the RFC3550 is '2'.
- **P** (Padding) is the padding indicator. Padding may be added at the end of the packet for the purpose of specific ciphering algorithms mandating specific bloc size.
- **X** (eXtension) indicates whether or not the header is followed by an extension.
- **CC** (CSRC Count) indicates the number of CSRC identifiers.
- **PT** (Payload Type) specifies either the coding type of user data (for a RTP packet) or the type of packet (for a RTCP packet).
- **Sequence number** is increased by one for each RTP data packet. This field allows the receiver to detect out-of-sequence or lost packets.
- **Time stamp** indicates the sampling corresponding to the first byte of the payload. This field is used by the receiver to remove the jitter.

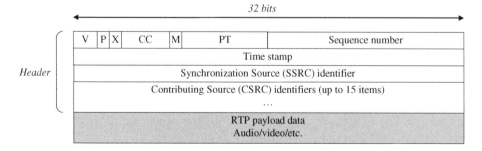

Figure 4.35 RTP packet format.

Table 4.4 Some examples of RTP Payload Type values.

PT	Name	Description
0	PCMU	G.711 speech codec at 64 Kb/s using μ-law
3	GSM	GSM speech codec
4	G.723	Dual rate (5.3 and 6.3 Kb/s) speech codec intended for H.324 videophone services
8	PCMA	G.711 speech codec at 64 Kb/s using A-law
12	QCELP	North-American IS-95 CDMA speech codec
34	H.263	Low bit rate video codec intended for H.324 videophone services
32	MPV	MPEG-1 and MPEG-2 video streams

- **SSRC** (Synchronization Source) is a random field identifying the source of the RTP flow.
- **CSRC** (Contributing Source) identify a source having contributed to the packet. This is typically used in the case of a conference call, when the speech inputs from multiple speakers have been mixed together before distribution.

As described above, the Payload Type (PT) field provides an indication on the coding of the payload data. The possible values of PT are specified in a separate document – RFC3551, 'RTP Profile for Audio and Video Conferences with Minimal Control'. From this field, the receiver knows exactly the audio and/or video codec applied to the data flow. In case of a RTP session being set up with SIP/SDP, the value of the RTP header Payload Type shall correspond to the 'RTP/AVP' value being indicated in the SDP session description and described above in this chapter.

For illustration purposes, Table 4.4 contains some typical values of PT, corresponding to the most well known coding schemes.

The Timestamp field of the RTP header is the information the receiver will use to evaluate packet arrival jitter. As opposed to the Sequence Number, which is increased by one at each packet, the Timestamp follows different rules, as it represents a time value associated with the first sample of the RTP data payload.

In the example of an audio codec with a clock rate of 8000 Hz, if a RTP packet is sent every 20 ms, then each packet will contain $160 = 8000 \times 20 \cdot 10^{-3}$ speech samples. Therefore, the value of the Timestamp will be incremented by 160 for each packet.

About RTCP
RTCP is the control protocol associated to RTP for real-time session operation. RFC3550 defines only five types of RTCP messages, each of them having a specific RTP header PT value:

- SR (Sender Report).
- RR (Receiver Report).
- SDES (Source Description) – this message contains information about the source of the flow, such as name, email address, phone number, etc.

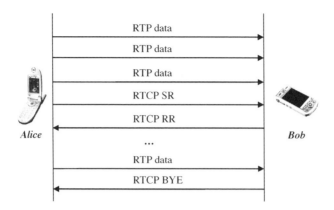

Figure 4.36 An example of a RTP session.

- BYE – this message is sent by an entity which leaves a RTP session.
- APP (APPlication) – this message format is left to application-specific features. This kind of message is used, for example, to support Talk Burst messages in PTT (Push-To-Talk) services. This service is further described in Chapter 6.

The Sender and Receiver Reports (SR and RR) messages are used to periodically exchange reception-quality feedback indications. These reports may contain lots of different information, such as the number of received or sent packets, the number of lost packets, an estimation of the variance of the jitter, etc. The aim of these reports is to allow the other party to adapt the session parameters based on the problem reported by the receiving entity. In the case of a broadcast or multicast session, this may help the session source to check that each participant is receiving the session in good condition.

RFC3550 does not impose any limitation on the RTCP traffic amount. However, it is recommended that RTCP traffic does not exceed 5% of the total session bandwidth.

For illustration purposes, Figure 4.36 is an example of RTP and RTCP messages exchanged between two terminals in the case of a unidirectional speech session from Alice to Bob. For simplicity, the figure does not display SIP and SDP signalling being exchanged to set up and terminate the session.

The RTP data frames contain encoded speech samples. From time to time, a Sender or Receiver Report is transmitted between the two parties. At some time, Bob decides to leave the session and a RTCP BYE message is sent to Alice.

An illustration of RTP and RTCP usage is provided in Chapter 6, through the Voice over IP and PoC (Push-To-Talk over Cellular) service description.

Reference documents about RTP

RTP IETF documents:

- RFC3550, 'RTP: A Transport Protocol for Real-Time Applications'
- RFC3551, 'RTP Profile for Audio and Video Conferences with Minimal Control'

Evolved UMTS Architecture

```
INVITE tel:+1-212-555-2222 SIP/2.0
Via: SIP/2.0/UDP [5555::aaa:bbb:ccc:ddd]:1357;comp=sigcomp;branch=z9hG4bKnashds7
Max-Forwards: 70
Route: <sip:pcscf1.visited1.net:7531;lr;comp=sigcomp>, <sip:scscf1.home1.net;lr>
P-Preferred-Identity: "John Doe" <sip:user1_public1@home1.net>
P-Access-Network-Info: 3GPP; cell-id-3gpp=234151D0FCE11
Privacy: none
From: <sip:user1_public1@home1.net>;tag=171828
To: <tel:+1-212-555-2222>
Call-ID: cb03a0s09a2sdfglkj490333
Cseq: 127 INVITE
Require: precondition, sec-agree
Proxy-Require: sec-agree
Supported: 100rel
Security-Verify: ipsec-3gpp; q=0.1; alg=hmac-sha-1-96; spi-c=98765432; spi-s=87654321;
    port-c=8642; port-s=7531
Contact: <sip:[5555::aaa:bbb:ccc:ddd]:1357;comp=sigcomp>
Allow: INVITE, ACK, CANCEL, BYE, PRACK, UPDATE, REFER, MESSAGE
Content-Type: application/sdp
Content-Length: 545
```
SIP header

```
v=0
o=- 2987933615 2987933615 IN IP6 5555::aaa:bbb:ccc:ddd
s=-
c=IN IP6 5555::aaa:bbb:ccc:ddd
t=0 0
```
SDP general parameters

```
m=video 3400 RTP/AVP 98 99
b=AS:75
a=curr:qos local none
a=curr:qos remote none
a=des:qos mandatory local sendrecv
a=des:qos none remote sendrecv
a=rtpmap:98 H263
a=fmtp:98 profile-level-id=0
a=rtpmap:99 MP4V-ES
```
SDP video media parameters

```
m=audio 3456 RTP/AVP 97 96
b=AS:25.4
a=curr:qos local none
a=curr:qos remote none
a=des:qos mandatory local sendrecv
a=des:qos none remote sendrecv
a=rtpmap:97 AMR
a=fmtp:97 mode-set=0,2,5,7; mode-change-period=2
a=rtpmap:96 telephone-event
a=maxptime:20
```
SDP audio media parameters

Figure 4.37 An example of an IMS SIP/SDP INVITE message.

4.4.5 A SIP/SDP IMS Example

As an illustration, Figure 4.37 describes an example of a SIP INVITE message initiated by a terminal and received by a P-CSCF, in the case of a point-to-point video telephony session set up between two subscribers. This message contains SIP parameters as well as a SDP part corresponding to session media parameters.

The message is represented in its uncompressed plain-text form; its length is around 1400 bytes (the INVITE is the largest SIP message exchanged during a SIP session setup) including the 28-byte overhead added by UDP/IP transport. Thanks to the SigComp compression method, the overall message size is approximately divided by two.

The SIP header part contains elements which have been mentioned in above sections, like:

- The **To** and **From** parameters which respectively identify the called and calling party.
- The **comp = sigcomp** field in the Via parameter, indicating the support of SIP signalling compression by the terminal.
- The **Route** parameter, which contains the address of the P-CSCF through which the SIP signalling shall be routed and the S-CSCF it is registered to.
- The **P-headers** which are IMS-specific extensions to SIP, carrying user identity and access network information.
- The **Content-Type** parameter, indicating the SIP message contains a SDP session description part.

The SDP part of the message contains a general part, applicable to the session and all its media flows. The '**t = 0 0**' parameter indicates that the session is permanent, starting immediately, with no time limit. Then, the SDP part describes two bidirectional media flows (the **sendrecv** attribute) – one for video and one for audio. Each SDP media part describes the proposed information coding schemes and the application protocol for data transport (RTP).

5
Life in EPS Networks

The objective of this chapter is to describe the different steps and mechanisms which are part of terminal and network life. To be more pragmatic, all these mechanisms are described from the viewpoint of the subscriber life, starting with the initial attachment at power-on to communication session setup and mobility in different conditions. In each, the interactions between the terminal and the network are described, as well as the related mechanisms which take place in the EPS network.

For that purpose, this chapter is split into four different parts:

- **Network attachment** – which focuses on the mechanisms in place to allow user initial registration.
- **Communication sessions** – which describes the different aspects of sessions, including session setup, Quality of Service negotiation, security procedures, etc.
- **Mobility in IDLE mode** – covering all procedures which take place when the terminal is inactive and moving through the network.
- **Mobility in ACTIVE mode** – which describes the most representative cases of user mobility during an active communication session.

5.1 Network Attachment

Network attachment is a key process by which the terminal registers to the network, so as to allow the subscriber to initiate or accept incoming communication sessions. Until this step is successfully completed, the user has no access to the network, except in case of emergency.

The network attachment is generally performed when the terminal is switched on, and relies on many different mechanisms within the network and the terminal, which are further described in this section.

Evolved Packet System (EPS)   P. Lescuyer and T. Lucidarme
Copyright © 2008 John Wiley & Sons, Ltd.

5.1.1 Broadcast of System Information

Broadcast of system information is an operation performed by the network. It is not specific to E-UTRAN networks, as similar processes exist, e.g. for UMTS, GSM and North-American CDMA systems, as well as IEEE Wireless Ethernet technologies. It is essential for terminals to understand which kind of network is present in terms of available technology, operators, and which channels and parameters shall be used to connect to the network in order to attempt a registration procedure.

Lots of different information can be broadcasted by the network so as to ease terminal life. However, the amount of broadcasted information shall not compromise the radio capacity. This trade-off has always been a subject of concern in standard committees, looking for solutions which allow the transmission of many parameters in the most efficient way.

Looking back to 2G/GSM, the broadcast channel was quite limited in terms of capacity (a few hundreds of bits per second) with very limited flexibility for extensions and the scheduling of information. In E-UTRAN networks, the system information broadcast mechanism is quite different, and makes use of two different schemes based on the BCH or DL-SCH transport channel.

The most critical information is broadcasted over the BCH transport channel, using fixed scheduling and transport format. The information broadcasted on this channel is called MIB (Master Information Block) and refreshes every 40 ms. The MIB contains key parameters, mainly related to the physical layer, such as:

- Downlink system bandwidth.
- Number of antennas.
- Reference signal transmit power.

In addition, less critical information is broadcasted on the DL-SCH transport channel, offering more flexibility in terms of the amount of information being broadcasted and the refresh period. The information broadcasted on this channel is structured in SU (Scheduling Units), where an SU consists of one or more SIB (System Information Blocks). The first of these SU is called SU-1; it is the most frequently broadcasted SU (every 80 ms) and has a specific purpose, as it describes the scheduling information (or periodicity) of all the other SUs. In addition, SU-1 contains key information for network access, such as:

- Current cell identity.
- TA code – or code of the Tracking Area the cells belongs to (details and purpose of Tracking Areas are described in the section about IDLE mode mobility).
- PLMN Identifier list – or the identities of the networks the cell belongs to (in some network-sharing deployment schemes, the cell may allow access to subscribers from different operators).
- Cell-barring status – which indicates if the terminal has permission to select this cell, e.g. for initiating a session.

Other SU contain other System Information, such as common and shared physical channel parameters, common radio measurement configurations, neighbouring cell list with associated thresholds (this is used for the cell reselection criteria evaluation, as described further in this chapter), etc.

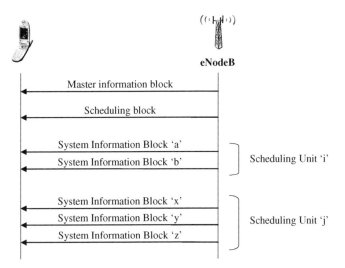

Figure 5.1 Contention-based random access procedure.

For illustration, Figure 5.1 describes the flow of messages related to the system information broadcast mechanism.

The aim of the *Scheduling Block* message is to provide the terminal with scheduling information (i.e. periodicity) of the Scheduling Units. Each Scheduling Unit is composed of different *System Information Block* messages.

5.1.2 Cell Selection

'Cell selection' is the process by which the mobile terminal chooses a suitable cell that it will camp on to perform initial registration.

In early cellular network deployment, cell selection used to be a fairly simple process, limited to the search of a cell in one frequency band and one unique radio technology. With the emergence of multi-layered networks combined with the availability of multi-mode terminals, this process is becoming more complex.

Basically, the cell selection criterion takes into account four types of information:

- **Network operator** (or PLMN, for Public Land Mobile Network) – A terminal equipped with a valid UICC card will naturally search for its home PLMN, or a cell which belongs to the network the user has subscribed to. However, there might be some cases in which the home operator is not available. When in a roaming situation, the terminal has to select another operator (possibly using a list of preferred visited operators that the user may have defined for service availability or charging cost reasons).
- **Technology** – the user (as well as the home operator) has the possibility to indicate and store in the USIM module a list of preferred technologies.
- **Radio criteria** – the received signal quality from the selected cell shall be good enough to allow user service. The parameters used in the cell quality-selection criteria (depending on the access technology) are broadcasted on the BCCH System Information.

- **Cell status** – this refers to the possible barring and reserved status of the cell defined by the operator.

About Cell Status
In its network, the operator has the flexibility to change the **cell status** for each of its cells, so as to allow or prevent the terminals to select and reselect some of the cells. This information is broadcasted by the network on the BCCH and shall be taken into account by all terminals in the cell-selection process.

In the normal case, for a cell being in service, the cell status is set to 'not Barred' and 'not Reserved'. The 'Barred' status means the terminal shall not select (or reselect) this cell, even for emergency cases. The 'Reserved for operator use' status means that the cell is reserved for operator use. Only one specific category of subscriber – the operator staff, corresponding to Access Classes 11 and 15 – may consider this cell for cell selection. Otherwise, all other categories of users, including public subscribers and emergency services, shall behave as if the cell is barred.

Such flexibility can be used, for example, during the early phases of network extension. Until the end of the network tuning phase, it is not desired that the terminal can select newly deployed cells so as to initiate or receive a service.

5.1.3 The Initial Access

In order to register to the network (and also on the occasion of a call setup), the terminal needs to go through the initial access procedure. This procedure makes use of the Random Access process, for the mobile to be able to setup a RRC connection with the network.

However, prior to this, the terminal needs to determine whether or not it can actually access the network through the cell it has selected – which is the purpose of Access Classes.

(i) About Access Classes
The concept of Access Class has been around since the beginning of 2G/GSM and is re-used in 3G/UMTS and EPS networks. The objective is to dynamically prevent or limit network access under specific conditions, such as:

- Exceptional network load.
- Emergency conditions – in such cases, priority should be given to public utilities and security services.

For this purpose, 16 Access Classes have been created, and each cell of the network broadcasts on the BCCH the list of classes which are barred from network access. The 0 to 9 class range corresponds to the public classes. Each subscriber is randomly allocated one of them, stored on the USIM card. In addition, some subscribers may also be allocated one of the specific purpose classes being defined by the standard for high-priority users. These classes allow a subscriber being a member of a specific corresponding category to initiate a call or answer an incoming one in situations where all public classes have been barred:

- Class 11 – reserved for PLMN use.
- Class 12 – for security services.
- Class 13 – for public utilities (e.g. water/gas suppliers).

- Class 14 – for emergency services.
- Class 15 – for PLMN staff.

Finally, class 10 has a specific purpose and corresponds to emergency calls. It allows the network to indicate if emergency calls are allowed on a per-cell basis. If class 10 is allowed, public class (from 0 to 9) terminals and even terminals without a valid USIM module can initiate an emergency call.

When the network load increases, the operator can then selectively bar one or several public classes from a given cell or group of cells. In the worst case, all public classes can be barred so as to save network resources for emergency or security services.

The Access Class restriction is a mechanism which does not substitute the cell status described in the previous section. The cell 'Barring' or 'Reserved' status only relates to the cell-selection (or reselection) process, regardless of terminal on-going activity. In contrast, the Access Class restriction is evaluated by the terminal at the occasion of access to the network and has no impact on the cell-selection and reselection processes. As a consequence, the fact that all public classes are barred in a given cell does not prevent a terminal from selecting such a cell.

Reference documents about terminal procedures in IDLE mode

3GPP technical specifications:

- 22.011, 'Service Accessibility'
- 23.122, 'Non-Access-Stratum (NAS) Functions Related to Mobile Station (MS) in Idle Mode'
- 36.304, 'User Equipment (UE) Procedures in Idle Mode'

(ii) About Random Access

Once the terminal has checked its Access Class is not restricted by the network, the Random Access procedure is the mechanism by which the terminal is attempting to access to the network while not having any other means to do so in the form of a dedicated communication channel. This procedure is used in many cases, which fall in three main categories:

- **Initial access from RRC-IDLE state:** This happens on many occasions of inactive terminals willing to set up a connection with the network. The most obvious user-related reason for doing so would be a terminal answering to an incoming session request (or paging) or a subscriber attempting to set up a session. In addition, Initial Access is also used in relation to pure signalling events, such as the initial registration (or attachment) performed when the terminal is powered on, or the Tracking Area Update procedures.
- **During handover procedure:** When the terminal arrives in the new target cell, it first needs to perform a Random Access before being able to resume the service which was in use in the old serving cell.
- **At uplink or downlink packet arrival in RRC-CONNECTED mode:** At some time, and while being in RRC-CONNECTED mode, the terminal may lose synchronization with the network or may not be allocated scheduled transmission resources in the uplink. This kind of situation may occur during a period of data transfer inactivity, or following a long DRX

cycle (as described in the section about data transmission). In such a case, a Random Access procedure is required, so that the terminal is able to resume packet data transmission or reception.

The terminals in those situations within a given cell will all compete between each other to gain access to the network. This is the reason why the Random Access procedure is possibly subject to collisions, failures and retries. In addition, the execution of the Random Access procedure is quite often critical. Excessive time spent in this process may lead to – depending on the situation – increased call setup time, degraded packet transfer delay performances and, at worst, call drop.

The E-UTRAN standard defines two types of Random Access procedures:

- The contention-based procedure – which is applicable to all events described above.
- The noncontention-based procedure – which only applies to handover and downlink packet arrival. This procedure is an optimized version of the 'contention-based' procedure in the case in which the eNodeB already knows the terminal identity.

(iii) The Contention-Based Random Access
Figure 5.2 describes the four steps of the contention-based Random Access procedure.

The 'Random Access Preamble' is sent on the RACH transport channel and contains a 5-bit random identity (which intends to identify uniquely the sending terminal on the RACH). The preamble sequence to be used is randomly chosen by the terminal among the available list broadcasted by the eNodeB on the BCH beacon channel.

The 'Random Access Response' answered by the network contains the random identity provided by the terminal in the preamble, a temporary C-RNTI and an initial uplink resource grant, which the terminal uses to transmit the 'Scheduled transmission' on the uplink shared transport channel. The Random Access Response's main purpose is to provide requesting terminals with an early resource allocation, so as to minimize the time to set up the RRC connection. This specific procedure was designed to comply with the very constraining requirement about terminal state change.

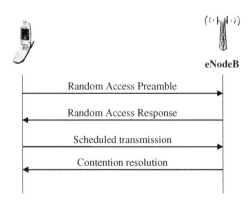

Figure 5.2 The principle of 'contention-based Random Access procedure'.

The C-RNTI, allocated by the eNodeB, is a Radio Network Temporary Identifier at the cell level. It is used to uniquely identify a terminal having an RRC connection in the Access Network. The 'Scheduled transmission' is sent using the simplest RLC TM (transparent) mode, meaning that the message is not segmented. In the case of initial access from RRC-IDLE state, the 'Scheduled transmission' sent by the terminal is actually an RRC Connection Request message. If the size of the early resource allocation permits, this initial message possibly contains a first NAS message, such as a Service Request, an initial attachment message or a Tracking Area update message.

The 'Contention resolution' message is not linked to the 'Scheduled transmission'. It aims at providing the actual contention resolution and identifying a winning terminal in case multiple terminals have requested access to the same resource at the same time. This message is sent early in the initial access process, in order to minimize latency for terminals not being elected which will try another attempt. Priority is, however, given to early allocation, so as to minimize the overall connection time for successful terminals.

(iv) The Noncontention-based Random Access
As described above, the noncontention-based procedure (Figure 5.3) is intended to Random Access in cases in which the eNodeB already has a reference for the requesting terminal. This is the reason why this procedure is applicable to inter-eNodeB handover cases, as well as downlink transmission resuming whilst the terminal is in RRC-CONNECTED mode.

At some time, the terminal is allocated a contention-free Random Access Preamble which does not belong to the list of preamble sequences broadcast on the BCH. In the case of handover, this information is provided by the target eNodeB (the one the terminal will access to) through the old serving eNodeB, using the RRC handover message.

(v) The Establishment of the RRC Connection
Figure 5.4 is an example of the use of the Random Access procedure to set up a RRC connection and send an initial NAS (Non Access Stratum) message to the EPC Core Network. This sequence may occur at the occasion of a user registration (described in the next section) or to initiate a data session setup.

In such a case, the terminal 'scheduled transmission' is actually a *RRC Connection Request* message which contains an initial NAS (Non Access Stratum) message intended to the MME.

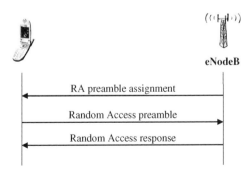

Figure 5.3 Noncontention-based Random Access procedure.

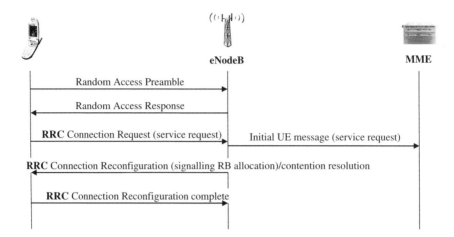

Figure 5.4 RRC connection setup and initial NAS message transmission.

The *RRC Connection Reconfiguration* is a multi-purpose procedure which performs the Random Access contention resolution, allows the eNodeB to establish, modify and release a RRC connection with a terminal, and possibly allocate signalling Radio Bearer resources.

When a Radio Bearer is allocated, the *RRC Connection Reconfiguration* message contains a Radio Bearer description so as to support the signalling messages further exchanged between the terminal and the eNodeB or the MME, e.g. for the purpose of the authentication procedure.

5.1.4 Registration

User registration is a mandatory process so that the subscriber can receive service from the network. The concept of being able to 'receive service' not only refers to the possibility for the subscriber to set up a voice or data session or any kind of multimedia session or service like 'Presence'; it also includes the possibility for the subscriber to be joined for a user-terminated session. Until the registration is successfully passed, the subscriber will be unable to exchange any kind of user data.

The only exception is emergency calls. Although being a useful feature, emergency calls are not considered a real service (in operator terminology), subject to registration and charging.

In short, the registration process serves four main purposes:

- Mutual user-network authentication – so that the terminal can trust the network it is connected to and also in order for the network to make sure it will get paid for the service used, etc.
- Allocation of temporary identity – the use of temporary identity allows maintaining the confidentiality of the user's private identity, also known as IMSI (International Mobile Subscriber Identity).
- User location registration – this allows the network to know the user's current location, and be able to page the terminal in case of an incoming mobile-terminated session request.
- Establishment of a default bearer – this is new to EPS networks and allows always-on connectivity for the end-user.

Figure 5.5 describes the signalling and procedures exchanged during the registration process.

The registration process is always triggered by the terminal itself when it is switched on, or if it loses completely the network coverage for some time and needs to register again.

Following the Random Access procedure (described in more detail in the 'Session Setup' section of this chapter), the terminal sends the Attach Request to the MME, which includes its identity. This identity may be either the subscriber's private identity (or IMSI) or a valid temporary identity (S-TMSI for S-Temporary Mobile Subscriber Identity).

Because the IMSI uniquely addresses each subscriber, it is seen as critical information from a security point of view and its transmission clearly has to be avoided as much as possible. By spying on and monitoring the IMSI, attackers could, for example, track a subscriber's location, movement and activity, determine user home country and operator. This is the reason why the

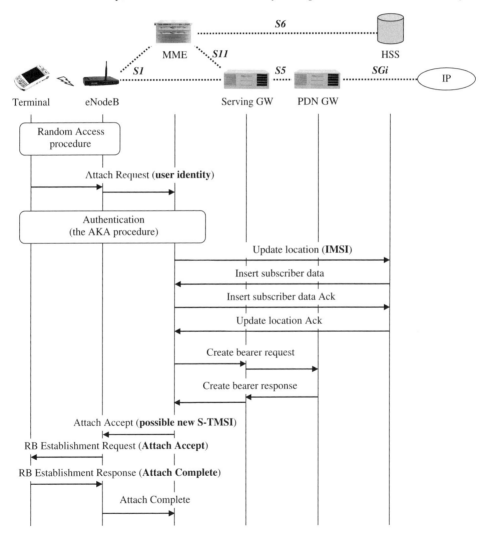

Figure 5.5 An example of subscriber registration.

NAS procedures make use of the S-TMSI temporary identity as much as possible instead of the IMSI. As described below, an S-TMSI is allocated by the network when the terminal has none, and its value is frequently renewed by the network.

Once the user identity is retrieved, the AKA process (Authentication and Key Agreement) is performed, so that network and terminal can mutually authenticate. More details on the AKA procedure, and EPS security concepts in general, is provided further in this chapter.

Because this is the very first attach performed at terminal power-on, the MME needs to update the HSS, so that the HSS can register the current MME corresponding to the subscriber (identified by its IMSI). In return, the HSS provides the MME with user subscription information (including Quality of Service limitations and access restriction), which will be further used by the MME to control and limit packet data context requests initiated by the terminal.

As part of the attachment procedure, the MME attempts to create a default bearer, which may be used by the terminal once the registration procedure is completed to initiate, for example, an IMS registration. The possibility to group registration and initial bearer establishment did not exist in GPRS or 3G/UMTS, as these two systems clearly separate the two procedures. The advantage of this evolution is that less signalling and processing time for both network and terminal is required.

Some definitions

- The **default bearer** is established when the terminal connects to the Packet Data Network. It remains established though the whole lifetime of the connection and provides the user with 'Always-On' connectivity.
- Any additional bearer established for the support of a specific service is called a **Dedicated Bearer**.
- The distinction between those two types of bearers in only meaningful for the Packet Core (meaning it is transparent to E-UTRAN), as each type of bearer has its own characteristics in terms of packet policing and Quality of Service.

The Serving and PDN GW which will support the packet bearer are chosen by the MME (similarly to the 2G/GPRS and 3G/UMTS architecture in which the SGSN is in charge of the determination of the GGSN). The default bearer is established using the *Create Bearer* GTP procedure, which implies both Serving GW and PDN GW.

Eventually, radio resources are set up over the air interface using the *Radio Bearer Establishment* procedure. To be more efficient, the request message sent by the eNodeB contains the *Attach Accept* message answered by the MME. This later may contain an S-TMSI allocated to the terminal in case it does not already have one, or to renew an old S-TMSI.

As a result of the whole registration procedure, the terminal gets an IP address (allocated by the PDN GW during the *Create Bearer* procedure and communicated to the terminal within the *Attach Accept* message) and full connectivity to the external IP network or IMS domain.

(i) About PDN GW Selection
The choice of the PDN GW is an important process supported by the MME during the subscriber registration process and is based on the concept of APN (Access Point Name) which is also used in 2G/GPRS and 3G/UMTS architectures. The role of the APN is to identify the gateway the subscriber will access to in order to get IP connectivity.

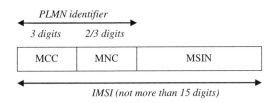

Figure 5.6 The structure of the IMSI (International Mobile Subscriber Identity).

For this purpose, the APN uniquely identifies a 2G/3G GGSN or a PDN GW, either from the home operator, or from the visited network in the case of roaming.

From a network deployment perspective, the operator may use this concept in a very simplified way by defining a single APN in its network, providing public IP connectivity to all its subscribers. Another possibility would be to allow multiple APN corresponding to, for example, an access to public Internet, access to IP operator-hosted services like streaming or Web-based information services, or a secured Intranet access.

During the session setup, the selection of the actual PDN GW is under the responsibility of the MME. This is a complex process which is able to address all the possible cases and to take into account information – or requests – from the subscriber as well as subscription information provided by the MME.

In addition, this process is backward-compatible with the APN selection process specified in the 2G/GPRS and 3G/UMTS standard, so as to allow IP connectivity through the EPC to subscribers under 2G and 3G radio access coverage.

(ii) About IMSI Structure

For information, Figure 5.6 describes the structure of the user IMSI private identity, which is common to 2G/GSM, 3G/UMTS and EPS subscribers.

The IMSI length is 15 decimal digit maximum. It is composed of three fields:

- The MCC (Mobile Country Code) – which uniquely identifies the country of the mobile subscriber's home operator.
- The MNC (Mobile Network Code) – which identifies the home operator of the subscriber within the country. Initially, this field was a two-digit field. It has been extended from the GSM specifications, to allow more operators in large countries like North America.
- The MSIN (Mobile Subscriber Identification Number) – which identifies the subscriber within the operator.

The concatenation of MCC and MNC codes forms the PLMN Identifier (Public Land Mobile Network), which uniquely identifies a network. The PLMN identifier is broadcasted on the BCCH, so as to inform the terminal about the operator the cells belongs to.

Reference documents about 3GPP network identities

3GPP technical specifications:

- 23.003, 'Numbering, Addressing and identification'

5.1.5 De-registration

The de-registration procedure is the counterpart of registration, described in the previous section. Once it is performed, the terminal has no more access to the network.

There are basically two kinds of de-registration procedures (or 'detach' procedures, as in the 3GPP terminology):

- The explicit detach – in which either the network or the terminal initiates the de-registration procedure. On the terminal side, this may be caused by the user switching its device off. On the network side, the reason may be the operator who decides to remove a customer from its network.
- The implicit detach – when terminal and network have lost communication for a certain period of time, each side implicitly considering that the registration is no longer active. The implicit detach is a necessary procedure to be avoided in the network to keep context and resources for subscribers whose terminal has disappeared from the network because of lack of coverage or any case in which the explicit detach could not be preformed.

Figure 5.7 describes an example of explicit detach initiated by the terminal.

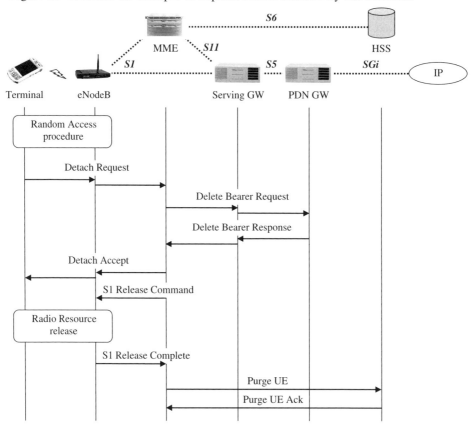

Figure 5.7 An example of terminal-initiated de-registration.

On receipt of the *Detach Request*, the MME initiates the bearer release procedure so as to release context and resources for the user tunnels supported by the Serving and PDN GW. The *Detach Accept* message is only sent back to the terminal if the cause of de-registration is not a terminal switch-off.

At the end, the release of radio-related resources and S1 resources is triggered by the *S1 Release* procedure initiated by the MME.

Once the resource release is done, the MME may keep the terminal-related context and subscriber information, so as to be able to perform a later attach from the subscriber without having to access the HSS. At some time, when the MME decides to delete the subscriber context, the *Purge* procedure is used to inform the HSS that it has deleted subscription data and that the corresponding subscriber is no longer attached to the network.

5.2 Communication Sessions

5.2.1 Terminal States

When presenting the evolved UMTS requirements, we briefly introduced the notion of 'terminal state'. In circuit applications, only two states are really useful ('connected' and 'disconnected'), reflecting the actual service state. However, when considering the various ranges of packet applications, intermediate states may have some interest, e.g. in phases where the session is set up but not active for a certain period of time. Such states would allow the network (including the Access Network and also the Evolved Packet Core) to manage resources in a more efficient way. This kind of flexibility cannot be provided by circuit-switched applications, as 'connected' implies guaranteed service capabilities in terms of bit rate and transfer delay.

For that purpose, two terminal state machines have been introduced within the standard, both maintained by the terminal and the network (Figure 5.8). The first machine is located at the RRC layer and maintained in the network by the eNodeB; the other is placed at the MM level and maintained in the network at the aGW level within the Packet Core.

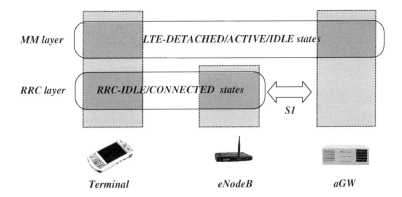

Figure 5.8 RRC and MM state machines.

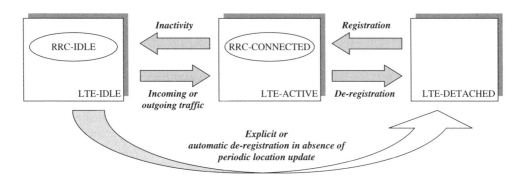

Figure 5.9 Evolved UMTS Terminal states and transitions.

(i) The Evolved UMTS Terminal States
The MM and RRC levels of state management do not exclude each other. As in Figure 5.9, the standard defines how they relate to each other and what the possible transitions are.

At the Packet Core level, three states are defined: LTE-DETACHED, LTE-ACTIVE and LTE-IDLE.

LTE-DETACHED corresponds to a state in which the mobile has been powered-on but is not registered to the network. It may be because it has not yet registered, or because the registration has failed in case no suitable network is available.

LTE-IDLE is a state in which the terminal is registered to the network, but not actually active. This corresponds to a low power consumption mode. In this state, the mobile location is known to the Packet Core domain at the Tracking Area level. In case of service setup or re-activation, the terminal is able to switch in LTE-ACTIVE mode within a very short period of time. In this state, the terminal mobility is ruled by the cell reselection algorithms and is therefore not controlled by the network. In LTE-IDLE mode, an EPS bearer may or may not be present between the network and the terminal. This allows a terminal in IDLE mode to resume a previously active data session without having to set up the EPS bearers again and renegotiate the associated Quality of Service attributes.

LTE-ACTIVE is the only real active state in which the terminal is exchanging data and signalling information with the network. This state is the only one where the terminal has a RRC connection being set up. In all other states, the terminal is not even known by the Access Network. In LTE-ACTIVE, the terminal location is also more accurate as the network knows its current cell, and the terminal mobility is ruled by the handover algorithms controlled by the network.

n principle, the terminal can de-register from any state. When in LTE-IDLE, de-registration can be either explicit, e.g. following a user action such as terminal power-off, or implicit. The aim of the implicit de-registration is to avoid maintaining context in the network for terminals which can no longer be reached. This can typically happen if the battery is suddenly removed, or if the mobile stays for a long period of time within a coverage hole. In such a case, the terminal is implicitly de-registered by the network, as no periodic location update is received.

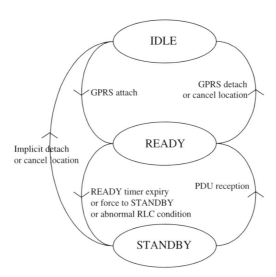

Figure 5.10 GPRS Mobility Management state machine (from 3GPP 23.060).

The Access layer terminal state view is slightly simpler, as the RRC only supports two states: RRC-IDLE (meaning that the terminal has no RRC connection being set up with any eNodeB) and RRC-CONNECTED which corresponds to the LTE-ACTIVE state.

(ii) Comparison with 2G/GPRS and UMTS
The evolved UMTS MM state model is actually very close to what has been defined for 2G/GPRS and UMTS Packet domain. The IDLE, READY and STANDBY 2G/GPRS Mobility Management states are functionally equivalent to the LTE-DETACHED, ACTIVE and IDLE states. As represented in Figure 5.10, the transitions between the 2G/GPRS states are also similar to those defined for evolved UMTS.

The only difference is that 2G and 3G/UMTS systems are actually supporting two state machines at the MM level – one for the PS (Packet) domain and the other for the CS (Circuit) domain. In Evolved UMTS, there is only one state machine, as all services are supported on one unique Packet Core network domain.

At the RRC level, the situation is a quite different. The UTRAN RRC layer introduced not less than four states (represented in by the grey box in Figure 5.11), reflecting the multiplicity of possible transport channels and combinations for data transmission over the radio interface. The intention was to define protocol states for transmission on dedicated channels (like CELL-DCH) and additional states to allow optimized resource management for connected users when the terminal is allocated shared resources (as in CELL-FACH) or no radio resource at all (as in CELL-PCH and URA-PCH). At the end, the standard looks quite flexible and gives lots of options for resource management optimization. The price to pay is a rather complex overall picture, not only to design, but also to operate.

From that perspective, E-UTRAN has been simplified, as all user data transmissions are supported by a single type of shared transport channel (putting aside the Multicast and

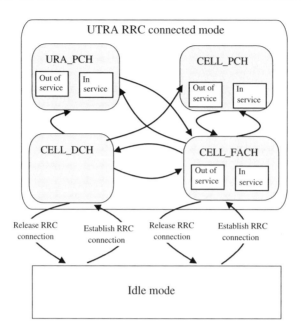

Figure 5.11 The 3G/UMTS RRC state machine (from 3GPP 25.331).

Broadcast transmissions). Therefore, the RRC state machine of E-UTRAN is much simpler, as it only proposes an idle and a connected state.

(iii) An Example of 'Evolved UMTS' State Transition
Figure 5.12 is an example of an MM state transition sequence which may happen when the mobile is powered-on or decides to register to the network once it finds a suitable cell.

Until it is successfully registered to the network, the terminal is in the LTE-DETACHED state. Once registered, the terminal may activate an EPS bearer (or use the default one), e.g. for IMS registration, or to update its 'Presence' information (the 'Presence' service is described in Chapter 6) and stays in the LTE-ACTIVE mode until inactivity detection occurs. The transition to LTE-IDLE has only the effect of releasing the RRC connection and the corresponding radio resources. All EPS bearers which were set up between the terminal and the MME are preserved,

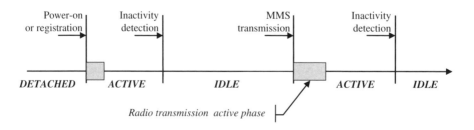

Figure 5.12 Example of MM state transition.

meaning that the terminal can resume all previously active sessions without having to renegotiate the session parameters and Quality of Service attributes.

When in the LTE-IDLE state, the user may start activity such as MMS transmission, which triggers a transition to the LTE-ACTIVE state. Once the data transmission is over, the terminal may move back to the LTE-IDLE state, based on inactivity detection criteria.

In any case, once the terminal is registered, it may be reached for in a user-terminated session in any of the LTE-ACTIVE or LTE-IDLE states.

5.2.2 Quality of Service in Evolved UMTS

(i) Concept of EPS Bearer
The definition of Quality of Service parameters and algorithms cannot be performed independently from the definition of information flows to which they will apply. For that purpose, the notion of 'EPS bearer' has been added to the standard.

The EPS bearer is an equivalent of the 'PDP context' being used in 2G/GPRS and 3G/UMTS standards. It is a logical concept which applies between the terminal and the PDN GW and aggregates one or several data flows transported between the two entities, as illustrated by Figure 5.13. An EPS bearer is actually composed of the three following elements:

- **An S5 bearer** – implemented by a tunnel which transports packets between the Serving and PDN Gateways.
- **An S1 bearer** – implemented by a tunnel which transports packets between the Serving GW and eNodeB.
- **A Radio Bearer** – implemented by a RLC connection between the eNodeB and the terminal. There is one RLC protocol machine per Radio Bearer.

The elementary data flows being transported by the EPS bearer are known as Service Data Flow (SDF). Each of them is characterized by the IP 5-tuple (source IP address, destination IP address, source port number, destination port number, protocol ID of the protocol above IP), which identifies both termination points (in the IP sense) as well as the application or service

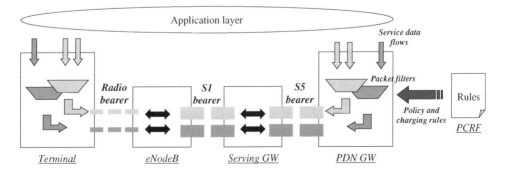

Figure 5.13 An example of two EPS bearers, from terminal to PDN GW.

being used. Practically speaking, a SDF may correspond to a connection to a Web or streaming server, or to a mailbox server.

From an Evolved UMTS perspective, an EPS bearer corresponds to one Quality of Service policy applied within the EPC and E-UTRAN. This means that all the SDF flows transported by the EPS bearer will be applied to the same packet scheduling algorithm, using the same priority, the same E-UTRAN RLC configuration, etc.

This notion is the EPS equivalent to the 'UMTS Bearer Service' of 3G/UMTS networks.

In addition to those mechanisms, and as described in Chapter 2, the PDN GW may enforce policy decisions (through packet filters) at the SDF level within an EPS bearer, based on rules provided by the PCRF (Policy and Charging Rules Function) node.

What is described in the rest of this section is:

- How the Quality of Service of an EPS bearer is characterized.
- How the EPS quality of service concept inter-works with 2G/3G 3GPP systems.

(ii) Quality of Service in EPS
An EPS bearer is characterized by the following parameters:

- **Allocation Retention Priority** (ARP) – this parameter refers to the priority used for the allocation and retention mechanisms. ARP is typically used for the allocation of the bearer resources at session setup or during handover mechanisms and also in the scope of resource pre-emption, for example, to decide which bearers need to be preserved in the case of a congestion situation. Once the bearer is established, the ARP has no impact on scheduling or packet-handling mechanisms.
- **Guaranteed Bit Rate** (GBR) – only applicable to bearers which require guaranteed Quality of Service for services such as voice or streaming.
- **Maximum Bit Rate** (MBR) – the MBR parameters help to set a limit on the data rate expected for the related service. In case the observed bit rate exceeds this limit, the EPS network can limit the effective rate by applying traffic-shaping functions.
- **QoS Class Identifier** (QCI) – which is used as a reference to a set of Access Network-related Quality of Service (QoS) parameters, for the transmission between the terminal and the eNodeB.

The purpose of the QCI, and associated parameters, is to provide a representation of QoS parameters to be shared between Core and Access parts of the network.

Each QoS class is associated with the following parameters:

- **Bearer Type** – this parameter indicates whether or not resources associated with the bearer need to be permanently allocated during the whole bearer lifetime, reflecting the difference between GBR (Guaranteed Bit Rate) and non-GBR bearers.
- **L2 Packet Delay Budget** (L2PDB) – This parameter describes the maximum time that packets shall spend transiting through RLC and MAC layers within the network and the terminal. In principle, this attribute is used to derive waiting queues and MAC HARQ operating parameters. For Guaranteed Bit Rate bearers, L2PDB indicates a maximum limit for packet transmission time. For nonGuaranteed Bit Rate bearers, this parameter does not

have the same strict meaning. It may be used, for example, as an input to the RLC/MAC queue-management system, in order to discard packets which have spent too much time in the waiting queue due to radio congestion.
- **L2 Packet Loss Rate** (L2PLR) – This parameter describes the maximum ratio of L2 packets which have not successfully delivered to the peer entity. As for the L2PDB, this parameter is intended for RLC and MAC HARQ configuration.

In addition to the bearer level parameters, the terminal is associated with another Quality of Service parameter: the **Aggregate Maximum Bit Rate** (AMBR). This parameter applies to nonguaranteed bit rate bearers only. Its purpose is to limit the overall bit rate of all bearers associated with this limit for a given Packet Data Network. This means that in case this limit is exceeded, the network has the possibility to apply traffic-conditioning algorithms for both uplink and downlink transmission, as for the MBR limit which is defined at the EPS bearer level.

GSM/GPRS and UMTS Quality of Service representations contain lots of parameters and do not limit the possible configuration which can possibly be defined. This results in a large number of possible parameter configurations which may lead to very different network implementation and behaviour on the field. The purpose of the QoS class is to limit this number of possible configurations by explicitly defining in the standard a list of 'label characteristics' or predefined sets of parameters which can fit to the various services that can be used.

Table 5.1 is an example of the possible QoS class definition for some GBR and non-GBR bearer types. In each case, some RT (Real-Time) and NRT (nonReal-Time) example services are given.

The Default Bearer (which is established when the terminal connects to the PDN network, at registration) can only be a non-GBR bearer type.

At the session setup, as part of the EPS bearer establishment, the terminal indicates associated requested Quality of Service attributes. These attributes are checked by the MME, as regards the user subscription rights provided by the HSS to the MME during the registration phase.

Eventually, an answer is provided to the terminal, possibly containing reduced values of the attributes, if the terminal request exceeded the user subscription.

Table 5.1 An example of label characteristics.

QCI	Bearer type	L2PDB	L2PLR	Example services
1	GBR	Low (<50ms)	Low (<10^{-6})	RT: Gaming
2	GBR	Medium (<100ms)	High (<10^{-3})	RT: Voice, Video (live)
3	GBR	High (<300ms)	Low (<10^{-6})	RT: Video (playback)
4	Non-GBR	Low (<50ms)	Low (<10^{-6})	NRT: SIP/SDP (IMS signalling)
5	Non-GBR	Medium (<100ms)	High (<10^{-3})	NRT: Web browsing RT: Interactive gaming
6	Non-GBR	High (<300ms)	Low (<10^{-6})	NRT: Bulk data transfer RT: Video (playback)

Table 5.2 Equivalence for PDP context QoS attributes.

R97/98 PDP attributes	R99 PDP attributes	EPS bearer attributes
Delay class	Traffic class	Bearer type
	Traffic handling priority	
Reliability class	SDU error ratio	L2PLR
	Residual bit error ratio	
	Delivery of erroneous SDU	
Peak throughput class	Max bit rate for uplink	MBR
	Max bit rate for downlink	
Precedence class	ARP	ARP
Mean Throughput	*Not applicable*	*Not applicable*
Not applicable	Max SDU size	*Not applicable*
Reordering required[b]	Delivery order	Delivery order[b]
Not applicable[a]	Transfer delay	L2PDB
Not applicable[a]	Guaranteed bit rate	GBR
Not applicable	*Not applicable*	AMBR

[a]Transfer delay in R99 only applies to real-time traffic classes (i.e. Conversational and Streaming). This is the reason why it has no R97/R98 equivalent, as GPRS is a best-effort service without guaranteed Quality of Service.

[b]In 2G/GPRS (respectively EPS), this attribute is not part of user-requested attributes; it is only present in SGSN and GGSN (respectively MME and User Plane Gateways).

When the MME requires E-UTRAN to set up the Radio Bearer, the Quality of Service attributes are translated by the eNodeB into radio resource allocation, scheduling priority, etc.

(iii) Comparison and Inter-Working with 2G/GSM and 3G/UMTS Systems
Table 5.2 describes the equivalence between the Quality of Service attributes defined in different versions of the standard:

- **R97/98 PDP attributes** refer to the Quality of Service attributes of PDP contexts for the 2G/GPRS part of the 3GPP standard up to Releases 1997 and 1998.
- **R99 PDP attributes** refer to the Quality of Service attributes of PDP contexts for 3G/UMTS as well as 2G/GPRS networks which implement Release 99 of the 3GPP standard.

From this table, it is interesting to consider how 3GPP Quality of Service has evolved from 2G technologies to EPS. 2G/GPRS standardization was initiated in 1996, in a wireless market which was mainly driven by circuit-switched applications. At that time, given the limitations of 2G/GSM radio interface technology, cellular packet data services were mainly thought of in terms of best-effort services with no guarantee of any kind in terms of transfer delay and bit rate. Later on, when UMTS and R99 were defined, lots of parameters were added to the Quality of Service representation. The intention was to allow all the possible flexibility for the network in terms of resource-allocation mechanisms, scheduling algorithms, etc., given all the circuit and packet service types known at that time and envisaged for the future. This level of flexibility was consistent with all the possible options allowed by the standard for the definition of dedicated transport channels in UTRAN.

Eventually, in the all-IP EPS world, Quality of Service representation is much simpler, based on fewer attributes and associated with predefined labels so as to limit the huge number of possible combinations and maintain consistency between network implementations from different manufacturers for a given type of service. The attribute set is actually limited to the minimum for resource allocation on shared radio channels, which is the only transmission scheme E-UTRAN allows. From this perspective, EPS looks more like a 2G/GPRS extension than an evolution of 3G/UMTS.

5.2.3 Security Overview

In the scope of EPS communication session management, it is important to spend some time on the security mechanisms put in place in the network and the terminal. In this chapter, two areas of security are envisaged:

- User to network security – which protects the network-to-terminal exchanges over the radio interface.
- Network domain security – which protects the interfaces between the EPS and IMS network nodes.

This section provides a brief description of the two, and subsequent sections give some more details about user-to-network security.

(i) User-to-Network Security
The overview of user security mechanisms in Evolved UMTS and IMS network parts is quite complex due to the multiple and imbricate mechanisms being defined between the terminal and network nodes. The overall simplified view is provided in Figure 5.14.

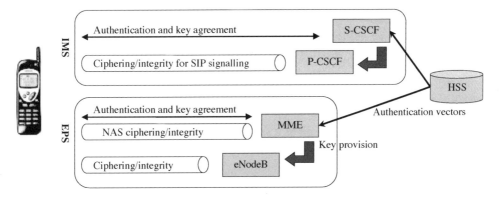

Figure 5.14 Overall and simplified security picture of user security architecture.

At the EPS level, data exchange between the terminal and the eNodeB is protected by ciphering and integrity mechanisms (this includes user data as well as RRC signalling). The security keys used as inputs to those mechanisms are provided by the MME to the eNodeB once the USIM (in the terminal) and the MME have been mutually authenticated. In addition, the NAS signalling is ciphered and integrity-protected between the terminal and the MME using separate keys. The latter did not exist in 3G/UMTS, which only provided one level of security protection over the radio segment between the terminal and the RNC.

Similarly, at the IMS level, the SIP signalling is protected by ciphering and integrity mechanisms, based on keys provided by the S-SCSF once terminal USIM and S-CSCF have been mutually authenticated.

The outcome of such a scheme is that some information sent over the radio interface benefit from multiple and combined security mechanisms. This is what will happen to SIP signalling exchanged between the terminal and the S-CSCF or an application server. When ciphering is applied at the P-CSCF, a ciphered SIP message will also be ciphered by the eNodeB before transmission over the radio interface.

In any case, the security protection and mechanisms performed at each level all derive from information provided by the HSS, or, to be more precise, the AuC (Authentication Center) part of the HSS.

EPS and IMS security mechanisms are then all based on user-specific secret information shared by the network and the subscriber module (USIM or ISIM) and make use of symmetric cryptographic algorithms. In short, 'symmetric cryptography' means that security algorithms (for example, ciphering and de-ciphering) are identical and using the same key, as opposed to 'asymmetric cryptography', which makes use of public and private keys for ciphering and de-ciphering.

It is interesting to note that, in contrast, most of the secured Web or IP-based services make use of certificates, such as recent radio technologies like WiMAX. The main reason for this situation was that EPS and IMS security was actually designed as an extension to 2G/GSM security concepts and architecture. Using security procedures based on certificates would have required deep architecture changes and the introduction PKI (for Public Key Infrastructure) for the creation and delivery of certificates.

The next two sections describe the user-security mechanisms in both EPS and IMS network parts in more detail.

(ii) Network Domain Security
Network Domain Security for IP (or NDS/IP) refers to as the user data and signalling exchange protection over the interfaces between network entities, either in the EPC (Packet Core) or E-UTRAN (Access Network). NDS/IP does not apply to terminal network data and signalling transmission, which is covered by 'user to network security'.

From a NDS/IP perspective, the network resembles Figure 5.15.

The overall network is composed of a single or multiple security domains, each domain being a subset of the network that is managed by a single administrative authority.

The SEG (Security Gateway) are located at the border of a security domain which concentrates all the traffic which enters or leaves the security domain. The NE (Network Entity) can be any kind of network nodes already presented and belonging to the E-UTRAN, EPC and IMS domains, such as an eNodeB, a MME, a S-CSCF, etc.

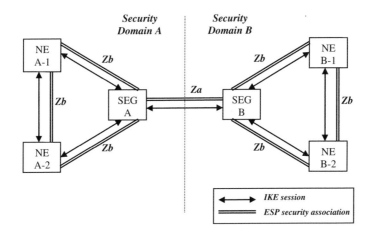

Figure 5.15 NDS architecture for IP-based networks (from 3GPP 33.210).

Zb applies between NE or between NE and SEG in a single domain and is under the sole responsibility and control of the operator.

In contrast, the Za interface connects two SEG of different security domains and is subject to roaming agreements between operators. As an example, E-UTRAN and EPC may be managed by different operators and therefore belong to different security domains so that the S1 interface will be mapped over a Za interface. Za may also be used between the EPC and IMS domains.

The purpose of NDS/IP is to provide security protection for sensitive information exchanged between network nodes. This includes user data, subscription information, authentication vectors and network data such as MM contexts, policy and charging information, as well as IMS-related information exchanged between CSCF nodes.

The NDS/IP framework provides three kinds of protection:

- Data origin authentication – to protect the receiving entity from packet injection from a rogue entity.
- Data integrity – to protect transmitted data to be modified.
- Data confidentiality – to prevent eavesdropping.

As a trade-off between security requirements and processing requirements on the hardware, not all the protections may be needed in all cases. For example, integrity and confidentiality protection is important over the S1 between eNodeB and MME, because of sensitive information exchanged over this interface such as user security keys and user identities.

However, user data integrity protection is not seen as critical enough to justify this mechanism to be implemented in all eNodeBs and MME equipment in the network. This is the reason why the user plane is only encryption-protected against eavesdropping.

For information, the NDS/IP mechanisms applicable to the interfaces of EPS networks are listed in Table 5.3.

Table 5.3 Summary of network domain security protection.

	Integrity/authentication	Encryption
S1 User plane	No	Yes
X2 User plane	No	Yes
S1 Control plane	Yes	Yes
X2 Control plane	Yes	Yes
EPC interfaces	Yes	Optional

From the NDS/IP point of view, network nodes are seen as pure IP nodes, regardless of their actual role within the network. Therefore, and not surprisingly, NDS/IP in 3GPP networks makes use of the classical security set of procedures and mechanisms defined by the IETF:

- The security between network elements is ensured through IPSec tunnels.
- The data authentication, integrity and confidentiality protection is provided by ESP (Encapsulating Security Payload) in 'tunnel mode'.
- The security keys are negotiated using the IKE (Internet Key Exchange) protocol.

(iii) About ESP

As described above, ESP is a complete security mechanism which provides three levels of protection and proposes for each of them a wide set of security algorithms. Figure 5.16 shows the effect of ESP protection in 'tunnel mode' in the example of a data packet with a TCP and IP header. In addition to TCP, ESP can actually accommodate any kind of transport layer, including UDP. In the 'tunnel mode', the whole initial packet is fully protected and encapsulated in a new IP header, as opposed to the ESP 'transport mode' which does not protect the initial IP header.

The ESP header only contains an index (which identifies the ESP Security Association) and a packet sequence number. The ESP trailer's purpose is to contain padding bits, depending on the encryption algorithms (block cipher algorithms require fixed block size). The ESP authentication data contain information to check the packet's integrity. The format and length depend on the chosen algorithm.

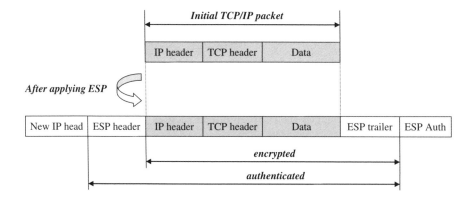

Figure 5.16 The effect of ESP protection in tunnel mode.

In the scope of 3GPP networks, the only mandatory algorithm for confidentiality is the recent AES (Advanced Encryption Standard). However, older 3DES (triple-DES for Data Encryption Standard) may also be used. The authentication and integrity protection can be provided using algorithms like SHA-1 (mandatory for 3GPP networks) or MD5.

> ***Reference documents about NDS/IP (Network Domain Security for IP)***
>
> SIP IETF documents:
>
> - RFC2401, 'Security Architecture for the Internet Protocol'
> - RFC2406, 'IP Encapsulating Security Payload (ESP)'
> - RFC2409, 'The Internet Key Exchange (IKE)'
>
> 3GPP technical specifications:
> - 33.210, '3G Security; Network Domain Security: IP Network Layer Security'

5.2.4 User Security in EPS

(i) Some Principles about Security in EPS

EPS systems propose three kinds of security protections, actually very similar to what exists in 3G/UMTS networks:

- Ciphering – which is the basic feature most of the wireless communication systems provide. Ciphering helps to ensure data confidentiality by providing protected information from being overheard. In EPS, ciphering not only applies to user data, but also to signalling messages. The latter is also important, as signalling messages convey in some cases user identity or network sensitive information that an attacker may use to gain unauthorized access to the network or breach network security protections.
- Integrity – this is the process by which the receiving entity is able to verify that signalling data have not been modified in an unauthorized way since it was sent by the sending entity. In EPS, integrity only applies to RRC and NAS signalling. The application-level signalling (such as, for example, RTCP, SIP and SDP) is not integrity-protected, as it is considered part of user data.
- Mutual authentication – which is used by the network to corroborate the subscriber's identity and, by the terminal, to make sure it is actually connected to an authorized serving network.

As described further in this section, EPS security procedures are based on security credentials and algorithms stored in the USIM module present in the terminal. EPS procedures have been designed in such a way that 3G/UMTS USIM cards will be able to access to an EPS network – ensuring backward compatibility to 3G/UMTS subscribers. This possibility is enabled by LTE-specific key derivation algorithms which will help to build E-UTRAN keys from 3G/UMTS CK (ciphering) and IK (integrity) keys – or the other way around.

However, 2G/GSM SIM access to EPS networks is not allowed because 2G security has been considered to be not robust enough (GSM network security is based on 64-bit encryption keys, whereas 3G/UMTS makes use of 128-bit encryption and integrity keys and 2G algorithms appear to be not so secured with time). In addition, and as explained below, the 2G/GSM AKA process lacks some important features which are seen as critical security threats in the scope of future E-UTRAN network deployments.

(ii) The AKA Procedure (Authentication and Key Agreement)
All the operations required for user security protections – i.e. security key derivation and mutual authentication – are performed during the AKA process (Authentication and Key Agreement). In short, the AKA process makes use of a challenge–response-based mechanism based on symmetric – or secret key – cryptography.

As in 2G/GSM and 3G/UMTS, the security procedures in EPS are all based and derived from a unique and permanent secret key, shared by both the subscriber module (or the USIM card) and the network AuC (Authentication Center) part of the HSS.

The AKA process used in EPS networks is the same as the one used in 3G/UMTS. The UMTS AKA was considered as secure enough for future EPS network deployment. This process is described in Figure 5.17 and differs from the 2G/GSM one by three major points:

- GSM does not provide mutual authentication, meaning that the USIM is not able to verify the identity and validity of the network entity it is connected to. This is prevented by the addition of the AUTN authentication token in the AKA procedure.
- GSM does not offer signalling protection (aside from the GSM ciphering, which protects – when activated – all data or signalling exchanged over the radio). This basically means the

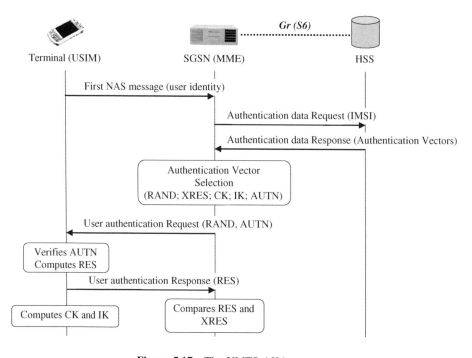

Figure 5.17 The UMTS AKA process.

signalling messages can be altered or simply replayed as such by rogue radio equipments. This is what integrity protection aims to prevent.
- GSM does not support anti-replay mechanisms, which means that an authentication vector (only composed of three elements in GSM) can be used more than once. The 3G authentication vector makes use of a sequence number which prevents the network (or rogue equipment) from re-using an eavesdropped authentication vector.

The UMTS AKA process is triggered by a connection or service request initiated by the terminal, and represented in Figure 5.17 by the 'First NAS message', which may be, for example, an *Attach Request* or a *Service Request* message. In most cases, the initial connection request is performed at the occasion of the terminal registration at power-on. The AKA process may, however, be performed on many occasions, such as a state change from IDLE to ACTIVE. This initial connection message contains the user identity which is used in the rest of the procedure.

Upon the user connection request, the SGSN (respectively, the MME) requests authentication information from the HSS over the Gr (respectively, S6) interface. The HSS answers with a set of one to five authentication vectors (AV), also called 'Quintets', as each of them contains the following five elements:

- **RAND** – the random challenge, which is one of the input parameters used to generate the four other elements of the vector.
- **XRES** – the Expected Result used by the network for USIM authentication.
- **AUTN** – the Authentication Token used by the USIM for network authentication.
- **CK** – the Ciphering Key.
- **IK** – the Integrity Key.

Using one of the vectors in the list, the SGSN (or MME) will actually engage the AKA procedure towards the USIM. This is done by sending to the terminal an authentication request containing the RAN and AUTN parameters.

Using the RAND and its own stored value of the K secret, the USIM is able to authenticate the network by verifying the value of the authentication token AUTN provided by the network. Then, the USIM generates a RES value, which is further checked by the SGSN (or MME) against the XRES expected result, so that the network can authenticate the USIM.

On the terminal side, the CK and IK security keys are computed in the same way as in the HSS, using the same algorithms and secret input values. In order not to compromise security, they are never exchanged over the radio interface.

(iii) The EAP-AKA Procedure
When the terminal attempts to access to the network using non-3GPP access (such as a WLAN access point), the AKA process presented above cannot be used. As presented in Chapter 2, the terminal in this case shall authenticate to the network through an AAA server (Authentication, Authorization and Accounting) which does not implement the protocol supported by the MME.

For that reason, a specific AKA protocol has been defined, known as EAP-AKA (Extensible Authentication Protocol). The EAP is an authentication and session key distribution framework defined by IETF which supports multiple authentication methods. The EAP protocol itself is quite similar to the UMTS AKA from a functional perspective.

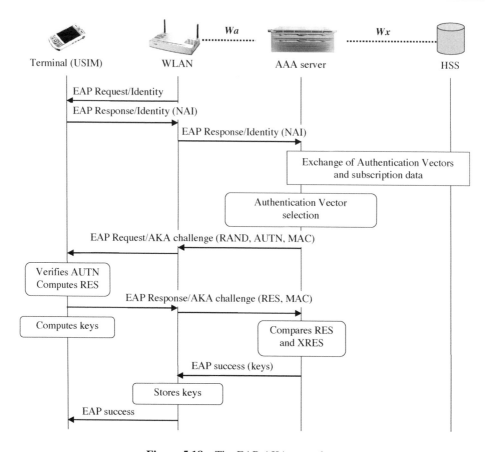

Figure 5.18 The EAP-AKA procedure.

EAP-AKA is a specific declination of the generic EAP, making use of security credentials and key derivation methods specified by the 3GPP. EAP-AKA therefore allows a terminal equipped with a USIM to authenticate with a 3GPP-AAA server using the IETF EAP protocol and the same credential and algorithms used in an E-UTRAN or 3G/UMTS network.

For information, Figure 5.18 describes the different steps of the EAP-AKA procedure.

Following a request from the WLAN access point, the terminal sends its identity in the form of a NAI (Network Address Identifier). This identifier, which follows the IETF generic 'username@realm' format, is built using the user IMSI, and the MCC and MNC parts of it. Figure 5.19 is an example of such a NAI (for information, the leading digit is not part of the IMSI.

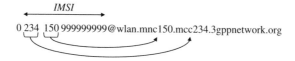

Figure 5.19 An example of NAI coding.

'0' indicates the NAI corresponds to EAP-AKA authentication, and '1' refers to SIM-AKA, which is another authentication process allowing WLAN access to 2G/GSM SIM-based credentials).

The MNC/MCC information is actually used by the WLAN to determine the relevant 3GPP AAA server which corresponds to the user. On reception of the *EAP Identity*, the AAA server retrieves a set of authentication vectors, similarly to the UMTS-AKA process, and selects one of them.

Then, the 3GPP AAA Server sends the *AKA Challenge* message to the terminal, containing the usual RAND, AUTN and a MAC (Message Authentication Code). As in the UMTS AKA, AUTN will be used by the USIM card to authenticate the network. If successful, the terminal generates a RES (allowing the network to authenticate the terminal) as well as the CK and IK keys.

When the terminal is successfully authenticated, the private session keys are transmitted from the AAA Server to the WLAN so as to protect further transmission between the terminal and WLAN Access Point.

There is actually no fundamental difference between the UMTS-AKA and EAP-AKA presented above. EAP-AKA is slightly more consuming in terms of signalling due to the fact that there is no equivalent to *EAP Success* in UMTS AKA, and also because the AAA server may optionally check again user identity before the challenge/response procedure.

Reference documents about EAP-AKA

IETF documents:

- RFC3748, 'Extensible Authentication Protocol (EAP)'
- RFC4187, 'Extensible Authentication Protocol Method for 3rd Generation Authentication and Key Agreement (EAP-AKA)'

3GPP technical specification:
- 33.234, 'Wireless Local Area Network (WLAN) Interworking Security'

(iv) The EPS Key Hierarchy

In order to apply the right security protection for the different information flows, a key hierarchy has been introduced in the standard, described in Figure 5.20.

K is the secret key permanently stored on the USIM and in the Authentication Centre AuC. It serves actually as the basis for all key-derivation algorithms in GSM, UMTS and EPS systems.

CK, IK is the pair of keys (respectively the Ciphering Key and the Integrity Key) derived in the AuC and in the USIM when the Security Association is set up.

K_{ASME} is an intermediate key derived by the terminal and in HSS from CK, IK during the AKA process. In the security specification document terminology, the ASME (Access Security Management Entity) is the network entity in charge of establishing and maintaining Security Associations with terminals based on keys received from the HSS. In EPS networks, ASME role is handled by the MME.

K_{eNB} is an intermediate key derived by terminal and MME from K_{ASME}. K_{eNB}, whose value is dependent on the eNodeB identity, is used in the eNodeB for the derivation of keys for RRC traffic and the derivation of keys for UP traffic.

Eventually, the five keys are generated, for integrity and confidentiality protection of the three different types of flows: NAS signalling (between terminal and MME), AS (RRC)

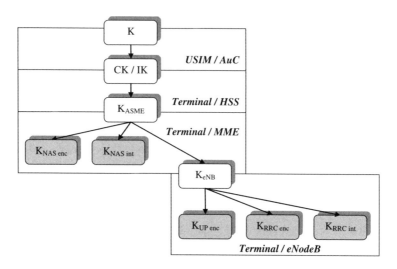

Figure 5.20 The key hierarchy in EPS.

signalling (between terminal and eNodeB) and User Plane data (between terminal and Serving GW):

- **K$_{NASenc}$** is derived by the terminal and the MME from K$_{ASME}$. It is used for the protection of NAS traffic with a particular encryption algorithm.
- **K$_{NASint}$** is derived by the terminal and the MME from K$_{ASME}$. It is used for the protection of NAS traffic with a particular integrity algorithm.
- **K$_{UPenc}$** is derived by the terminal and the eNodeB from K$_{eNB}$. It is used for the protection of User plane traffic with a particular encryption algorithm.
- **K$_{RRCenc}$** is derived by the terminal and the eNodeB from K$_{eNB}$. It is used for the protection of RRC signalling traffic with a particular encryption algorithm.
- **K$_{RRCint}$** is derived by the terminal and the eNodeB from K$_{eNB}$. It is used for the protection of RRC signalling traffic with a particular integrity algorithm.

The function of each key is summarized in Figure 5.21 for the downlink flows, from network to terminal. As the security algorithms in 3GPP are symmetrical, similar processes are applied for uplink flows.

As described in Chapter 4, E-UTRAN integrity protection and ciphering are both applied by the PDCP layer, located within the eNodeB.

In the 3G/UMTS standard, and in the case of handover from GSM to UMTS, the GSM 64-bit Kc ciphering key is converted to 128-bit IK and CK keys, which is felt secure enough in the context of 3G-UMTS networks.

However, in the case of handover from GSM to E-UTRAN, and as stated above, the GSM security keying system is not considered as future-proof. For that reason, when a terminal moves from GSM to E-UTRAN, the security context needs to be re-established and new keys have to be determined by the E-UTRAN target system.

For the other direction (from E-UTRAN to 2G or 3G), the 2G/GSM or 3G/UMTS are derived from EPS keys, allowing secure handover to those systems.

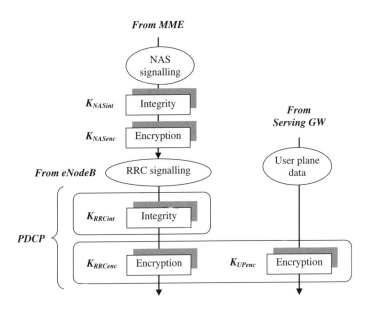

Figure 5.21 Correspondence between security keys and information flows in the network.

In 3G/UMTS, the Security Association (which corresponds to the CK and IK ciphering and integrity keys) is identified by a KSI (Key Set Identifier). KSI is allocated by the network during the authentication process and stored in the USIM. It is used by the network to know which are the keys stored by the terminal and they can be re-used at subsequent connection requests. This allows the start of ciphering on a new connection without authentication.

EPS network security makes use of a similar mechanism. In the EPS case, KSI identifies K_{ASME} from which all the session keys are derived.

(v) About E-UTRAN Ciphering and Integrity Algorithms
In the early versions of 3G/UMTS, only one set of Encryption and Integrity algorithms were defined: UEA1 (for UMTS Encryption Algorithm 1) and UIA1 (for UMTS Integrity Algorithm 1), both derived from the KASUMI block cipher algorithm developed in the scope of 3GPP. This algorithm is defined in 3GPP Technical Specification 35.202, 'Specification of the 3GPP Confidentiality and Integrity Algorithms; Document 2: Kasumi Specification'.

From Release 7 of UMTS, a new set of algorithms has been introduced: UEA2 and UIA2, both derived from the SNOW 3G stream cipher algorithm and defined in 3GPP Technical Specification 35.216, 'Specification of the 3GPP Confidentiality and Integrity Algorithms UEA2 & UIA2; Document 2: SNOW 3G specification'. This specification does not provide much detail about the algorithm itself, as SNOW 3G was actually specified by the SAGE (Security Algorithms Group of Experts) ETSI committee.

The intention behind UEA2 and UIA2 introduction was to provide an alternative to the KASUMI set, in case KASUMI should be broken.

Regarding E-UTRAN, it was decided to base confidentiality and integrity protection on the UAE2/UIA2 set inherited from 3G/UMTS. As an alternative, it was also decided to specify the use of AES (Advanced Encryption Standard) for encryption and an AES-based algorithm for integrity protection.

AES is one of the most popular and worldwide used symmetric cryptography algorithms providing block cipher encryption. It became an official standard in 2002.

5.2.5 User Security in IMS

The IMS domain applies two kinds of user security procedures:

- The IMS AKA (Authentication and Key Agreement) – which provides user and S-CSCF mutual authentication.
- The IMS SA (Security Association) – which provides security protection for SIP signalling between the terminal and the P-CSCF.

(i) The IMS AKA Procedure

The IMS AKA aims at providing mutual authentication between the ISIM (the IMS application present on the terminal UICC card) and the IMS domain. In addition, IMS AKA also provides the means to agree on session keys on the network and terminal side, so as to build the IMS Security Association between the terminal and the P-CSCF and protect SIP signalling with confidentiality and integrity. The IMS AKA is a mandatory process and shall be performed before the subscriber can get any sort of IMS services.

As described in Figure 5.22, the IMS AKA is very similar to the UMTS AKA and is based on the same algorithms to derive the keys, verify the AUTN and generate the XRES for mutual authentication. The main difference is that IMS AKA makes use of the SIP signalling (instead of 3GPP-specific messages) and different input parameters like an IMS-specific long-term key (the equivalent of K) combined with the user's IMS private identity.

For simplicity, the figure does not represent P-CSCF and I-CSCF nodes, whose role in the IMS AKA process is limited to SIP signalling routing and identification of the S-CSCF.

As in the UMTS AKA, the authentication vectors are provided from the HSS to the S-CSCF, following a request from the terminal to register to IMS. Once the S-CSCF has chosen one of the authentication vectors, it sends a SIP *401 Unauthorized* response which contains the RAND random challenge, as well as the AUTN, which the ISIM will use to authenticate the network.

Once the terminal and network are mutually authenticated, the S-CSCF sends a SIP *200 OK* to the terminal through the P-CSCF. Those keys are removed from the answer and stored by the P-CSCF, so as to further build the Security Association with the terminal.

(ii) The Security Association (SA) between the Terminal and the P-CSCF

The aim of the Security Association built between the terminal and the P-CSCF is to protect the SIP signalling exchanged between these two entities. The Security Association provides two levels of protection:

- SIP signalling Integrity protection – which is mandatory.
- SIP signalling Ciphering – which is optional.

As the terminal and the P-CSCF are two IP nodes, the Security Association is actually an IPSec tunnel and makes use of ESP (Encapsulating Security Payload) in 'transport mode'.

Life in EPS Networks

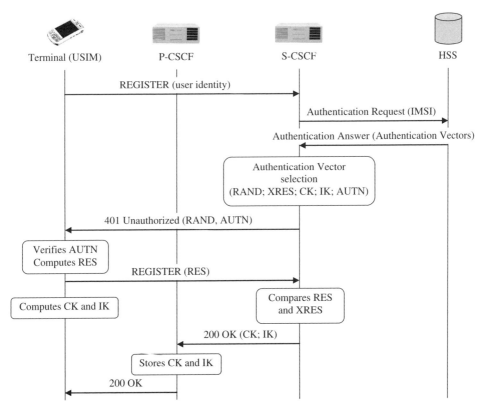

Figure 5.22 The IMS AKA process.

Reference documents about IMS and EPS Security

3GPP technical specifications:

- 33.102, '3G Security: Security Architecture'
- 33.203, '3G Security: Access Security for IP-Based Services'

5.2.6 Session Setup

This section describes the different phases involved in a communication-session setup. From this point, it is assumed that the subscriber terminal is switched on and successfully registered to a PDN network based on the procedures described above in this chapter.

Because of the Always-On intrinsic nature of EPS, session setup in an Evolved UMTS network is fairly different from the traditional circuit-switched call setup. For that reason, the two following types of procedures are described in this section:

- Service Request – this procedure is used by the terminal to request the network for a Radio Bearer. This corresponds to an IDLE to ACTIVE terminal state transition because the user is resuming a data session or activating a new service.

- Dedicated Bearer activation – which corresponds to a service activation while the terminal is in ACTIVE mode.

(i) Paging

Paging is the mechanism by which the network can inform a registered subscriber about an incoming call (or mobile-terminated call). Before being able to page a terminal (or send paging information on the PCCH logical channel), the network needs to know the actual terminal location. In E-UTRAN and EPC networks, terminal location management is enabled by the 'Tracking Area Update' mechanism, further detailed hereafter in section 5.3, dedicated to 'Mobility in IDLE Mode'.

In any case, when the terminal receives a paging message, it sets up a connection in the same way as in a mobile-originated call setup.

In order to avoid the terminals listening to the paging channel all the time, a DRX mode (for discontinuous reception) is proposed, configured by the network and which is similar to the one which exists in 2G/GSM and 3G/UMTS networks. When this mode is used, the terminal only decodes the paging channel at some specific periods of time. On the network side, paging indications are sent to terminals according to those periods, also called 'paging groups'.

(ii) Service Request

Figure 5.23 describes the different steps of a user-initiated service request. This sequence of messages occurs when, for example, the terminal user in IDLE mode is resuming a Web session, initiating a voice call or activating any kind of service, including the whole set of IMS services. This whole process actually results in a transition from the IDLE state to the ACTIVE state.

The network-triggered Service Request looks exactly the same. The only difference is that the Service Request is actually triggered by the reception by the terminal of a paging containing its identity.

As part of the 'Always-On' EPS philosophy, when the terminal changes to IDLE mode, all the EPS bearers are 'preserved' in the Core Network, with their Quality of Service characteristics. This means that when the terminal becomes ACTIVE again, there is no need to re-establish the bearers in the Serving or PDN network. Only the E-UTRAN part of the bearer (also referred to as the 'Radio Bearer') needs to be set up again.

To initiate the service, the terminal sends a *Service Request* NAS message to the MME using the Random Access procedure described above. On the S1 interface, for the transmission between the eNodeB and the MME, this message is encapsulated in the *Initial UE Message*. The *Service Request* message contains the user's temporary identity (or S-TMSI) as well as an indication about the requested service type (like 'data' or 'answer to a paging message').

When needed, the NAS authentication procedure takes place, based on the AKA procedure presented above.

Once the AKA procedure is performed, the terminal and the MME are mutually authenticated and share a common (and secret) session key system. The MME can therefore possibly set up or change NAS signalling integrity and ciphering protection, using the *NAS security Mode Command* message. The KSI_{ASME} parameter identifies the set of NAS keys to be used for those algorithms.

Once the NAS security is effective, the MME performs the *Initial Context Setup* procedure, which aims at creating the EPS bearer(s) and terminal context within the eNodeB. In addition,

Life in EPS Networks

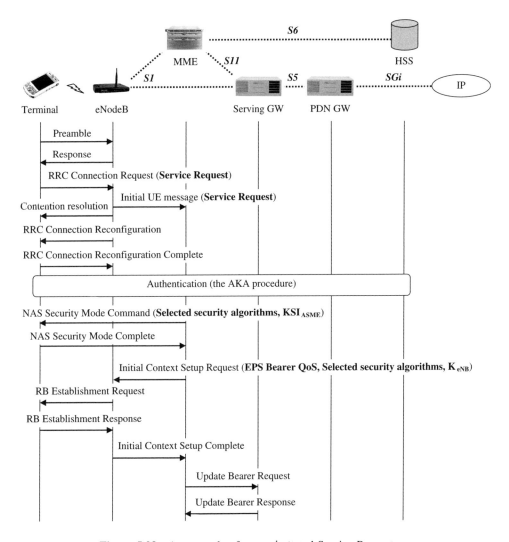

Figure 5.23 An example of a user-initiated Service Request.

this procedure also aims at establishing the necessary resources on the Radio and S1 interface bearers. On this occasion, the *Initial Context Setup* message contains NAS elements such as the Quality of Service attributes associated to the EPS bearer(s) as well as integrity and ciphering algorithms chosen by the MME and K_{eNB} – the set of keys which will be used by the eNodeB at the PDCP level for integrity and encryption. Access Network-level security is started as part of the Radio Bearer (RB) establishment procedure.

At the end of the this procedure, all the bearers (including the default EPS bearer and possibly other Dedicated Bearers) which were active between the network and the terminal before the last transition to IDLE mode are available and usable again.

Eventually, the MME updates the user-related bearer for the Serving GW, which is then able to transmit downlink data to the terminal. There is no signalling exchange between the Serving and PDN GW due to the fact that this segment of the EPS bearers was preserved in the EPC while the terminal was IDLE.

When the whole procedure is over, in order to cope with the user's application requirements, a new additional EPS bearer – or Dedicated Bearer – may be created with specific Quality of Service characteristics.

(ii) Dedicated Bearer Activation

The Dedicated Bearer activation procedure is used when the terminal or the network is activating a new application service while in ACTIVE mode. Therefore, to activate a new bearer in this condition, there is no need to set up a RRC connection or perform the AKA process.

In Figure 5.24, it is assumed that application-level signalling is exchanged between the terminal and its peer (which could be an application server or another terminal), possibly based on SIP and SDP, as presented in Chapter 4.

At some point, the PDN GW – which plays the role of PCEF (Policy and Charging Enforcement Function) – determines the Quality of Service (QoS) characteristics for the new EPS bearer and initiates the creation of the bearer. This action can be triggered by the PCRF (Policy and Charging Rules Function) node, not represented in the figure.

Further on, associated resources are created on the S1 and radio interface using, respectively, the S1 *Bearer Setup Request* procedure and the *RB Establishment Request* RRC

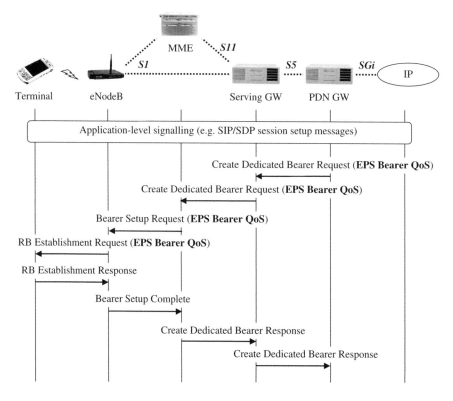

Figure 5.24 An example of Dedicated Bearer activation.

message. The QoS parameters – part of the SM layer (Session Management) – are transferred from the MME to the terminal via those messages so as to limit the amount of signalling.

5.2.7 Data Transmission

(i) DRX in Connected Mode
Discontinuous reception (or DRX) is a key point for terminal battery saving. While being active (or in the LTE-ACTIVE state), there may be some period of time during which the terminal has no need to maintain radio reception and associated processing capability. This is typically the case for some nonreal-time packet applications, like Web browsing, Instant Messaging or Push-To-Talk. Existing WLAN technologies like WiMAX also support similar mechanisms, known as 'Power Save Mode'.

The 'DRX in connected mode' is actually a kind of trade-off between the ACTIVE and IDLE states. This mode is only applicable to terminals in the ACTIVE state and does not introduce any additional complexity in the 'terminal state' model already presented in this chapter. When it was specified in the standard, the objective of the DRX mode was to allow the terminal to maintain its connection with the network while consuming the same amount of power as in IDLE mode.

Power saving is not the only interest of the DRX mode. It is also good to know that in such a mode, the terminal can resume its activity much quicker than in IDLE mode, as a consequence of the fact that less signalling is needed accordingly. Another side effect is that the paging load on the network is reduced, since the terminal position is known at the cell level in ACTIVE mode.

However, the price to pay for the DRX mode is the need to maintain in the eNodeB a full communication context (and associated memory resources) for each terminal being in such a mode. If we assume that all terminals in a network are possibly using this mode, this can lead to significant resource usage. Besides, for fast-moving mobiles, the DRX mode may not be appropriate. Fast terminals involve frequent cell change and cell update procedures, which spoil the benefit of DRX in terms of power saving and increases the amount of radio signalling.

Figure 5.25 shows an example of the DRX mode in E-UTRAN, using periodic 'DRX cycles' composed of 'On' periods of time, during which the terminal will decode the downlink physical channel, and 'Off' periods, during which the terminal receiver is turned off.

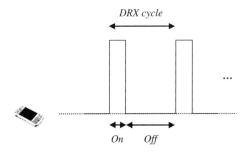

Figure 5.25 The DRX mode.

The DRX mode is terminal-specific (meaning that each terminal has its own DRX mode-handling and configuration) controlled by the MAC layer, based on parameters provided by the eNodeB.

The length of the DRX cycle is flexible and can be as long as the DRX mode for IDLE modes (related to the reception of the Paging channel).

5.3 Mobility in IDLE Mode

5.3.1 Cell Reselection Principles

As in 2G/GSM or 3G/UTRAN, the purpose of cell reselection is to ensure that the terminal in IDLE mode (meaning not active and not engaged in an on-going service) is camped on the best cell in terms of signal strength and quality. In wireless networks, cell reselection is a necessary process, mainly because of terminal mobility, but also because of the fluctuation in the radio environment, which implies variations in the signal strength and interference level, even for a still or slow-moving terminal.

As a basic principle (still being used from 2G/GSM networks), the cell reselection criteria are evaluated by each terminal, using radio measurements performed by the terminal itself (like received beacon channel levels) and parameters (such as threshold values) provided by the network and part of the System Information. For a multi-mode terminal in an environment which provides multiple access types, the terminal needs to evaluate the reselection criteria using measurements from the different available frequencies and access technologies.

The fact that the network controls the value of parameters used in the reselection criteria allows the network to drive terminals in IDLE mode towards the most relevant network layer when appropriate. Some examples are provided below:

- **Network sharing** – When the coverage of a country is shared between different operators (which typically occurs in low-density areas), cell reselection proposes mechanisms to direct the terminal to the appropriate network operator at a network-sharing border
- **Network load condition** – Depending on the load condition of certain layers or access technologies of the network, cell reselection helps to give a better precedence to appropriate cells.
- **Private networks and home zones** – Similarly, cell reselection may be used to give precedence in the criteria evaluation to private/home cells.

5.3.2 Terminal Location Management

Location management is a critical function for cellular networks. When being active, the mobile location is known at the cell level, as the network needs to quickly react to terminal cell change, allocate new resources in the new cell and release old, unused resources in the previously serving cell. In this case, the terminal mobility is driven by handover procedures, described further in this chapter.

For all mobiles not being active (or being in 'IDLE mode'), location management is still an important item, as the network needs to know the current terminal location at any time in case of mobile-terminated session setup or push services. However, IDLE mode procedures do not require the network to know each terminal location with a high degree of accuracy (such as the cell level). For that reason, the concept of Tracking Area (TA) has been introduced.

Basically, a TA is defined as a set of contiguous cells. The identity of the TA the cell belongs to, or TAI (Tracking Area Identity), is part of the system information broadcast on the BCCH. As in the 3GPP definition, Tracking Areas do not overlap each other. When the network needs to join the terminal, a paging message is sent in all the cells which belong to the Tracking Area.

The dimensioning of TA is a typical network-engineering issue, which results from a trade-off between network signalling load and radio paging load:

- If TA are too small, terminal moves will result in a large number of TA update procedures and high signalling load. This issue can be worked out by increasing the size of the TA.
- On the other hand, if the TA are too large, all the cells within the TA will have to cope with a high traffic load on the Paging channel. Since the Packet Core does not know the idle terminal location with more accuracy than the TA, one single mobile-terminated call will generate a paging message in each cell of the TA in which the terminal is located.

In practice, Tracking Areas are dimensioned according to the estimation of IDLE mode terminal density. In hot-spot, or low-speed, dense urban areas, TA are usually small so as to limit the paging load. In contrast, in rural or low-speed, dense areas, TA size can be increased without compromising the network signalling load.

There are actually three cases in which the current terminal TA is signalled to the Core Network:

- At initial registration, the terminal communicates to the Core Network its current Tracking Area.
- When the terminal changes zones, as a result of subscriber move within the network, the new TA is updated in order to keep the Packet Core network updated.
- In addition, the current TA is periodically updated (or refreshed), even if it does not change, so that the Packet Core network does not keep alive a context for a terminal which is no longer reachable in the network. This can happen if the terminal fails to de-register or suddenly runs out of coverage.

(i) Multi-TA Registration
As an enhancement to UMTS, and in order to allow further optimization of the signalling load exchanged between the network and terminal in IDLE mode, the standards leave the possibility for the terminal to be registered into multiple Tracking Areas. In this situation, the terminal does not perform any TA update as long as it remains under the coverage of the Tracking Areas it was registered to (like TA1, TA2 and TA3 in Figure 5.26), with the exception of periodic TA update. The multi-TA registration mechanism helps to reduce the number of TA updates that the networks has to process for terminals located at the edge of tracking areas.

The list of TA that the terminal is registered to is communicated by the network during the TA update process. The terminal considers it is registered to the whole TA list until it enters a TA which does not belong to the list, or gets an updated list from the network, e.g. on the occasion of a periodic TA update.

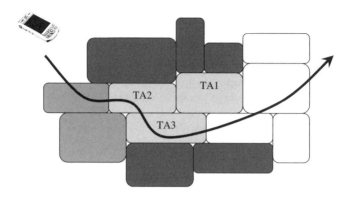

Figure 5.26 An example of multi-TA registration.

(ii) Comparison with 2G/GPRS and UMTS

The concept of location area, such as the Tracking Area, is not new to Evolved UMTS, as it was introduced at the beginning of 2G/GSM systems. Later on, when 2G/GPRS and UMTS were introduced in the standard, this principle became more complex. In UMTS, as presented in Figure 5.27, no less than four types of areas are being used:

- **LA** (Location Area), which is the type of area supported by the Circuit Core network domain MSC/VLR.
- **RA** (Routing Area), which is the equivalent of the LA for the Packet Core network domain. RA is defined in such a way that a LA may include one or more RA.
- **URA** (UTRAN Registration Area), which is a registration area for the use of the UMTS Access Network. URA was introduced to provide flexibility in UTRAN terminal location

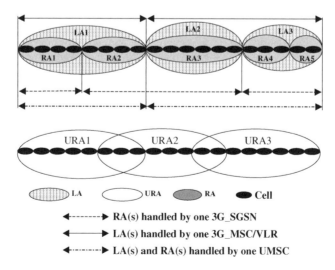

Figure 5.27 Relationship between UMTS localization areas (from 3GPP 23.221).

management, in connection with the protocol states which were introduced in the UTRAN RRC layer (see section on terminal states). As it is managed by the Access Network, URA has no relation with the Core Network LA and RA.
- **Cell**, which provides the best accuracy localization information.

LA and RA are quite similar to the concept of TA, as being a nonoverlapping group of cells. However, the URA concept has no equivalent in E-UTRAN. The possibility of defining overlapping URA was introduced as a way to decrease the signalling load impact of 'URA update' procedure, similarly to the 'Multi-TA Registration' concept presented above.

From the perspective of terminal location management, Evolved UMTS has been simplified, as there is only one type of Core Network Domain (the Evolved Packet Core) and no registration area has been defined for the Access Network – like the 3G/UMTS URA. As described further in this chapter, this will also have an impact on RRC state management simplification.

5.3.3 Tracking Area Update

Figure 5.28 describes an example of a Tracking Area update (TAU). In this case, the terminal changes both MME and Serving GW nodes. A similar procedure also applies when the terminal is moving between EPS and the 2G or 3G network. In such a case, the terminal moves between two different types of zones: a TA on the EPS side, and a RA (Routing Area) on the 2G or 3G side. However, the principles for such a procedure are not that different from the intra-EPS Tracking Area Update described hereafter.

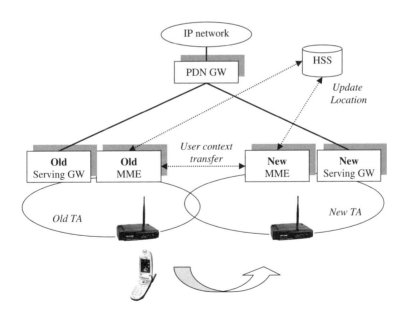

Figure 5.28 An example of a Tracking Area Update.

In short, the following operations need to be performed during a TAU:

- Bearer path update – when a bearer is available, as a consequence of the Always-On IP connectivity, the bearer path in the EPC (corresponding to the initial default bearer, or any bearer the subscriber would have asked for) needs to be updated. This implies that the PDN Gateway is updated with the reference of the Serving GW in charge of the new TA, and a new bearer needs to be created between the terminal and new Serving GW.
- User context transfer from old to new MME – this allows the new MME to get the subscriber context information from the old MME. This context includes information such as user IMSI and subscription data.
- HSS database update – at the end, the HSS is updated with the terminal's new serving MME identity and IP address.

Figure 5.29 shows the messages exchanged between the network entities so as to achieve the operations listed above. Not all Tracking Area Updates generate as much signalling. In the case that the new and old TA are served by the same MME and Serving GW, the overall procedure is much simpler and only limited to the TA update between the terminal and the MME, without involvement of the PDN GW and the HSS.

At some time, following a cell reselection, the terminal detects that it has entered into a TA which does not belong to the list of TA it is registered to. This triggers the initiation of a TA Update Request message sent to the new MME which serves the current eNodeB. This message contains two key parameters: the S-TMSI (which identifies the subscriber) and the old TAI. This will help the new MME to identify the old serving MME.

The new MME can then retrieve the user information from the old MME, using the *MME Context Request* procedure, which contains the S-TMSI and the old TAI. This context contains the user's IMSI, user's subscription information as well as a set of authentication vectors. The new MME can therefore run the AKA procedure in order to authenticate the terminal and protect all subsequent signalling exchanges over the radio interface.

Once the TA update is accepted by the new MME, the bearer path needs to be updated. This operation is under the control of the new MME (using the *Create Bearer Request* message) and relayed by the new Serving GW to the PDN GW.

Following this phase, the HSS is updated with the new terminal location (*Update Location* procedure) and the old MME is informed by the HSS that the subscriber has been successfully located within another MME. This later triggers the old bearer release between the old Serving GW and the PDN GW.

Eventually, the new MME informs the terminal about the successful outcome of the whole procedure. The *TA Update Accept* message may contain a new S-TMSI allocated by the new MME, which is acknowledged by the terminal using the *TA Update Complete* message.

5.4 Mobility in ACTIVE Mode

This section aims at presenting how EPS networks support mobility cases for active terminals engaged in communication sessions. As opposed to IDLE mode, active terminal mobility (also called handover) is completely under the control of the network. The decision to move as well as the choice for the target cell and technology (when applicable) is made by the current serving eNodeB, based on measurements performed by the eNodeB itself and the terminal. In addition,

Life in EPS Networks

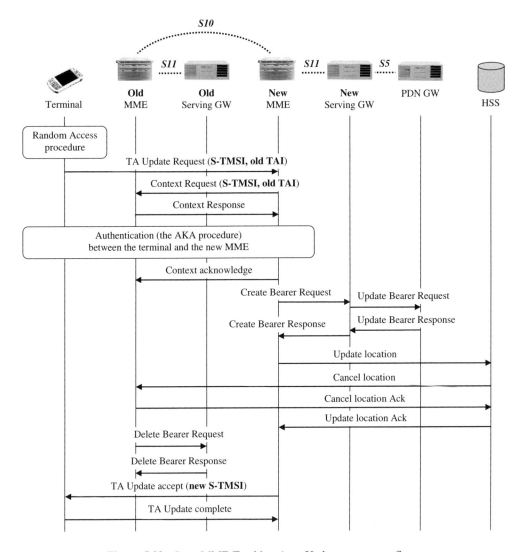

Figure 5.29 Inter-MME Tracking Area Update – message flow.

ACTIVE mode mobility requires some specific features to be supported and implemented by the network so as to limit interaction on user experience and preserve the on-going service.

In this domain, the E-UTRAN handover cases follow (as much as possible) two main principles inherited from 2G/GSM and 3G/UMTS systems:

- **Make before break.** In all the cases, the resources and context in the target nodes (whatever the target technology is) are reserved before the actual handover (or change of radio equipment which serves the terminal) is performed. This ensures that the interruption time is kept to a minimum, since the time for resource reservation in the target nodes is not predictable – if it does not fail ...

- **Packet data forwarding.** Due to the nature of the E-UTRAN radio interface, the amount of packets stored in radio equipment before scheduled transmission over the radio may not be negligible. For that reason, some mobility cases make use – when applicable – of packet-forwarding mechanisms between source and target nodes so as to limit packet loss during the overall handover.

There may be many drivers for terminal mobility in ACTIVE mode. The most common one is related to degrading radio conditions due to, for example, increasing interference or terminal mobility. In such conditions, not doing the handover to a suitable cell in a timely manner will most probably lead to a call drop. However, there may be some other reasons to perform a handover, corresponding to less critical conditions, which may depend on operator policy and network engineering constraints. For example, this may include:

- Traffic load balancing between network layers using different frequencies or radio access technologies.
- Handover for service reasons, depending on the service being used or requested by the end-user.
- Network sharing. When local agreements have been set up between operators, a roaming subscriber may be handed over a cell from its home network when available.

The next sections provide some more details on a subset of mobility cases considered as being the most representative, from the simplest to the most complex:

- **Intra-E-UTRAN mobility with X2 support** – the basic one.
- **Intra-E-UTRAN mobility without X2 support** – a refinement of the previous case, when direct communication between the source and target eNodeB is not available.
- **Intra E-UTRAN mobility with EPC node relocation** – a more complex case involving more support of Packet Core nodes.
- **Mobility between 2G/3G packet and E-UTRAN** – an example of inter-technology packet handover.
- **Voice Call Continuity between 2G/Circuit and E-UTRAN** – this case combines a change of service domain and technology so as to ensure legacy 2G/GSM circuit voice continuity.

For illustration purposes, the next sections present some message exchange flows. In general, they all are composed of two parts:

- The preparation phase, which corresponds to the handover decision (made by the source nodes) and the resource reservation (in terms of radio, terrestrial interfaces and possibly memory context and processing capability) in the target nodes.
- The execution phase, which is the handover execution itself, including the synchronization to the target radio nodes and resource release in the old serving nodes.

5.4.1 Intra-E-UTRAN Mobility with X2 Support

Figure 5.30 shows the general architecture of an intra E-UTRAN mobility case, which is the simplest case of radio mobility in ACTIVE mode. In this example, the whole procedure

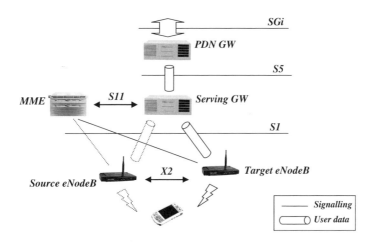

Figure 5.30 Overview of intra E-UTRAN mobility with X2 support.

benefits from the availability of the X2 interface between the source and target eNodeB, so that the involvement of the MME and Serving GW in the handover process itself is at a minimum. In addition, as described below, the X2 interface allows packet loss limitation thanks to buffered packet forwarding from source to target eNodeB.

The only impact on EPC nodes relates to the update of the signalling and user plane connectivity. As the terminal is moving from one node to the other, the new eNodeB needs to build an S1 connection with the MME which is in charge of the user session, and also needs to build a new tunnel for user data transmission with the Serving GW. Once the handover is completed, the old resources and connections on the radio and S1 interface (represented using dotted lines) are released. In any case, the handover is completely transparent to the PDN GW, which keeps tunnelling user data to and from the same Serving GW.

Figure 5.31 describes in more detail the different steps and signalling messages which are part of the handover procedure.

When in ACTIVE mode, the handover decision is made by the source eNodeB, based on measurements reported by the terminal and also possibly made by the eNodeB itself.

Once the decision is made, the source eNodeB sends a *Handover Request* message over the X2 interface to the target eNodeB, which allocates all needed resources to accept the incoming terminal and associated bearers. When done with the allocation phase, the target eNodeB answers with a *Handover Request Ack* message which encapsulates the *Handover Command* content eventually sent to the terminal by the source eNodeB. On reception of the *Handover Request Ack*, the source eNodeB forwards all buffered downlink RLC SDUs that have not been acknowledged by the terminal to the target eNodeB. Those packets will be stored by the target eNodeB until the terminal is able to receive them.

Once the terminal is synchronized with the target eNodeB, it sends a *Handover Confirm*, which triggers the transmission of the *Path Switch* procedure to the MME. The role of the *Path Switch Request* message is to inform the MME about the successful completion of an intra E-UTRAN handover performed via the X2 interface and request a path switch of the user plane data towards the new eNodeB. On reception of this message, the MME is now aware that the

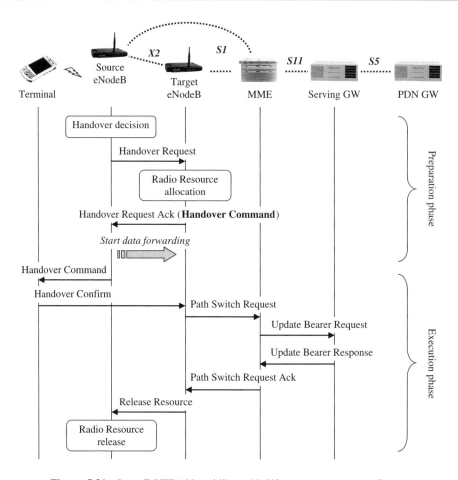

Figure 5.31 Intra-E-UTRAN mobility with X2 support – message flow.

terminal has successfully changed eNodeB and can therefore update the Serving GW about the new data path (the *Update Bearer* exchange). Once the *Handover Confirm* is received, the target eNodeB can transmit over the radio the buffered packets for the downlink.

Eventually, the Release Resource is sent by the target eNodeB over X2, which has the effect of releasing old resources allocated in the Source eNodeB.

5.4.2 Intra-E-UTRAN Mobility without X2 Support

In some cases, it may happen that the X2 interface is not available between eNodeBs. This may result from network equipment failure, or simply from the fact that the operator is not willing to deploy X2 connectivity between eNodeB for cost reasons.

In such a case, the network architecture picture is the same as the previous case. However, the overall handover process is much more complex, as there is no direct communication between source and target eNodeB. As a consequence, the MME is no longer transparent to the handover process, as it acts as a signalling relay between the two eNodeBs.

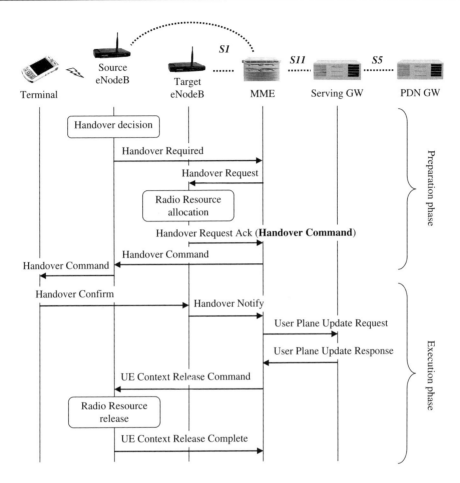

Figure 5.32 Intra-E-UTRAN mobility without X2 support – message flow.

The procedures and messages being exchanged between entities in this case of handover are described in Figure 5.32. Although looking different, the main principles are actually very similar to the previous case.

Instead of being sent directly to the target eNodeB, the request for handover is transmitted from the source eNodeB via the MME, using the *Handover Required* and *Handover Request* S1 messages. Similarly, once the resources have been allocated in the target eNodeB, the answer is sent back to the source eNodeB using the *Handover Request Ack* and *Handover Command* S1 messages. As in the 'X2 support' case, this answer message contains information related to the target cell radio resource.

Finally, once the handover is completed, the resources in the source eNodeB are released under the MME control, once the MME receives the *Handover Notify* message informing that the handover procedure is successful from the E-UTRAN perspective. In parallel, the Serving GW is updated about the new data path towards the new eNodeB.

The main difference from the 'X2 support' previous case is that no data-forwarding is performed between the source and target eNodeBs. As a consequence, all the RLC PDU being buffered at the source eNodeB level will be lost. The impact on user perception will depend on the application and the corresponding protocol stack being used.

For all nonreal-time applications (like Web browsing) which rely on secured end-to-end transport layers like TCP, such a handover may induce a delay in end-to-end information transmission, but no actual loss of data due to data-recovery mechanisms implemented at the OSI layer 4 transport level.

However, for real-time applications based on unsecured transport layers like UDP (for example, streaming or voice), the handover will result in a loss of data frames, with a possible impact on user quality of experience.

5.4.3 Intra-E-UTRAN Mobility with EPC Node Relocation

In this mobility case, the target eNodeB has no connectivity with the current MME and Serving GW. For that reason, the terminal mobility will also imply a relocation of Evolved Packet Core nodes (Figures 5.33 and 5.34). From the terminal and eNodeB perspective, this handover is not different from the previous 'no X2 support' case. The only real difference relies on the fact that the session needs also to be handed over from one MME to the other. In practice, this is performed by transferring the user communication context from the source MME to the target MME using the S10 interface. In addition, the PDN GW needs also to be updated, so as to maintain User plane connectivity.

If there is X2 connectivity between the source and the target eNodeB, packet forwarding can be applied so as to limit packet loss during the handover.

Depending on the network engineering choice, there might be other simpler cases of mobility with EPC node relocation. As Serving GW and MME are separate nodes, it may happen that a user mobility case implies a change of MME with no change of Serving GW.

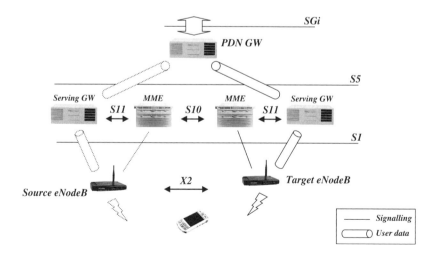

Figure 5.33 Overview of intra-E-UTRAN mobility with EPC relocation.

Life in EPS Networks

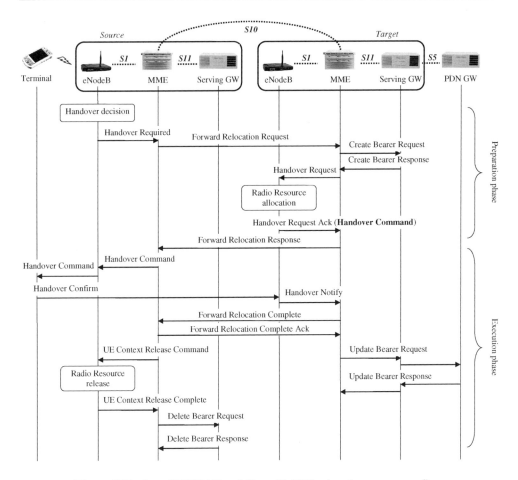

Figure 5.34 Intra-E-UTRAN mobility with EPC relocation – message flow.

Similarly, the source and target eNodeBs may be connected to the same MME, but different Serving GW.

Although looking more complex than the previous example (Intra E-UTRAN mobility without X2 support), this mobility case uses the same principles. The main difference is in the fact that the source and target MME are different nodes, which requires the transfer of the user context (containing the user IMSI, user subscription information, authentication vectors as well as on-going allocated EPS bearers) between the two MME using the *Forward Relocation Response/Request* exchange. In addition, a new User plane bearer is created between the PDN GW (which is the User plane anchor point for the session) and the new Serving GW.

During the execution phase, once the handover is complete from the Access Network perspective, the new MME informs the old one about the successful outcome using the *Forward Relocation Complete*, so that the old radio resources and bearer path can be released. In addition, the bearer path is updated using the Update Bearer procedure, so that the PDN GW can transmit the downlink packet to the relevant new Serving GW.

At the end (not represented in the diagram), if the terminal determines that the new cell belongs to a Tracking Area it is not registered to, a TA Update procedure is performed towards the new MME. As a consequence, the HSS is updated accordingly.

5.4.4 Mobility between 2G/3G Packet and E-UTRAN

As a basic feature, EPS networks are able to support seamless mobility to and from 2G and 3G packet systems. Figure 5.35 describes an example of such a mobility case, for a terminal moving from a E-UTRAN access towards a 3G/UTRAN target cell.

For simplicity, the target 3G RNC and BTS nodes are represented as one box, connected to the target SGSN towards the standard UMTS Iu interface. In the case of mobility towards a 2G/GPRS system, the picture would actually be quite similar, as the SGSN node exists in both 2G and 3G packet core architecture.

As represented in the figure, the Serving GW acts as a sort of User plane anchor point. The control plane for NAS signalling (for session setup and control) is moved over the S3 interface from the serving MME to the target SGSN, which is the standard point for terminating this protocol in 2G and 3G packet architecture. Regarding the User plane, a new tunnel is built between the Serving GW and the target SGSN over the S4 interface so as to ensure packet transmission continuity.

Since the Serving GW still remains in the data path, the PDN GW is not involved in the mobility procedure. There is, however, one little exception to that. Due to the change in radio access technology, the Serving GW may inform the PDN GW about the handover, mainly for

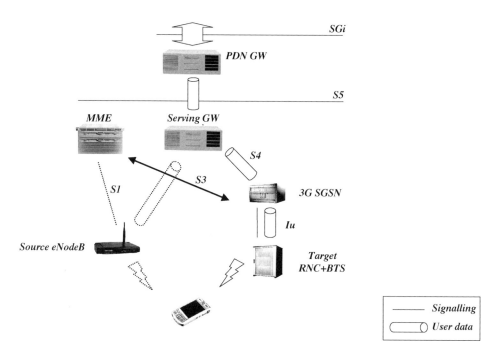

Figure 5.35 Overview of E-UTRAN to 2G/3G mobility.

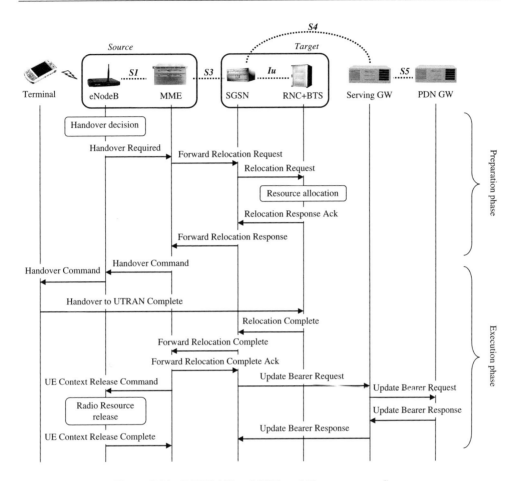

Figure 5.36 E-UTRAN to 2G/3G mobility – message flow.

charging purpose. This gives flexibility to the charging system to apply different rate and billing procedures, depending on the access system technology and associated specific Quality of Service representation.

Once the handover is completed, the old resources and connections on the radio as well as S1 user and signalling interface (represented using dotted lines) are released.

Figure 5.36 describes the messages and procedures involved in such an E-UTRAN to 2G or 3G mobility. Handover in the other direction is not further described, as it actually makes use of very similar principles and procedures.

When the handover decision is made, the session context (including session-related EPS bearers and associated Quality of Service attributes) is moved from the source MME to the target SGSN using the *Forward Relocation* procedure, as in the 'Intra E-UTRAN mobility with EPC nodes relocation' mobility case. This procedure is actually an extension of the existing Forward Relocation procedure which applies in the case of inter-SGSN mobility within 2G or 3G networks.

On this occasion, the MME translates the EPS Quality of Service attributes (presented above in this chapter) into their 2G or 3G equivalent, in the form of PDP context attributes.

E-UTRAN to 2G/3G mobility may support data forwarding, from the eNodeB to the target SGSN, so as to avoid that all packets still stored at the eNodeB may eventually be sent to the terminal. Data forwarding is always requested by the eNodeB (and reflected by the content of the *Handover Required* message). The 3GPP standard proposes two types of data forwarding:

- Direct forwarding – in which buffered data are sent directly from the eNodeB to the target SGSN. All necessary information (such as IP address and Tunnel identifier) is part of the *Forward Relocation Response* message.
- Indirect forwarding – in which buffered data are transmitted to the target SGSN via a Serving GW.

The handover execution phase is then triggered by a *Handover Command* message. When applicable, this message contains all information for the eNodeB to be able to forward buffered data, either in direct or indirect mode.

Once the terminal is synchronized on the target BTS and the handover considered as completed from the Access Network point of view, a *Forward Relocation Complete* is sent from the SGSN to the MME. This signal is used as an indication that resources in the old serving E-UTRAN and MME nodes are no longer useful and can be released. Simultaneously, the target SGSN updates the bearer path towards the Serving GW using the *Update Bearer* procedure.

6

The Services

The aim of this chapter is to describe some of the typical services which can be run on top of Evolved UMTS networks. Most of these services are not specific to EPS networks. Therefore, this chapter focuses on how the services would be supported over EPS, in trying to emphasis what are the added value and specific aspects of service deployment over such packet-based new networks.

In this chapter, the following services are described:

- Push-to-talk Over Cellular (PoC) – which can be seen either as an enhancement of text-based Instant Messaging or an alternative to the traditional voice service.
- Presence – which is sometimes considered as a companion service to PoC or Instant Messaging.
- Multimedia broadcast and multicast.
- Multimedia telephony – which includes Voice over IP as well as video telephony.

6.1 The Role of OMA

The availability of attractive end-user services is one of the main conditions for the success of a communication technology. However, being attractive is not the only condition for mass-market adoption and success. It is also critical that services are designed with a specific focus on inter-operability, not only between devices and service platforms, but also when moving across access technologies, network operators and country borders.

This becomes obvious when considering that several hundreds of handset device models have been made available worldwide for 3G technologies (either cdma2000 or UMTS), with a commercial life which does not exceed 12–18 months.

In the area of service inter-operability, the Open Mobile Alliance (OMA) plays a central role. The OMA is responsible for the delivery of a set of open technical specifications for application and service frameworks. The OMA is a global organization, which includes a large list of members, mainly product and service manufacturers; most of them are also represented in standard organizations such as the 3GPP or 3GPP2.

Evolved Packet System (EPS) P. Lescuyer and T. Lucidarme
Copyright © 2008 John Wiley & Sons, Ltd.

The OMA activity is oriented towards the main existing or upcoming service definition and evolution. Here is an extract of the list of those applications:

- Multimedia Messaging Service (MMS). MMS provides the possibility to exchange rich content (audio, video, text) between subscribers.
- Digital Rights Management: the OMA has defined a framework to enable secured packaging and content distribution to subscribers.
- Push-to-talk Over Cellular (PoC), which is basically a walkie-talkie type of communication service.
- Presence Service, which allows a subscriber to make its Presence information available to other service users.
- WAP, which was formerly handled by the WAP Forum. The WAP Forum has actually been consolidated into the OMA and no longer exists as an independent organization.
- Mobile Location Service.
- Instant Messaging.

In the next chapters, some of these services are described more in detail.

6.2 Push-to-talk Over Cellular

Push-to-talk Over Cellular (PoC) is very similar to the well known walkie-talkie communication service. It provides a one-to-one or one-to-many speech service to a group of people in half-duplex mode, which means that only one participant can talk at a given time, while the others are listening. Like traditional walkie-talkie services, a session participant requests the floor by pressing a 'talk button' on its terminal. As the participant stops talking, he releases the talk button, allowing another participant to request the speech channel.

However, there are some basic differences with analogue or digital low-cost solutions which everyone can buy and easily deploy, due to the fact that PoC is not only supported by the terminal, but also relies on a full network infrastructure. The added value of PoC can be summarized by the two following points:

- PoC provision is not limited to a geographical area. The fact that PoC is supported on top of IMS architecture makes it available to every network subscriber attached to the network, regardless of its current location within the network coverage and access technology.
- In a PoC session, access to the speech channel is under the control of a PoC server and is not fully left on human being behaviour policy. This offers many possibilities to define evolved algorithms for speech access contention resolution. For example, a PoC server may queue and serve requests based on user priority, enforce specific service policy (such as imposing a maximum speech burst duration), or allow pre-empting the current talker in case of emergency speech access.

PoC is much more efficient than conventional speech services in terms of resource usage. The service is, however, not quite the same. Because PoC mandates each user to request the floor before talking, there is not the same level of interactivity between users and the quality of experience may also not be as good as conventional speech services. Because PoC is actually not a real-time conversational service, it is more tolerant to jitter, packet loss and overall

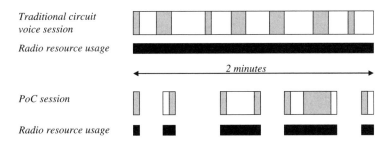

Figure 6.1 Comparison between traditional voice and PoC resource usage.

packet transmission delay. As described later in this chapter, this is obviously not the case for multimedia conversational services.

As an example, Figure 6.1 compares two types of voice sessions. In the case of a legacy voice session, there is almost no silent period of time. In order to guarantee the Quality of Service, resources are reserved for the whole duration of the call and only released when the call is torn down. In the case of a PoC session, there might be some silent periods during which resources may be released by the network. PoC offers the possibility to the network to reserve resources 'on demand', each time the floor is requested, and release them if nobody is talking anymore.

Because it is quite an efficient service in terms of consumed radio and network resources, this service is very popular for professional services and is provided on top of most PMR (Private Mobile Radio) systems. This service is also becoming popular on public networks, as an evolution or voice declination of the text Instant Messaging service.

It is worth noting that PoC is independent of the access technology, provided the available bandwidth and latency figures are good enough. For example, PoC could be provided over EDGE/GPRS 2G systems. However, user experience will definitely benefit from very low E-UTRAN latency and the ability to wake up from standby states. The high-level requirements and capability of E-UTRAN will enable optimized strategies for network manufacturers to deliver efficient resource-management schemes for PoC services.

To be efficient, PoC depends on two companion services, also defined by OMA:

- **The Group Management service.** As PoC provides a one-to-many communication mode, the Group Management service is an easy way to predefine groups so that PoC sessions can be initiated very easily.
- **The Presence service.** This service will indicate to any PoC user if other people from the group it is part of can be reached or are still part of the PoC session.

As described below, PoC makes use of the existing IETF protocol suite. The PoC sessions are managed using SIP protocol, and session signalling and bearer transport are performed through RTP/RTCP. On top of RTCP, PoC has defined its own extension – known as TBCP (Talk Burst Control Protocol) – for the purpose of speech channel management. This part is further described in this section.

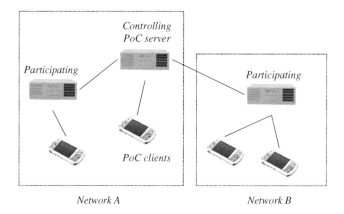

Figure 6.2 Relationship between PoC entities.

And, finally, this section gives some figures about the signalling and application bandwidth that a typical PoC session would require.

6.2.1 Service Architecture

Figure 6.2 describes the PoC service architecture and shows the signalling and bearer path between PoC entities. The key part of the PoC service is based on a **PoC server**. However, for each session, there may be two types of PoC servers: one unique Controlling server, and possibly one or many Participating PoC functions. The PoC controlling server is mainly in charge of managing the SIP session with each of the PoC session participant and managing the access to the floor. In addition, the Participating function may provide local charging information or may support local policy enforcement (such as access control or session barring) in case the PoC session spans over more than one area or network.

In terms of data transmission, the PoC service does not work in multicast or broadcast mode. Speech packets received by the server from the participant having access to the speech channel are forwarded to all other participants in unicast mode, even if several participants to the same session are localized in the same radio cell or group of cells.

More details about PoC interfaces and associated functions are provided in Figure 6.3. When engaged in a session, each subscriber is linked to the server through the POC-3 interface, which supports both speech frame and PoC-related signalling – also known as Talk Burst Control protocol – for session joining, leaving and floor management. As described later in this section, POC-3 is based on RTP for speech transport and RTCP for signalling exchange.

POC-1 and POC-2 interfaces support the SIP signalling to set up and tear down the PoC session. From an IMS and PS Core domain perspective, a PoC session is seen as a classical IMS packet session, set up using SIP signalling through the S-CSCF that the subscriber is currently registered to. In order to support session data and signalling (including SIP and PoC signalling) transport, a PDP context (or an EPS bearer) over the Core and Access network is established, from the terminal to the GGSN or PDN GW in the case of an EPC Core network.

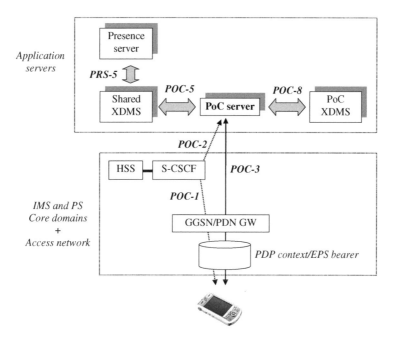

Figure 6.3 PoC service architecture.

About XDMS Servers

The PoC server is in relation to two additional servers:

- The PoC XDMS (XML Document Management Server), which maintains information about the PoC groups.
- The Shared XDMS, which is a multipurpose XDMS server. In the scope of the PoC service, this server provides Presence information about the members of a PoC group. More details are provided about the 'Presence' service in the next section.

In the OMA architecture, the role of XDMS servers is to manage user-specific service-related information in the form of XML (eXtensible Markup Language) documents. 'Managing documents' is a global concept which includes document creation, modification, retrieval and deletion. XDMS is not specific to PoC or Presence services, as it is a general framework which is able to handle information associated to other kinds of existing future services.

In a given network, there might be different types of XDMS servers, depending on the type of information to manage. However, they all rely on the same concepts:

- The use of XML as a common standard or format to manipulate information. Briefly, XML is a kind of language or set of rules to structure data. From a visual perspective, XML looks very similar to HTML, making use of tags, attributes and the well known '<' and '>' separators.

```
<?xml version="1.0" encoding="UTF-8"?>
<group xmlns="urn:oma:params:xml:ns:list-service"
 xmlns:rl="urn:ietf:params:xml:ns:resource-lists"
 xmlns:cr="urn:ietf:params:xml:ns:common-policy"
 xmlns:xsi="http://www.w3.org/2001/XMLSchema-instance">

  <list-service uri="sip:myconference@example.com">
    <display-name xml:lang="en-us">Friends</display-name>
    <list>
      <entry uri="tel:+43012345678"/>
      <entry uri="sip:hermione.blossom@example.com"/>
    </list>
    <max-participant-count>10</max-participant-count>
    <cr:ruleset>
      <cr:rule id="a7c">
        <cr:conditions>
          <is-list-member/>
        </cr:conditions>
        <cr:actions>
          <join-handling>true</join-handling>
          <allow-anonymity>true</allow-anonymity>
        </cr:actions>
      </cr:rule>
    </cr:ruleset>
  </list-service>

</group>
```

Figure 6.4 An example of PoC description using XML.

- XDMS proposes a client–server model. The application protocol used to support requests and answers is HTTP (Hypertext Transfer Protocol), as for Web servers.

Figure 6.4 is an example coming from OMA specifications of XML-structured data managed by the 'PoC XDMS' server. This server is in charge of managing PoC-specific information, including **PoC Groups** and **PoC User Acess Policy**.

The URI 'sip:myconference@example.com' represents the PoC group identity. The <**list**> element contains the PoC group members. In this example, there are only two members; one is represented using a legacy telephone number, and the other is addressed through its SIP URI. As specified by the <**max-participant-count**>, this group cannot accept more than 10 participants.

In addition to group attributes, the XML PoC group syntax allows defining a set of rules for each group. In this example:

- <**join-handling**>set to 'true' means that the request initiated by clients to join a PoC session will be accepted by the PoC server.
- <**allow-anonymity**>set to 'true' means that the PoC Server accepts anonymous access to the PoC session.

As mentioned above, POC-5 and POC-8 make use of HTTP protocol to exchange information between the PoC server (which plays the role of a XDMS client) and the two

XDMS servers it is connected to. HTTP is the well known application-level protocol defined by IETF for information transfer, as the basis of many Web-based services.

Any XDMS client willing to retrieve information from the server will make use of the GET method defined in HTTP, as for Web-page retrieval. Similarly, any client willing to create a group will use the PUT method.

6.2.2 PoC Protocol Suite

(i) Overview
Figure 6.5 represents the stack of protocols being used for PoC sessions between each client and the PoC server. This protocol stack is in charge of ensuring three main functions:

- **PoC session establishment and termination.** As for all IMS applications, this is ensured by SIP and SDP layers. SDP (Session Description Protocol) is used to describe all the media flows and speech codec intended for the PoC session. As for VoIP, a 3GPP PoC session can possibly make use of any of the speech codecs being defined by the 3GPP documents.
- **Audio frame transmission and feedback control.** Speech block transmission is performed using RTP (Real-Time Protocol), as PoC data are sensitive to transfer delay variation. In addition, both participants and PoC servers may send RTP-quality feedback reports using RTCP Sender and Receiver Reports.
- **Floor control.** All features associated to the access of the speech channel are supported by a PoC-specific protocol layer know as TBCP (Talk Burst Control Protocol). Built on top of RTCP, the TBCP messages are actually defined as application-specific RTCP messages.

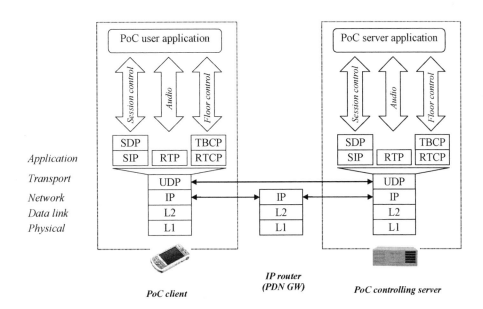

Figure 6.5 The PoC protocol stack.

All information (including SIP signalling or RTP-related data) are supported over a UDP transport layer, in unacknowledged mode.

(ii) The TBCP Messages

The main purpose of the TBCP layer (Talk Burst Control Protocol) is to arbitrate access to the floor for all the PoC session participants. The TBCP name comes from the PoC OMA terminology which defines a 'Talk Burst' as being an audio flow issued by a client who has the permission to send it. The actual Talk Burst control is performed through a small set of messages. For illustration purposes, some of them are briefly described hereafter:

- **Talk Burst Request**: this request is issued by a participant willing to access the floor. This request contains a 'priority level' which the user can use to notify the server about the importance of the request.
- **Talk Burst Granted/Deny**: this is used by the server to notify the requesting client that its request has been either accepted or rejected. In case the Talk Burst was granted, the server may notify a 'stop talking' timer in order to limit the length of the Talk Burst.
- **Talk Burst Release**: this is sent by a client to notify the server that it has finished sending a Talk Burst.
- **Talk Burst Idle/Taken**: this is used by the server to notify the participants whether the floor is free or not.
- **Talk Burst Revoke**: this is used by the server to pre-empt an on-going Talk Burst. This may be used, for example, for emergency purposes. This message provides a cause value to the client, indicating whether the pre-emption was done because the Talk Burst was too long or because a higher-priority request has arrived.

Figure 6.6 describes the generic format of a TBCP message, showing that they actually are defined as application-specific RTCP messages.

The TBCP messages contain the following fields:

- **Subtype**: this field identifies the type of TBCP message.
- **PT** (payload type): this field has the value 204, which indicates that the RTCP packet is application-specific.
- **SSRC**: this field identifies the PoC entity sending the message. In PoC, each client assigns itself a unique SSRC.
- **Application name**: this field is ASCII-coded. In PoC the value is assigned to 'PoC1'.

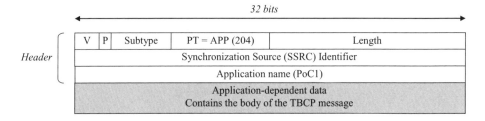

Figure 6.6 TBCP message format.

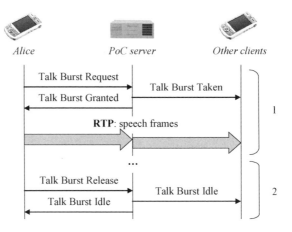

Figure 6.7 An example of Talk Burst control.

(iii) An Example of TBCP Signalling Exchange
In order to illustrate TBCP protocol operation, Figure 6.7 describes an example of Talk Burst management.

1. In the first phase, it is assumed that the Talk Burst is idle. Alice sends a *Talk Burst Request* to the PoC server, which decides to accept. As a result, Alice receives a *Talk Burst Granted* message while the others receive a Talk Burst Taken message.
2. When Alice stops talking and releases the speech button, her terminal sends a *Talk Burst Release* message to the PoC server. As nobody is granted the floor anymore, the PoC server sends a *Talk Burst Idle* message to all participants of the session.

6.2.3 An Example of PoC Session Setup

This section describes an example of PoC session setup with manual answer (meaning that the invited participants need to answer manually to the invitation). In this example, it is assumed that there is only one PoC server controlling the session. Figure 6.8 describes the signalling exchange between Alice (initiating the session) and Bob (a participant being invited to the session).

For simplification, the preliminary steps are not described. When Alice decides to initiate the PoC session, it is assumed that a PDP context or an EPS bearer for signalling support is already established between her terminal and the GGSN, and all participants are IMS-registered to their S-CSCF.

1. In the first phase, Alice sends an INVITE message to the PoC server in order to initiate a session. The message contains a SDP part, which describes the session media parameters, and an XML-encoded part describing the participants whom Alice is willing to contact. Like any

end-to-end SIP signalling, the INVITE message is forwarded by the server to Bob (as one of the recipients) through Alice and Bob's S-CSCF servers.
2. The '180 ringing' is a SIP notification sent to Alice indicating that Bob's terminal is ringing, and the '200 OK' message is sent as Bob accepts the session. In the mean time, Bob's terminal has requested a specific bearer in order to support the PoC session data transfer (which includes both RTP data frames and TBCP signalling).
3. As the SIP session is established end-to-end, the PoC server grants Alice for the floor and indicates to Bob that the Talk Burst is not free. At that time, speech frames can be sent from Alice to all the participants.

(i) What about PoC Bearers, Transport Connections and Media Flows?
The OMA standard is quite flexible as regards to the resources to be established for PoC session support. In terms of PDP context or EPS bearer, the standard does not mandate for any specific solution to be applied. The network has the flexibility to either set up a unique context for both PoC session SIP signalling and data (RTP and RTCP) transport or allocate separated bearers in the case of specific Quality of Service requirements.

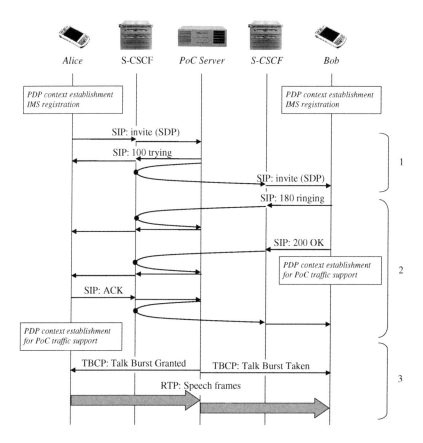

Figure 6.8 An example of PoC session setup – manual answer.

The Services 291

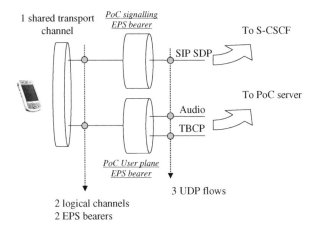

Figure 6.9 An example of control and user flow mapping.

Figure 6.9 describes an example of the bearers being used for a PoC session, as seen by the terminal and the network (including the Access and Core network views).

At the Transport OSI level, there are three UDP flows being set up to support the SIP session signalling, and the audio and TBCP messaging. Those three flows may be supported over two separated bearers. The first one supports the SIP signalling with Quality of Service requirements corresponding to an interactive service. The other, supporting the User plane data, has more constraining Quality of Service requirements, as for streaming or conversational services.

These two logical traffic channels may then be supported by transport channel levels having different configurations and characteristics. For example, the SIP signalling logical channel may be supported by a E-UTRAN RLC AM (Acknowledged Mode) transport channel, whereas the audio channel is supported by a RLC UM (Unacknowledged Mode) transport channel.

(ii) What about the Data Rate Induced by PoC Services?
This paragraph proposes an estimation of the requirements of PoC service in terms of signalling and pure user data bandwidth, as a comparison to what classical voice service requires.

In the following example, we consider a 1-minute PoC session with two participants only. Five-second Talk Bursts are sent alternatively by each participant, so that each participant is six times talker, six times listener during the session. For simplification, it is assumed that no Sender or Receiver Reports for quality feedback are exchanged during the session.

Also for simplification, it is assumed that the TBCP messages have a mean length of 16 bytes (which includes the RTCP header and the TBCP message part).

In this example, all speech frames are encoded using the 3GPP AMR 12.2 full-rate speech algorithm which delivers around 32 bytes every 20 ms. These bytes are further encapsulated by a 12-byte RTP header. As mentioned in RFC3267, 'RTP Payload Format for AMR and AMR-WB', a RTP frame may contain more than one AMR speech frame. This has the virtue of reducing the amount of protocol overhead relative to the speech data, at the expense of an increased latency.

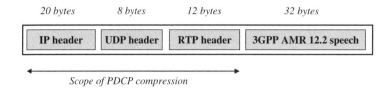

Figure 6.10 An example of a speech packet.

All sent packets (including the media packets or the RTCP ones) are encapsulated by UDP (adding an 8-byte overhead) and IPv4 (adding a 20-byte overhead), as illustrated in Figure 6.10.

Table 6.1 describes the number of bytes sent and received by a given participant during the 1-minute session.

The first obvious remark is that although the channel is used in half-duplex mode (one talker at a time), the channel occupation is about the same as a full-duplex speech service. This is due to the fact that the IPv4/UDP/RTP encapsulation adds 40 bytes of overhead on top of the 32 bytes of encoded speech.

This basically means that PoC cannot be a commercially viable service on wireless networks without a minimum level of protocol header compression. Thanks to RoHC, the compression layer of Evolved UMTS presented in Chapter 4, the amount of overhead can be efficiently reduced. The figures in the table assume that the 40-byte IPv4/UDP/RTP overhead is reduced to 6 bytes, which is the common value observed under reasonable radio conditions. The resulting bit rate is halved, which is more in line with traditional voice-over circuit occupation.

The use of a reduced-rate speech codec is another way to reduce the service bandwidth requirements, at the expense of lower speech quality.

Finally, it is worth noting that, in the scope of the example above, the TBCP protocol does not add much overhead to the number of bytes being sent. In the case of a 5-second Talk Burst, the TBCP overhead is less than 0.5%.

6.2.4 Charging Aspects

As usual, the standard does not mandate any specific charging scheme to be applied, but provides manufacturers and network operators with all the mechanisms to define flexible mechanisms for PoC charging.

Table 6.1 PoC session example data rate estimation for one participant.

	Without RoHC compression	With RoHC compression
Sent as listener	0	0
Received as listener	108528	57120
Sent as talker	108528	57120
Received as talker	528	120
Resulting bit rate UL or DL	14.5 Kb/s	7.6 Kb/s

Table 6.2 Extract of PoC CDR.

PoC CDR	
Start Timestamp	PoC session starting time
End Timestamp	PoC session ending time
SDP session description	Describes the PoC session media components and their parameters
Originator	Describes the asserted SIP identity of the PoC session originator
Number of participants	The number of invited parties of the PoC session
List of participants	Contains the addresses (in the form of a SIP URI or E.164 phone number) of the parties of the session
List of Talk Burst exchanges	Contains detailed information about the number of Talk Bursts, Talk Burst bearer volume, sum of Talk Burst times
...	...

The network entity in charge of producing charging data is the PoC controlling server. The server can distinguish the session owners among all the participants and apply specific charging rules to each of them. Besides, as the PoC controlling server has responsibility for arbitrating the speech channel, receiving and forwarding all the Talk Bursts, it can provide lots of detailed information for the purpose of PoC charging. In addition, the PoC participating server may also collect local charging information, e.g. in case the PoC session spans over multiple networks.

The charging architecture is similar to what has been presented in Chapter 2. PoC-related CDR (Charging Data Record) are sent by the PoC controlling or participating servers to the CDF (Charging Data Function) through the Rf interface, using the Diameter IETF protocol.

Table 6.2 is an extract of the information that the CDR may contain. In case PoC Participating nodes are also involved in the session, the CDF may contain more than one CDR. Ultimately, the CDF will consolidate all the information it receives before sending the final report to the BD (Billing Domain).

Reference documents about PoC

3GPP technical specifications:

- 23.279, '3GPP Enablers for OMA Push-to-Talk over Cellular (PoC) Services: Stage 2'
- 32.272, 'Push-to-Talk over Cellular (PoC) Charging'

OMA specifications:

- OMA-AD-PoC, 'Push to Talk over Cellular (PoC): Architecture'
- OMA-TS-PoC-ControlPlane, 'PoC Control Plane'
- OMA-TS-PoC-UserPlane, 'PoC User Plane'
- OMA-TS-PoC-XDM, 'PoC XDM Specification'

6.3 Presence

The Presence service's purpose is to allow a subscriber to make its Presence information available to other users of Application Servers in the network. By extension, the Presence service is not limited to user availability information. During the past few years, there have been lots of new extensions brought to this service, coming from IETF activities or OMA specifications. These may also include lots of different information being defined by IETF documents, or by the OMA Presence specifications. These extensions provide richer information about the subscriber status, such as:

- The current activity (travelling, sleeping, ...).
- The current location (at home, in office, ...).
- The mood (afraid, hungry, ...).
- Time offset (specifies the current time zone of the person).
- Contact address (an address in the form of a URL to which the user can be joined).
- The services the user has access to, e.g. PoC, Instant Messaging, ...
- The list of devices and network connectivity the user has access to (802.11, GPRS, IMS, ...).

The rest of the section describes the service architecture as well as the relations with other parts of the Evolved UMTS network, and an example of Presence service usage.

6.3.1 Service Architecture

Figure 6.11 is a simplified overview of the Presence service architecture, as described in OMA specifications. The service itself relies on a central node – the Presence Server – in charge of

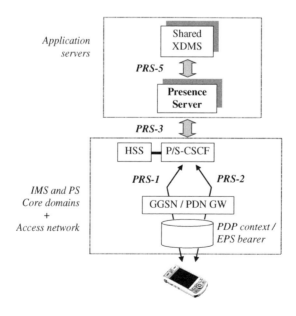

Figure 6.11 Presence Service architecture.

storing and distributing Presence information. From the IMS perspective, the Presence Server is seen as a SIP Application Server. The Presence Server is linked through the PRS-5 interface to the Shared XDMS server, already presented in the section about the PoC service. The Shared XDMS server is in charge of managing user-specific service-related information, including the Presence information provided by the Presence Server.

On the other side, as the OMA Presence Server is build on top of IMS architecture, the Presence Server is also linked through the PRS-3 interface to IMS SIP servers such as the S-CSCF. As regards the Presence service, the role of the S-SCSF is only to route SIP signalling to and from the server.

The end-user terminal involved in the Presence Service may play two different roles, which are described in the specifications as 'Presence Source' or 'Watcher'. Those two roles do not exclude each other, as a given terminal may be simultaneously a Presence Source for himself and a Watcher to somebody else's information.

When being a 'Presence Source', the terminal provides its Presence information to the Presence Server via the S-CSCF and the PRS-1 interface. When being a 'Watcher', the terminal subscribes to Presence information and receives notifications from the server through the PRS-2 interface. Although PRS-1 and PRS-2 are presented as two different functional interfaces, they may be supported by the same PDP context (or EPS bearer) between the terminal and the GGSN (or PDN GW).

6.3.2 An Example of a Presence Session

(i) Presence Information Format

The Presence information in the OMA specification is structured in three components:

- Person: this set of information contains attributes related to the user itself.
- Service: this set describes the services that the user has access to.
- Device: this describes which physical devices are associated to each of the supported services. This part also indicates the kind of connectivity that the device has access to.

The Presence information is distributed between Source, Watcher and Server using XML-generic format. Figure 6.12 is a simplified example from the OMA Presence specification, describing what Presence information looks like. This example includes the following elements:

- <status> and <willingness> indicate the availability and willingness of the user to receive a PoC service request, PoC being the service listed further in the XML description.
- <registration-state> and <barring-state> parameters indicate that the user is registered to PoC and that no incoming or out-going service communication barring is active.
- A <contact> information is also part of the Presence information, to be used by other parties willing to contact the user through the PoC service.
- The <timestamp> is an indication of the time the Presence information has been issued.

The format defined by OMA is mainly based on RFC3863, 'Presence Information Data Format (PIDF)' and other IETF extensions.

```xml
<?xml version="1.0" encoding="UTF-8"?>
<presence xmlns="urn:ietf:params:xml:ns:pidf"
 xmlns:rpid="urn:ietf:params:xml:ns:pidf:rpid"
 xmlns:op="urn:oma:xml:prs:pidf:oma-pres"
 entity="sip:someone@example.com">
   <tuple id="a1232">
     <status>
       <basic>open</basic>
     </status>
     <op:willingness>
       <op:basic>open</op:basic>
     </op:willingness>
     <op:registration-state>active</op:registration-state>
     <op:barring-state>terminated</op:barring-state>
     <op:service-description>
       <op:service-id>org.openmobilealliance:PoC-session</op:service-id>
       <op:version>1.0</op:version>
     </op:service-description>
     <contact>sip:someone@example.com</contact>
     <timestamp>2005-02-23T12:14:56Z</timestamp>
   </tuple>
</presence>
```

Figure 6.12 An example of XML-encoded Presence information.

(ii) Presence Signalling Procedures

The Presence service makes use of three SIP methods, all defined by IETF documents:

- SUBSCRIBE: used by a Watcher to subscribe to Presence information. In the SIP message header, the Watcher indicates the user's identity whose events he wants to subscribe to.
- PUBLISH: used by a source to provide its Presence information to the server. The PUBLISH message is actually a SIP message which encapsulates XML-coded Presence information.
- NOTIFY: used by the Presence server to notify Watchers about Presence information change or availability. Like the PUBLISH message, Presence information is also XML-coded. The information may, however, be different from the one provided by the source, as the Server may process the information such as filtering or composing information from different user devices.

Figure 6.13 describes how these methods are used in the scope of a Presence session.

As for IMS SIP signalling, the Presence messages are subject to header compression based on SigComp. Although OMA strongly suggests that Presence messages should be compressed, it is still an option to do so. The decision is actually left to the endpoints (e.g. Watchers and Presence Sources) and shall be enforced by IMS CSCF nodes.

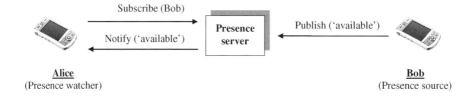

Figure 6.13 An example of a Presence session.

In terms of transmission requirements, Presence does not require a large number of resources. In most cases, XML-encoded Presence information and associated SIP headers sent to or received from each subscriber should not be much larger than 1000 bytes.

From a Core and Access network perspective, Presence requires the user (either Watcher or Presence Source) to be IMS-registered and have a PDP context or EPS bearer established while the service is active. However, there is no need for radio physical resource to be allocated from the beginning to the end of the session.

In terms of RRC and MME states, the terminal needs to be RRC-connected in order to publish or be notified about a Presence state change. When in RRC IDLE mode, the terminal needs to connect to the network before any kind of activity related to the Presence Service (as for any other service). In the case of an incoming Notify to be sent to an IDLE mode terminal, the network has to send a paging message which will cause the receiving terminal to connect to the network.

6.3.3 Charging Aspects

Thanks to SIP signalling, the intermediate S-CSCF node has the possibility of monitoring and counting any kind of event associated to a SIP message. This includes, as an example:

- Presence service registration.
- Subscription to Presence information published by another user.
- Presence information retrieval to the XDMS server.
- Reception of Presence information notifications.

This gives flexibility for the operator to implement any kind of charging algorithm, based on the different events.

Reference documents about Presence

IETF documents:

- RFC3856, 'A Presence Event Package for the Session Initiation Protocol (SIP)'
- RFC3863, 'Presence Information Data Format (PIDF)'
- RFC3903, 'SIP Extension for Event State Publication'

3GPP technical specifications:

- 22.141, 'Presence Service: Stage 1'
- 23.141, 'Presence Service: Architecture and Functional Description: Stage 2'
- 24.141, 'Presence Service Using the IP Multimedia (IM) Core Network (CN) subsystem'
- 33.141, 'Presence Service: Security'

  OMA specifications:

- OMA-AD-Presence-SIMPLE, 'Presence SIMPLE Architecture Document'
- OMA-TS-Presence-SIMPLE, 'Presence SIMPLE Specification'

6.4 Broadcast and Multicast

This section describes the broadcast and multicast service, as specified by the 3GPP documents. This service is known as MBMS (Multimedia Broadcast and Multicast Service).

From a general perspective, the benefit of broadcast and multicast on networks is that multiple subscribers can receive the same data at the same time, sent only once on each link. For the radio interface, the obvious benefit is that in a given cell, the radio resource cost is limited to what is needed for one transmission in a given cell, for the sake of radio interference and capacity. The gain on terrestrial interfaces is not also not negligible, as the interfaces between the content source and the radio equipment will also benefit from broadcast techniques.

Back to history, GSM networks already offers a text-based broadcast service known as SMSCB (Short Message Service Cell Broadcast). The SMSCB is limited to low bit rate data, transmitted to all subscribers in a given set of cells over a shared broadcast radio channel. The capacity of the GSM broadcast channel is quite limited, as it only allows sending a block of 88 bytes of text every eight 51-multiframe (approximately 2 seconds), which provides a maximum bit rate of around 350 bits per second.

As described below, MBMS has a much broader set of rates and applications, allowing multimedia types of data to be delivered.

MBMS shall not be confused with 'IP Multicast' – the multicast solution proposed by IETF. IP Multicast can, however, be supported by wireless networks but, in this case, it does not allow sharing common radio resources for subscribers located in the same radio cell.

6.4.1 Some Definitions

The MBMS 3GPP service is actually composed of two distinct services: broadcast and multicast. There are many common areas between broadcast and multicast. However, although supported in the same way over the radio interface, the two services shall be distinguished, as there is a key difference between them:

- Broadcast service may be received by any subscriber located in the area in which the service is offered.
- Conversely, multicast services can only be received by users having subscribed to the service and having joined the multicast group associated with the service.

In the MBMS scope, broadcast and multicast services are unidirectional point-to-multipoint transmissions of multimedia data. These two services may be used to broadcast, for example, text, audio, picture, video from a single source (which is called the BM-SC for Broadcast Multicast Service Centre) to:

- Any user located in the service area (in the case of a broadcast service).
- Only members of a multicast group (in the case of a multicast service).

Broadcast only requires the service to be enabled by the user on its terminal. For this reason, the operator cannot apply charging rules to the end-user, as the network does not know which subscribers have received the service and for how much time the users have been receiving it. For such a service, only the broadcast service providers can be charged, possibly based on the amount of data broadcasted, size of service area or broadcast service duration.

Multicast is subject to service subscription, and requires the end-user to explicitly join the group in order to receive the service. Because it is subject to subscription, the multicast service allows the operator to set specific user charging rules for this service.

6.4.2 Typical Applications

This section gives some examples of typical applications suitable for MBMS. The list is not exhaustive, but tries to explore the possible service range, along with the corresponding rates and real-time characteristics.

(i) Audio and/or Video Streaming
This set of applications includes all kinds of news or advertisement message distribution. This could be, for example, a weather report, an announcement for time-limited discount in a store or pay-per-view video access. As for streaming applications, such a service puts some Quality of Service constraints on the network in terms of bandwidth, packet transfer delay and jitter. The typical application bit rate requirements is from 32 Kb/s to more than 300 Kb/s, depending on the quality and media type.

(ii) Audio and/or Video Downloading
MBMS could also be used for multimedia content distribution, for off-line use. As opposed to streaming, the provision of real-time transfer services is not required. However, enough bandwidth shall be provided to the broadcast/multicast server especially for large content downloading.

(iii) File Downloading
A less obvious application of MBMS is the possibility to efficiently download data to large populations of terminals. This could be, for example, for the purpose of software update (user terminal application update of new codec or plug-in downloading) or virus list update downloading.

Here, again, we see the interest of multicasting versus broadcasting, as it allows restricting the service to the set of end-users which have actually subscribed to it.

(iv) Still Image and Text Distribution
This kind of application is closer to what GSM SMSCB already provides. The requirements in terms of bandwidth are similar to what a classical messaging service would mandate, i.e. a maximum bit rate of around 10 Kb/s, with no specific real-time constraint.

6.4.3 Service Architecture

The MBMS service architecture is based on the Packet Core domain, and is therefore compatible with Enhanced Packet Core, as well as 2G/GSM or 3G UMTS Packet Core nodes like the SGSN and GGSN.

MBMS is, however, not part of the standard features supported by the Packet Core, as it requires specific additional capabilities from the relevant nodes for the purpose, for example, of broadcast or multicast session establishment, or multicast authorization.

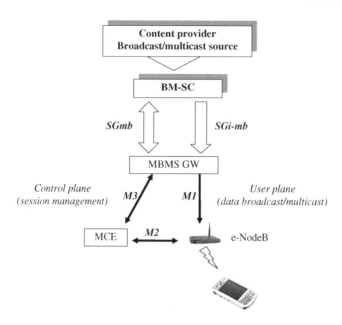

Figure 6.14 The MBMS logical architecture.

In the scope of EPS networks, it was decided to create two additional logical network entities: the MCE and the MBMS GW, as presented in Figure 6.14.

The MCE (Multi-cell/multicast Coordination Entity) is a new logical entity, responsible for allocation of time and frequency resources for multi-cell MBMS transmission. The MCE actually does the scheduling on the radio interface. The MCE is a logical node which may be integrated as part of the eNodeB (in which case, the M2 interface becomes an internal eNodeB interface).

The MBMS GW (MBMS Gateway) is the entry point of incoming broadcast/multicast traffic. Its role is to broadcast the packets to all eNodeBs within a service area, as well as MBMS session management (like *Session Start* and *Session Stop*). It is also in charge of collecting charging information relative to the distributed MBMS traffic for each terminal having an active MBMS session. In 2G/GPRS and 3G/UMTS networks, this role is ensured by the GGSN.

The Broadcast Multicast Service Centre (BM-SC), already present in 2G and 3G MBMS architectures, is the functional entity in charge of providing the service to the end-user. For that purpose, the BM-SC serves as an entry point for content providers or any other broadcast/multicast source which is external to the network.

The BM-SC is in charge of the following main functions:

- Membership function, which is responsible for providing authorization for terminals requesting to activate an MBMS service.
- Session and transmission function, which is responsible for the scheduling of broadcast and multicast sessions.

- Service announcement function; MBMS session announcement is triggered by the BM-SC but can be performed through different mechanisms such as SMS (Short Message Service) or any kind of Push service.
- Security function, which ensures integrity and confidentiality protection of MBMS data.

From the BM-SC, two interfaces are defined: the SGmb and SGi-mb.

The **SGmb** interface supports all the MBMS signalling procedures between the BM-SC and the MBMS GW. This includes MBMS bearer signalling for setting up and releasing context at MBMS session establishment and termination. The SGmb also supports user-related signalling, e.g. for Multicast session authorization, or user session joining or detach.

From a protocol perspective, the SGmb interface is based on 3GPP-specific extensions to the Diameter protocol defined at the IETF in the document RFC3588, 'Diameter Base Protocol'.

The **SGi-mb** interface supports the MBMS traffic plane. This also introduces some interfaces:

- The **M1 interface**, associated to the MBMS data (or User plane) makes use of IP multicast protocol for the delivery of packets to eNodeBs.
- The **M3 interface** supports the MBMS session control signalling, e.g. for session initiation and termination.
- The **M2 interface** is used by the MCE to provide the eNodeB with radio configuration data.

Figure 6.15 represents the architecture of the EPS User plane for the MBMS service. The MCE is not represented here, as it is not part of the MBMS data-transmission path. The top-level layer represents the MBMS application protocol; more details on this part are provided later in this section.

The TNL (Transport Network Layer) boxes represent the IP-based transmission protocols used on network interfaces. The TNL on the M1 interface is based on 'IP Multicast', so as to allow point-to-multipoint delivery of MBMS data for multi-cell transmission.

The SYNC protocol, used over the M1 interface, has a specific purpose. Its main role is to allow content synchronization in the case of multi-cell transmission. SYNC carries additional

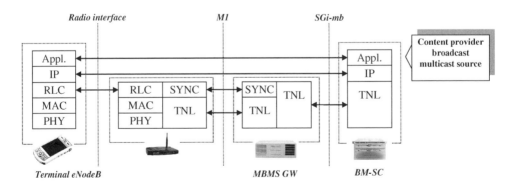

Figure 6.15 MBMS User plane architecture.

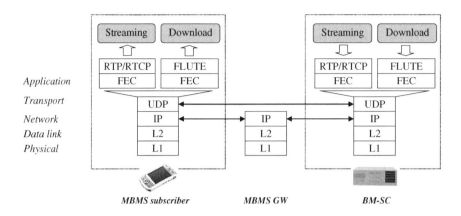

Figure 6.16 The MBMS data transport protocol suite (simplified view).

information so as to allow eNodeBs to identify transmission radio frame timing and detect packet loss.

(i) The MBMS Protocol Suite

Figure 6.16 describes the MBMS protocol suite for MBMS data transfer. There are actually two options for data transfer, depending on the type of service.

For streaming services, MBMS data transfer is based on RTP/RTCP (Real Time Protocol/ Real Time Control Protocol), described in Chapter 4. This type of protocol allows the receiving end to regenerate packet data timing information which may be altered by transmission jitter. A FEC (Forward Error Correction) block is appended to each RTP data packet to allow error detection and possibly correction at the receiving end. It allows the receivers to reconstruct missing portions of the source block.

For download services (e.g. for still image, 3GPP audio/video file or binary data transfer), MBMS makes use of the FLUTE protocol (File Delivery over Unidirectional Transport), as defined in the RFC3926 IETF document. FLUTE is a unidirectional file-transfer protocol specifically designed for multicast transmission and massively scalable distribution, although it may also be used for unicast (point-to-point) file distribution.

FLUTE provides a generic format for describing files to be transferred. This format, known as FDT (File Delivery Table), specifies the name, size and type (among other parameters) of file to be transferred). In addition, because FLUTE is inherently designed for multicast traffic, it does not require feedback or even connectivity from the receiver to the sender. As for streaming services, data-transfer reliability is ensured using a FEC scheme integrated into FLUTE.

Both streaming and download application services are eventually supported over a UDP/IP transport scheme. In addition to those protocols, specific to data transfer, MBMS makes use of other existing layers:

- The 3GPP NAS (Non Access Stratum) protocol for packet session management has been modified, as MBMS requires the terminal to open a 'MBMS context'. This new type of 'packet context' is equivalent to the classical 'PDP context' for an IP multicast address.

The Services   303

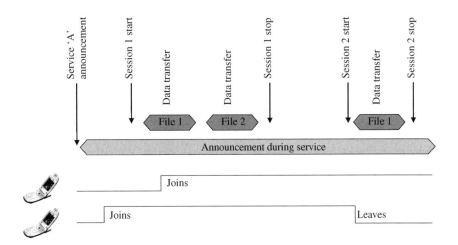

Figure 6.17 Timeline of an MBMS service.

- The existing IETF IGMP (Internet Group Management Protocol) is also part of MBMS signalling protocols, as it is used by subscribers for joining or leaving MBMS sessions.
- The IETF SDP (Session Description Protocol), described in Chapter 2, is used as a basis for the description of MBMS sessions.

(ii) Some Vocabulary
MBMS specifications make a difference between **service** and **session**.

A **session** is a period of time during which broadcast or multicast data are transmitted, either in one set or using several data files. As in Figure 6.17, there might be several consecutive sessions of data transfer for one single **service**. In principle, the service is announced up to the end of the last session of the service. From a receiver point of view, mobiles can join or leave the service at any time, even during a data-transfer session.

6.4.4 MBMS Security

When compared to point-to-point communications, multicast service security faces some new specific challenges. In the MBMS context, ciphering has a different purpose, as MBMS subscribers have no real interest in broadcast or multicast privacy. However, encryption is a key feature for operators and content providers, as it is the only way to prevent unauthorized customers from receiving the data for free.

MBMS security mechanisms like ciphering need to be designed in a way which is suitable for point-to-multipoint communications. In addition, those mechanisms shall prevent any bypass attempts from subscribers, by exchanging keys or passwords, for example. This is the reason why MBMS security mechanisms are quite specific, involving complex key structures and frequent key updates.

As described Figure 6.18, the MBMS key structure is based on three different types of keys:

- The MUK (MBMS User Key).

- A set of MSK (MBMS Service Key) per MUK.
- A set of MTK (MBMS Traffic Key) per MSK.

As opposed to most point-to-point communications, MBMS security is not applied on a specific part of the transmission path (like the radio segment) but is rather performed end-to-end, between the BM-SC and the subscribers. The basis of MBMS security is that the multicast data are protected by a symmetric key (the MTK), which is a secret key shared by both the sender (the BM-SC) and the receivers (the MBMS subscribers). In order not to compromise security, MTK generation and distribution follow a complex process, illustrated in Figure 6.18.

At first, each terminal builds its own unique secret MUK, derived from CK (Ciphering Key) and IK (Integrity Key), both stored on the terminal USIM. Those references are also stored within the HSS, so that the network can also build its own reference of the MUK. The MUK is further used by the terminal to retrieve the MSK provided by the BM-SC during the authorization process.

MSK and MTK key sets are the actual keys to secure MBMS data; they are both generated by the BM-SC. Each MTK corresponds to a unique data flow (either a streaming or FLUTE file download flow) and is used to cipher the data block for this flow. The set of MTK is transmitted to the authorized subscribers by the BM-SC in a secured way, protected by the MSK.

As a way of enhancing security, it is then possible for the BM-SC to define two different sets of MSK (and associated MTK) and update the MTK during a service or a session by indicating to the authorized receivers the new MSK to use.

MSK and MTK are usually updated in a subscriber's terminal using unsolicited 'push' procedures from the BM-SC. The protocol used to perform this update is MIKEY (Multimedia Internet Keying) – a key-distribution protocol defined by the IETF and documented in RFC3830.

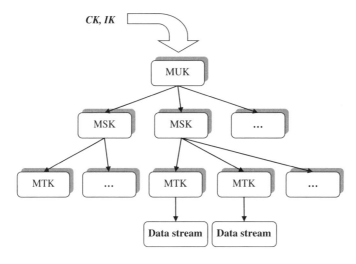

Figure 6.18 MBMS key structure.

The Services 305

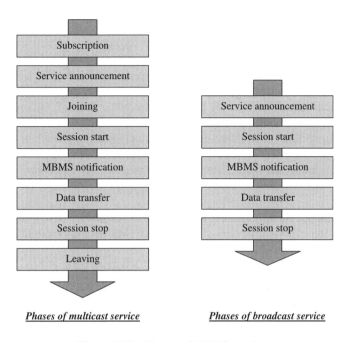

Phases of multicast service   *Phases of broadcast service*

Figure 6.19 Phases of MBMS services.

6.4.5 The MBMS Service Steps

Figure 6.19 describes the various steps of multicast and broadcast service provision. This illustrates the fact that multicast is actually a superset of the broadcast service, as it requires user subscription and access authorization procedures to prevent nonsubscribers from accessing the data.

The **Subscription** phase role, specific to multicast, is to build the relationship between the end-user and the service or content provider. As a result, the end-user is allowed to receive the multicast service. One of the consequences of the subscription phase is that the USIM of the subscriber is updated with MBMS-specific information, such as MBMS service keys and user keys, so that the user will be able to join upcoming multicast sessions and be able to decipher the data received.

The **Service announcement** phase aims at informing end-users about the available MBMS services.

During the **joining** phase, specific to multicast, the subscriber indicates its willingness to listen for a specific service, so that it can be charged for the multicast data.

The **Session start** phase is the actual start of the MBMS session. Network resources to support the MBMS bearers are reserved and established at that time.

The **MBMS notification** phase is used to inform users about the upcoming MBMS sessions.

During the **Data transfer** phase, the MBMS data are transferred to the terminals.

At **Session stop**, network resources previously allocated are released by the network.

At the end, the **Leaving** phase, corresponding to the joining, is an explicit indication to the user that it is no longer willing to follow the session and is not receiving anymore.

In the rest of the chapter, more details are provided on the Service announcement phase.

Service Announcement

Service announcement is the procedure by which the end-user is informed about the available MBMS services. Service announcement is part of the features supported by the BM-SC. There might be several ways to achieve this, using, for example, services like SMS, MMS, WAP Push service or even an existing MBMS bearer. For that reason, the standard does not further specify the signalling and mechanisms for doing the announcement, but only the format of information that needs to be sent to the subscribers.

Service announcement involves the delivery of 'metadata fragments', or blocks of information which contain the details of the services. To be more efficient, the data may possibly be sent compressed, using the well known GZIP algorithm. As examples, these blocks of information include:

- Service description: As specified by the 3GPP, MBMS service description is done using SDP (Session Description Protocol), similarly to an IMS session description. Figure 6.20 is an example of such a description.
- Associated service delivery procedures: This XML-encoded set of information is related to MBMS transmission auxiliary procedures, e.g. for file post-delivery repair, or reception reporting. All these procedures have to be initiated by the MBMS terminal receivers.
- Service protection method: This XML-encoded set contains information for service key management and related security procedures necessary for information decryption.
- MBMS data error control related information: MBMS streams make use of a generic Forward Error Correction (FEC) mechanism. This allows the receiver to detect and possibly repair data-transmission errors.

| *Session parameters* | ```
v=0
o=ghost 2890844526 2890842807 IN IP4 192.168.10.10
s=3GPP MBMS Streaming SDP Example
i=Example of MBMS streaming SDP file
u=http://www.infoserver.example.com/ae600
e=ghost@mailserver.example.com
c=IN IP6 FF1E:03AD::7F2E:172A:1E24
t=3034423619 3042462419
b=AS:77
a=mbms-mode:broadcast 1234
a=source-filter: incl IN IP6 * 2001:210:1:2:240:96FF:FE25:8EC9
a=FEC-declaration:0 encoding-id=1
``` |
| *1st medium parameters (video stream)* | ```
m=video 4002 UDP/MBMS-FEC/RTP/AVP 96
b=TIAS:62000
b=RR:0
b=RS:600
a=maxprate:17
a=rtpmap:96 H264/90000
a=fmtp:96 profile-level-id=42A01E; packetization-mode=1
``` |
| *2nd medium parameters (audio stream)* | ```
m=audio 4004 UDP/MBMS-FEC/RTP/AVP 98
b=TIAS:15120
b=RR:0
b=RS:600
a=maxprate:10
a=rtpmap:98 AMR/8000
a=fmtp:98 octet-align=1
a=FEC:0
``` |

Figure 6.20 An example of MBMS session description.
```

The picture is an example of an MBMS streaming session description based on SDP. As usual in SDP, the description is split into several parts, the first on presenting session-level parameters, and the subsequent ones relate to parameters at the medium level (in this example, there are two media – one for the audio stream, and one for the video stream).

This description contains all the required parameters for a subscriber to be able to join the session and decode the information:

- The sender IP address (an IPv6 in this example) is described by the '**a = source-filter**' parameter
- The session start and end times are specified by the SDP timing field '**t =**'
- The session bearer mode is described by the '**a = mbms-mode:broadcast**'. In this line, '**1234**' is the session Temporary Mobile Group Identity (TMGI) which the subscriber refers to when it is joining the service
- The data codec and transport protocol are specified for each of the data bearers (in this example, the video stream is H264 encoded, whereas the audio is encoded as AMR 8KHz).

6.4.6 The E-UTRAN Aspects of MBMS

E-UTRAN is quite flexible and offers many possible options for MBMS service deployment. In MBMS, the operator has the possibility of reserving a frequency layer to MBMS transmissions. In this case, the cells belonging to this layer only offer MBMS service. In those **dedicated** cells, there is no support for unicast (or point-to-point) service.

In contrast, when no specific frequency is reserved for MBMS, **mixed** cells provide simultaneous unicast and MBMS services.

In parallel, there may be two types of MBMS data transmission in E-UTRAN:

- Single-cell transmission – in this case, MBMS data are only provided and available over the coverage of one single cell.
- Multi-cell transmission – in this case, the MBMS data sent in the different cells are tightly synchronized. This allows the receiving terminal to recombine the signals received from various cells and improve the signal-to-noise ratio, as compared with conventional point-to-multipoint transmission.

More information about the radio physical aspects of MBMS are provided in Chapter 3.

6.4.7 Charging Aspects

MBMS charging mechanisms have two objectives. First, they aim at collecting charging information related to mobile subscribers receiving MBMS services. This is more applicable to the multicast service (which require a user-authorization phase) than the broadcast service, as the end-user may join a broadcast session or service without the support of explicit signalling. In addition to end-user charging, information related to content or service providers delivering content using MBMS are also collected.

MBMS charging architecture follows the global 3GPP charging architecture presented in Chapter 2. The BM-SC is actually considered as an Application Server from the CDF.

Table 6.3 Extract of MBMS CDR.

MBMS subscriber CDR	
Served IMSI	Contains the IMSI of the subscriber
Bearer services description	Describes the MBMS SDP session media parameters
Duration	Describes the time period to which the CDR applies
TMGI	Describes the Temporary Mobile Group Identity allocated to the MBMS service
...	...
MBMS content provider CDR	
Served PDP address	Describes the IP multicast address which was used for the service
Recipient address list	Contains the addresses of the recipients registered to receive the service
Service area	Describes the area over which the MBMS bearer service was distributed
Service type	Indicates whether the CDR corresponds to Broadcast or Multicast service
...	...

As for other Application Servers, the BM-SC is in charge of sending charging information to the CDF over the Rf interface, and using the Diameter protocol. From its side, the CDF generates CDR (Charging Data Record), which are further used as the basis for the actual charging. Because the MBMS requires both subscribers and content providers to be charged, two different formats of CDR have been defined. An example of the information fields of these two CDR types are described in Table 6.3.

In principle, the CDR shall allow the network operator to charge based on the size of the broadcast/multicast area, the type of Quality of Service required (e.g. in terms of bit rate), the amount of data actually being sent, and the number of recipients.

MBMS charging is triggered by typical events, parts of MBMS session establishment and termination, such as 'Session Start' and 'Session Stop' messages (actually related to network resource allocation and release), and successful user authorization.

Reference documents about MBMS

3GPP technical specifications:

- 22.146, 'Multimedia Broadcast/Multicast Service (MBMS): Stage 1'
- 22.246, 'Multimedia Broadcast/Multicast Service (MBMS) User Services: Stage 1'
- 23.246, 'Multimedia Broadcast/Multicast Service (MBMS): Architecture and Functional Description'
- 26.346, 'Multimedia Broadcast/Multicast Service (MBMS): Protocols and Codecs'
- 33.246, '3G Security; Security of Multimedia Broadcast/Multicast Service (MBMS)'
- 32.273, 'Multimedia Broadcast and Multicast Service (MBMS) Charging'

IETF documents:

- RFC3830, 'MIKEY: Multimedia Internet KEYing'
- RFC3926, 'FLUTE: File Delivery over Unidirectional Transport'

6.5 Voice and Multimedia Telephony

This section deals with voice and general-purpose real-time conversational multimedia services in Evolved UMTS networks.

Video telephony in not new to Evolved UMTS networks, as it is already proposed as a service by 3G networks using two alternatives:

- **The circuit-switched solution**, based on the ITU H.324M standard (M for 'Mobile'). H.324M is actually the low bit rate variant of H.324, using 3GPP voice-coding algorithms specially designed for wireless transmission.
- **The packet-based solution**, using IETF protocols for signalling and data transport like the 3GPP IMS solution.

In the scope of Evolved UMTS networks, only this latter alternative will be described in this chapter. In its most general definition, an IMS multimedia session may include any combination of the following three media types: voice, video and text.

While voice and video media flows are very constraining from real-time and requested resource perspectives, text transmission only requires a low bit rate bearer, as the transmission rate does not exceed 30 characters per second. Unlike text-messaging services, typed character flow is simply time-sampled so that no specific action is needed from the user to request transmission. In principle, the characters typed on a keyboard or drawn on a screen on the sending terminal should be rendered in real time on the display of the receiving terminal.

The media flows being set up for a multimedia session may not necessarily be arranged in duplex mode. This means that, for example, video transmission may only be configured from terminal A to terminal B, but not in the other direction.

6.5.1 About Circuit and Packet Voice Support

(i) Circuit Voice in 2G Networks
2G/GSM Cellular network was actually thought of as an extension of public land-line telephone networks, also known as PSTN (Public Switched Telephone Network), and was therefore specifically optimized for voice transport.

For transmission over PSTN, audio sound is digitized at an 8-kHz sample rate using 13-bit PCM (Pulse Coded Modulation) samples. Each sample is then compressed to 8 bits using a logarithmic compression system (the 'µ-law' in North America and Japan and 'A-law' in Europe and most of the rest of the world). Eventually, one voice channel occupies a full-duplex 64-Kb/s channel.

A side effect of the 8-KHz sampling value is that any voice frequency above 4 KHz is distorted or completely lost, still allowing voice recognition but definitely not high-grade quality.

The traditional PSTN voice processing was never envisaged to be used as such on public cellular networks, not only because of the too high requested bandwidth, but also because the voice-coding process was not designed for high bit error rate radio transmission.

This is the reason why a specific voice-coding scheme was introduced in all cellular networks, most of them being based on LPC (Linear Prediction Coding). Those coding

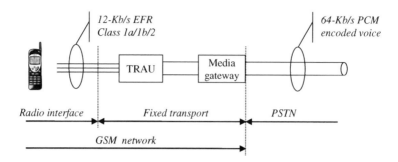

Figure 6.21 Voice transport in 2G/GSM networks.

algorithms allow dividing the voice rate by a factor of four (and even more for low bit rate coders) whilst preserving the specific characteristics of sounds generated by the vocal tract (the throat and mouth). As in GSM, this voice-coding operation is handled by the TRAU (Transcoding and Rate Adaptor Unit). In the example of Figure 6.21, voice is encoded into a 12-Kb/s EFR (Enhanced Full Rate) voice stream.

As a result, voice transmission over the GSM radio interface benefits from quite a high degree of optimization, which can be summarized by the four following items:

- UEP (Unequal Error Protection): As a result of speech coding algorithms, the TCU produces blocks of data or speech frames containing three classes of bits. Class 1a, as the most critical bits, are protected by both cyclic and convolutional codes, whereas class 1b bits are only protected by the convolutional code. Class 2 bits, being less critical, are sent unprotected over the radio interface. This mechanism, known as UEP (Unequal Error Protection), allows optimizing the amount of protection and redundancy to be added to user data sent over the radio interface, for the sake of efficient resource usage.
- Very limited overhead: In GSM speech handling, the access network is fully aware of the voice channel characteristics in terms of codec and voice frame structure. For that reason, the speech frame overhead over the radio interface is reduced to the minimum and limited to a few bits.
- High-quality low bit rate voice-coding algorithms: Thanks to evolved signal-processing algorithms, the GSM radio interface can transmit voice with almost no impact on voice quality as compared to PSTN, using only 25% of the bandwidth required for a PSTN voice circuit.
- Discontinuous transmission: Discontinuous transmission (also known as DTX) has been introduced in 2G/GSM voice processing, considering the fact that in a normal conversation, participants are actually not talking at the same time. When silent periods are detected by the terminal, SID (Silence Descriptor) frames are sent instead of full frames; these SID frames are used by the other terminal to generate local background comfort noise. As silent terminals are able to reduce the amount of information they transmit, DTX helps to increase battery life. In addition DTX also reduces the average interference level over the air interface, leading to better spectrum efficiency, although full-duplex radio resources are allocated for the whole duration of the call in circuit voice systems.

(ii) From Circuit to Packet Voice

As described above, circuit speech processing over the radio used to be a very optimized process. As a comparison, VoIP was very often perceived as a technology not suitable for radio resource-constrained public cellular networks, mainly because VoIP speech generates lots of protocol overhead. The encapsulation of speech frames into the full set of RTP/UDP/IP protocols is a source of a significant amount of signalling in addition to user data.

In addition, VoIP speech does not take advantage of UEP (Unequal Error Protection). The codec information is negotiated end-to-end, so that intermediate nodes (like a radio base station) are not aware of the details of the frame format. When the IP-encapsulated speech packet to be transmitted to the terminal arrives, the base station has no indication about the type and format of the information to be transmitted.

However, thanks to the introduction of HSDPA and other high-speed transmission schemes into wireless access, some new techniques have been studied and evaluated so as to reduce the gap between circuit and IP packet voice:

- PS voice handling can benefit from high-speed channels and availability of fast HARQ mechanisms.
- Efficient and error-resilient header compression schemes like ROHC dramatically reduce the VoIP overhead contribution.

Figure 6.22 describes how voice packets are processed over an EPS network interconnected with the legacy PSTN. The Media Gateway is in charge of voice transcoding, between the IP world and the PCM transport. Putting aside ROHC compression (which reduces the RTP/UDP/IP overhead), the speech packets are unmodified, from the Media Gateway to the terminal. The HARQ process is implemented only on the last segment of the communication, over the radio interface between the terminal and the eNodeB.

CS allows guaranteed delay and bit rate offered to applications for full-duplex communications, which is a good point for real-time conversational services. However, voice does not require strict CS bearer, at least for the two following reasons:

- Communication is rarely real full-duplex mode, as only one participant is speaking at a time in most cases.
- The actual voice rate is not constant but implies some low bit rate silent periods.

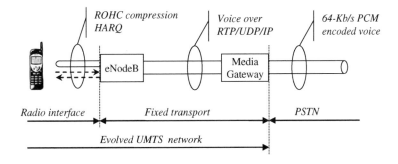

Figure 6.22 Voice transport in Evolved UMTS network.

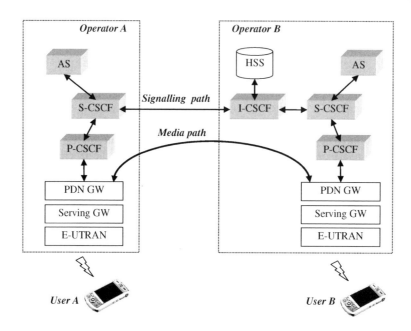

Figure 6.23 The architecture of a mobile-to-mobile IMS-based multimedia service.

Finally, the user bit rate of a voice service is far from being constant, so that an All-IP transport can obviously be a suitable solution to voice, even on the radio interface.

6.5.2 Service Architecture

Figure 6.23 describes the architecture of a mobile-to-mobile multimedia service based on IMS, in the case of a subscriber from an Operator A calling a subscriber from an Operator B.

In the control path, the SIP signalling used to set up, possibly modify and tear down, the session goes through the various IMS SIP servers already introduced in Chapter 2. The P-CSCF is the first SIP proxy server seen by each of the subscribers and the S-CSCF is actually the SIP server that each terminal is subscribed to. The role of the I-CSCF is to interrogate the HSS database in order to retrieve the actual S-CSCF that the called party B is currently registered to.

In addition to session setup signalling support, the CSCF nodes as well as the multimedia Telephony AS (Application Server) are also involved in the support of associated supplementary services, which are described in a subsequent session.

Aside from the control path, the user data packets may be routed through a complete separated media path, as in Figure 6.23.

On the EPS side, both signalling and user data are transported in EPC via PDN and Serving GW, as well as the E-UTRAN Access Network.

Figure 6.24 describes the protocol stack supported by an IMS multimedia terminal, for both the User and the Control planes. As part of the Control plane, IMS-based multimedia sessions rely on SIP and SDP for end-to-end session setup, media parameter negotiation and configuration. On the User plane, user data (being audio, video or text) are transported by a RTP/UDP/IP stack, as for the PoC service. The RTCP (RTP Control Protocol) is used to

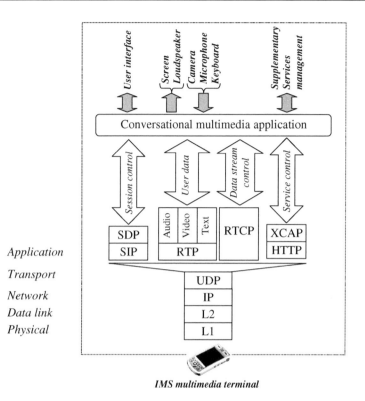

Figure 6.24 The IMS Multimedia Telephony protocol stack.

report statistic and feedback to the peer entity. In addition, RTCP is also used for end-to-end speech Codec control.

Multimedia telephony services may be associated to a set of supplementary services, providing support for features such as multi-user conference or call forwarding to a voice mailbox if no answer. Depending on the service, the support of a specific telephony Application Server may be required, so as, for example, to store and enforce end-user preferences for call forwarding. In addition, the subscriber has the possibility of activating, de-activating or interrogating the status, depending on what each individual service allows. In the IMS-based architecture, this is provided by a specific XCAP (XML Configuration Access Protocol).

The XCAP protocol is based on the flexible XML (eXtensible Markup Language) to describe information exchanged between the client (or subscriber terminal) and the server on the network side. XCAP uses the well known HTTP methods PUT, GET and DELETE to access and modify the information stored in the server.

6.5.3 About Information Coding

(i) Speech Coding
The speech-encoding algorithms of Evolved UMTS are based on those defined for 3G/UMTS and listed in Table 6.4. The list on the left contains all modes defined for 3G/UMTS speech

Table 6.4 AMR and AMR-WB source codec bit rate (from 3GPP 26.071 and 26.171).

Codec mode	Source codec bit rate	Codec mode	Source codec bit-rate
AMR_12.20	12.20 kbit/s (GSM EFR)	AMR-WB_23.85	23.85 kbit/s
AMR_10.20	10.20 kbit/s	AMR-WB_23.05	23.05 kbit/s
AMR_7.95	7.95 kbit/s	AMR-WB_19.85	19.85 kbit/s
AMR_7.40	7.40 kbit/s (IS-641)	AMR-WB_18.25	18.25 kbit/s
AMR_6.70	6.70 kbit/s (PDC-EFR)	AMR-WB_15.85	15.85 kbit/s
AMR_5.90	5.90 kbit/s	AMR-WB_14.25	14.25 kbit/s
AMR_5.15	5.15 kbit/s	AMR-WB_12.65	12.65 kbit/s
AMR_4.75	4.75 kbit/s	AMR-WB_8.85	8.85 kbit/s
AMR_SID	1.80 kbit/s	AMR-WB_6.60	6.60 kbit/s
		AMR-WB_SID	1.75 kbit/s

services. It is interesting to note that the list not only contains the legacy 2G/GSM EFR (Enhanced Full Rate) codec, but also some originated from other 2G technologies, such as the IS-641 coder from the TDMA IS-136 standard, as well as the EFR codec from the PDC Japanese system.

The list on the right contains all the WB (Wide Band) modes. WB coders were introduced into the wireless standard in order to improve the voice quality of the traditional 8-KHz sampling process. The WB coders rely on a 16-KHz up-sampling process, so that the speech bandwidth is increased from the typical land-line 3400 Hz to 7000 Hz. As a result, the voice sounds more natural, with quality being a bit closer to what face-to-face conversation would provide.

The WB modes do not apply to PSTN communications because of the limitations of the 8-KHz PCM coding still largely in use for land-line networks. However, for terminal-to-terminal voice communications and provided the terminals engaged in the session support these modes, AMR-WB can contribute to significantly improve the voice quality, at the cost of an increased transmission bit rate.

The SID (Silence Descriptor) modes correspond to SID frames being sent when no voice activity is detected. Those frames are used by the receiving terminal to generate background comfort noise so as to avoid the feeling of communication interruption on the part of the listener. The bit rate is calculated assuming that SID frames are continuously transmitted.

For any voice call, a fixed codec among the lists above can be chosen for the whole duration of the call. This leaves the possibility for the operator to propose service grades in which premium customers would be served by a higher source rate codec and benefit from higher quality, or to optimize system capacity by, for example, allocating lower source rate as the network load increases.

Another possibility is to allocate a list of possible codecs for the speech call and benefit from the codec adaptation mechanisms, known as AMR (Adaptive Multi-Rate). The main interest of AMR is to allow a trade-off between speech quality and channel protection as the radio transmission bit error rate varies. When the transmission quality is good enough, characterized by high values of C/I (Carrier to Interference ratio), high source codec bit rate, like AMR_12.20, is preferable in order to privilege voice quality.

The Services 315

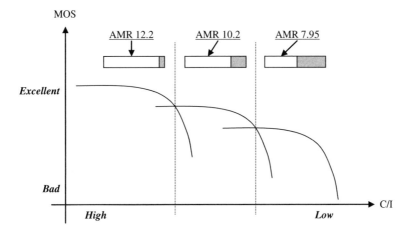

Figure 6.25 AMR codec adaptation.

In contrast, when the radio transmission quality degrades, received frames contain more errors, leading to degraded voice quality. This can be compensated for by AMR, through the use of a lower bit rate source codec, like AMR_10.2, and an increased amount of channel protection, aimed at correcting the transmission errors.

This adaptation is illustrated in Figure 6.25, showing the best performing codec, as regards C/I radio conditions. In this diagram, voice quality is expressed in terms of MOS (Mean Opinion Score). MOS is calculated by averaging the results of subjective tests in which a group of listeners are asked to rate the audio quality of test sentences. The MOS rating varies from 1 (bad) to 5 (excellent).

Each of the curves shows how the corresponding codec performs accordingly to the radio quality. The radio block size is represented by a rectangle, the greyed part being the part dedicated to the channel protection.

(ii) Video Coding
IP multimedia video coding makes use of the following list of algorithms:

- H263 – this was a mandatory video-coding component for H.324 services and actually the only mandatory one for IMS.
- MPEG4.
- H264.

H.263 video coding, initially released in 1996 by ITU, was targeted to low bit rate communications. Numerous additional coding tools have been brought to this initial version in 1998 and 2000 so as to enhance picture fidelity, coding efficiency, error resilience, reduce encoder and decoder complexity, etc.

In order to limit option combinations and bound implementation complexity, several profiles have been created, each of them corresponding to a set of coding options to be implemented. In the scope of 3GPP networks, only two of them are expected to be supported by terminals:

Table 6.5 Some H.263 modes of operation.

Format	Picture size	Picture rate (Hz)	Max bit rate (Kb/s)
QCIF	176 × 144	15	64
QCIF	176 × 144	30	128
CIF	352 × 288	15	128
CIF	352 × 288	30	384

- **Profile 0**, which is the mandatory baseline profile and does not contain any optional modes of operation.
- **Profile 3**, which is the optional profile suited to interactive and streaming wireless transmission. This profile has been defined in order to provide enhanced coding efficiency performance and enhanced error resilience for wireless transmission.

The basic H.263 picture format is known as CIF (Common Intermediate Format), which has 288 lines of 352 pixels. All the other four formats are built using CIF dimensions, like QCIF (Quarter CIF), whose size is one-quarter of CIF, and 4CIF, which is four times the CIF size.

H.263 video coding was initially defined for up to 64-Kb/s transmissions. However, depending on the picture size and rate, it may happen that H.263 delivers a much higher bit rate. This is illustrated by Table 6.5.

H.264 video coding, initially released in 2003, is also known as AVC (Advanced Video Coding). It is to be considered as an evolution of the previous H.263 generation. The aim of H.264 was to be able to cope with the growing need for higher compression techniques for the benefit of generic audiovisual applications such as video telephony, video conferencing, television broadcasting and Internet streaming.

Like H.263, H264 proposes a set of profiles corresponding to different levels of complexity and service grade. For 3GPP IMS multimedia telephony, it is expected that only the baseline profile is supported by terminals.

(iii) Real Time Text Coding

Text transmission is based on ITU T.140, which specifies the 'Protocol for Multimedia Application Text Conversation'.

All the transmitted characters shall comply with ISO 10646-1, which is one of the standards for multiple octet character sets. The objective of this standard is to support all the characters from the widest range of alphabets. In the T.140 scope, all the characters sent shall comply with the two octet version of the ISO 10646-1 standard.

Reference documents about information coding

3GPP technical specifications:

- 26.071, 'AMR Speech Codec: General Description'
- 26.171, 'Adaptive Multi-Rate–Wideband (AMR-WB) Speech Codec: General Description'
- 26.114, 'IMS Multimedia Telephony: Media Handling and Interaction'

6.5.4 About Supplementary Services

The concept of supplementary services was initially related to legacy circuit-switched-based infrastructure. It refers to services which are not provided by default to subscriber, but can be subscribed to in addition to other services. The supplementary services were introduced first in the wireline telephone networks, and were also part of wireless 2G/GSM network standards from the beginning.

The supplementary services encompass a very large set of services and functions, some of them very well known and popular, like 'call forwarding to a voice messaging system', or 'calling line identity presentation'. As a possible source of revenue for the operators, supplementary services are, in most cases, subject to registration and de-registration, as well as activation and de-activation from the end-user.

As an illustration, Table 6.6 presents some of the supplementary services available for 2G/GSM and 3G/UMTS circuit-based services.

As a basic principle, supplementary services have been defined and implemented, regardless of the access technology, using a specific set of messaging protocols to allow all possible actions for the end-user. However, when moving to IMS networks based on a completely different signalling and infrastructure, the key question that operators and subscribers would ask is

Table 6.6 Some CS-based supplementary services.

CFU	**Call Forwarding Unconditional**
	Allows a called mobile subscriber to have the network send all incoming calls to another directory number, such as another fixed or wireless terminal, or a voice-messaging service
CFNRy	**Call Forwarding on No Reply**
	Same as CFU, except that the call forwarding is only performed if the called subscriber does not answer the incoming call
HOLD	**Call Hold**
	Allows a mobile subscriber to interrupt communication on an existing active call and then, subsequently, if desired, re-establish communication
CW	**Call Waiting**
	Allows a mobile subscriber to be notified of an incoming call whilst already engaged in another active or held call. Subsequently, the subscriber can either accept, reject or ignore the incoming call
AoCI	**Advice of Charge – Information**
	Allows providing the terminal with the information to produce an estimate of the cost of the service used
BAOC	**Barring of All Outgoing Calls**
	Allows the subscriber to have barring of all outgoing calls, except emergency calls
CLIP	**Calling Line Identification Presentation**
	Provides the called subscriber with the possibility of receiving the line identity – or telephone number – of the calling party
CLIR	**Calling Line Identification Restriction**
	Allows the calling party to prevent presentation of the line identity to the called party

'How is it possible to preserve and implement those services so as to maintain customer service grade and operator revenues?'

To answer this question, a new set of SIP-based services has been defined, which aims at providing the same support as the supplementary services. All this activity was handled by the TISPAN technical body (Telecommunication and Internet Converged Services and Protocols for Advanced Networking), which is part of the ETSI; all the results have been taken as a basis for IMS-based multimedia services. However, as there may be some differences in the way some features are supported, the TISPAN services are more often referred to as 'PSTN simulation services'.

The services defined by TISPAN to emulate the PSTN supplementary services have been grouped into several categories, which more or less cover all of them:

- **OIP/OIR** (Originating Identity Presentation/Restriction)– see below.
- **TIP/TIR** (Terminating Identity Presentation/Restriction)– see below.
- **CDIV** (Communication Diversion)– this category contains equivalents to all call-forwarding supplementary services like CFNRy (Call Forwarding on No Reply) and CFB (Call Forwarding on Busy user).
- **HOLD** (Communication HOLD)– see below.
- **CB** (Call Barring)– this category contains equivalents to all call-barring PSTN services like BAOC.
- **MWI** (Message Waiting Indication)– this service enables the network to indicate to a subscriber that there is at least one message waiting in its mailbox.
- **CONF** (Conference)– this service enables a subscriber to set up a session involving several users simultaneously.
- **ECT** (Explicit Call Transfer)– this service enables a subscriber involved in a communication to transfer that communication to a third party, as a opposed to CDIV services, which only apply to the service setup phase.

In general, IMS emulation of supplementary services makes use of SIP protocol extensions, SDP protocol (when the media bearers are involved in the service) as well as support when required from the telephony Application Server (which maintains, for example, the subscriber busy of barring status) or CSCF IMS control nodes.

In the rest of the section, some of these categories are described in more detail, to illustrate how IMS-based emulation of supplementary services is performed.

(i) OIP/OIR (Originating Identity Presentation/Restriction); TIP/TIR (Terminating Identity Presentation/Restriction)
OIP and OIR services deal with possibility of presenting or restricting the identity of the originator to the called party. Similarly, TIP and TIR services allow displaying or restricting called party identity to the calling party. Those four services are the IMS equivalents of CLIP, CLIR, COLP and COLR circuit-switched supplementary services.

To allow subscriber identity presentation or restriction, the service makes use of SIP extension header fields to convey useful information and requires specific actions from terminals and S-CSCF nodes to add useful information when needed. These extensions, briefly presented in Chapter 4, are defined as part of SIP IMS specifications:

- The **From** header field.

- The **P-Preferred-Identity** header field.
- The **P-Asserted-Identity** header field – which represents the IMS subscriber identity once it has been authenticated.
- The **Privacy** header field.

(ii) HOLD (Communication HOLD)

The HOLD supplementary service enables a user to suspend an established session and resume it at a later time. This feature is typically used when a subscriber receives an incoming call while being engaged in an active session. If nothing specific is done, the incoming call is eventually forwarded to the voice mailbox as the called party does not answer (and provided the corresponding Call Forwarding Service was configured accordingly).

Otherwise, if the HOLD service is active, the subscriber may put the on-going active session on hold, accept the incoming call, and resume the held session when the second one is over.

In the IMS world, and as illustrated in Figure 6.26, this service is provided through the SDP protocol whose role (in the scope of the HOLD service) is to suspend or resume the media streams of the session.

In the case of a full-duplex voice session, when a user presses the 'hold' button, its terminal will update the media stream attribute from 'sendrecv' (which is the normal value for an active conversational service) to 'sendonly' (meaning that the held terminal shall not transmit anymore on its media bearers) using a SIP INVITE message. In addition, the HOLD requesting terminal will also locally mute, so that no media are sent and played out at the held terminal side. This INVITE message is acknowledged by the peer terminal using a SIP 200 OK containing the new media attribute, as seen by terminal B.

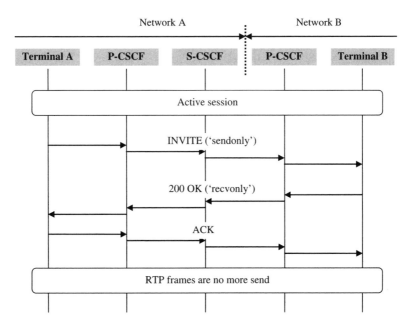

Figure 6.26 An example of a HOLD service.

For the HOLD service, all the signalling is transmitted through P-CSCF and S-CSCF IMS nodes, as any other SIP messages that would be exchanged between the two terminals. Further on, when the initial session is resumed, the media stream attribute is changed back to 'sendrecv'.

This behavior is not specific to IMS multimedia communication, as it was initially described in the document RFC3264, 'An Offer/Answer Model with the Session Description Protocol (SDP)', as the normal procedure to put a unicast media stream on hold.

Reference documents about supplementary services

3GPP technical specifications:

- 22.004, 'General on Supplementary Services'
- 24.173, 'IMS Multimedia Telephony Service and Supplementary Services'

ETSI TISPAN technical specifications:

- 181002, 'Multimedia Telephony with PSTN/ISDN Simulation Services'
- 183004, 'Communication Diversion (CDIV)'
- 183005, 'Conference (CONF)'
- 183006, 'Message Waiting Indication (MWI)'
- 183007, 'Originating Identification Presentation (OIP) and Originating Identification Restriction (OIR)'
- 183008, 'Terminating Identification Presentation (TIP) and Terminating Identification Restriction (TIR)'
- 183010, 'Communication HOLD (HOLD)'
- 183011, 'Anonymous Communication Rejection (ACR) and Communication Barring (CB)'
- 183029, 'Explicit Communication Transfer (ECT)'

6.5.5 Multimedia Services in EPS Systems

The major issues for supporting a conversational real-time service over a packet network are jitter (or transmission delay variation) and packet loss. Packet-based radio systems like E-UTRAN will experience the same radio conditions as radio systems based on dedicated resources (like 2G/GSM or 3G/UTRAN). However, shared radio systems like EPS are more sensitive to traffic load, so that scheduling algorithms have to be designed in a very specific way, so as to ensure that latency and jitter values do no compromise the end-user's quality of experience.

The jitter is not a problem by itself, as long as it remains within an acceptable range. As in most of the existing IP-based real-time applications, it can easily be compensated for by using a buffer at the receiving terminal, so that the packet timing variation can be regenerated. This is exactly what real-time transport protocols like RTP, presented in Chapter 4, can propose. Each RTP data packet contains a Timestamp which aims to reconstruct the exact data time sequence. However, jitter compensation for a conversational service shall not compromise the overall transmission delay.

In addition, packet data schedulers have to cope with additional points which may affect real-time resource allocation:

- **Sender Reports and Receiver Reports (SR and RR)**– these messages are part of the RTCP layer, sent from time to time. These reports are sent periodically and the induced traffic load should not exceed 5% of the total session bandwidth, as requested by the IETF RTP document.
- **ROHC Full packet header transmission**– At some time, or when a transmission error occurs, ROHC compression and decompression machines need to be resynchronized, as the decompressor may not be able to successfully retrieve compressed information. Resynchronization is achieved by the compressor by sending a full uncompressed header. This is unpredictable, as it depends on radio transmission quality as well as ROHC implementation.

Glossary

This includes definitions of all acronyms and technical terms used.

16 QAM (16 Quadrature Amplitude Modulation). 16 QAM is a 16-state phase and amplitude modulation scheme used in HSDPA radio interface.

3GPP (3rd Generation Partnership Project). 3GPP is a cross-country organization initially in charge of the definition of the 3G UMTS standard.

3GPP2 (3rd Generation Partnership Project 2). 3GPP2 is a cross-country organisation in charge of the definition of the 3G cdma2000 standard and its evolutions.

8PSK (8 Phase Shift Keying). 8PSK is an eight-state phase-modulation scheme used in EDGE radio interface.

AAA (Authentication, Authorization and Accounting). AAA is a generic model for IP network access control, initiated and developed by IETF.

AES (Advanced Encryption Standard). AES is a robust symmetric encryption algorithm.

AF (Application Function) AF is the generic name for the network element which supports applications that require dynamic policy and/or charging control. For 3GPP IP access types, the AF is supported by the P-CSCF.

AKA (Authentication and Key Agreement). The AKA process is a challenge–response-based mechanism which aims at mutual network/terminal authentication and security key distribution.

AMR (Adaptive Multi-Rate). The AMR speech coder is a single integrated speech codec with multiple source rates, allowing a trade-off between speech quality and channel protection as the bit error rate varies.

ARQ (Automatic Repeat Request). ARQ is a method for repetition of erroneous or lost packets.

AS (Access Stratum). The AS is the set of features associated to the radio interface and Access Network.

AuC (Authentication Centre). The AuC is one of the HSS components.

AUTN (Authentication Token). AUTN is the parameter used by the USIM to authenticate the network.

AUTN. is part of the authentication vector (AV).

AV (Authentication Vector). The AV is a set of five elements (RAND, AUTN, CK, IK, XRES) which enables the EPC to run the AKA process.

BCH (Broadcast Channel). The BCH is a downlink E-UTRAN transport channel associated to the BCCH.

BCCH (Broadcast Control Channel). The BCCH is a downlink E-UTRAN logical channel supporting broadcast of system information.

BLAST (Bell Lucent Layered Space Time coding). BLAST is a MIMO technology consisting in placing different symbols in different respective antennas.

BLER (Block Error Rate).

CCCH (Common Control Channel). The CCCH is an uplink E-UTRAN logical channel for signalling transmission until a RRC connection is available.

CDF (Charging Data Function). The CDF is in charge of building CDR which content and format are specified by 3GPP standard documents.

CDMA (Code Division Multiple Access). CDMA is a medium-access method in which different users are allocated different code sequences.

CDR (Charging Data Record). A CDR is a collection of information (connection time, allocated resources, etc.) for use in billing and accounting.

CGF (Charging Gateway Function). The CGF is a gateway between the Core Network nodes and the Billing Domain. It supports CDR storage and management (opening, closing).

CK (Ciphering Key). CK is the secret key used to ensure transmission confidentiality between the terminal and the eNodeB. CK is part of the authentication vector (AV).

CQI (Channel Quality Indicator).

CRF (Charging Rules Function). The CRF is a PS Core Network element introduced in UMTS for IMS flow-based charging.

CP (Cyclic Prefix).

CS (Circuit Switched domain). The CS domain is the subset of the 2G/GSM and 3G/UMTS Core Network domain dedicated to the support of packet-based services.

CSCF (Call Session Control Function). The CSCF is a SIP-based server in the IMS architecture for establishing, terminating or modifying IMS sessions.

DAB (Digital Audio Broadcast). Digital radio broadcast based on OFDM technology.

DCCH (Dedicated Control Channel). DCCH is a bi-directional E-UTRAN logical channel for the transmission of control information.

DFT (Discrete Fourier Transform). DFT is a version of the continuous Fourier Transform using digital samples.

DFT-SOFDM (Discrete Fourier Transform Spread Orthogonal Frequency Division Multiplex). DFT-SOFDM is a method of modulation allowing the decrease of transmitted signal PAPR.

DL (Downlink). Communication link from Base Station to mobile.

DL-SCH (Downlink Shared Channel). The DL-SCH is a downlink E-UTRAN transport channel for the transmission of user and control data.

DTCH (Dedicated Traffic Channel). DTCH is a bi-directional E-UTRAN logical channel for the transmission of traffic information.

DVB (Digital Video Broadcast). DVB is a digital television broadcast system based on OFDM technology.

E-UTRAN (Evolved-UTRAN). E-UTRAN is the Access Network evolution for 3GPP 3G standards.

E-DCH (Enhanced Dedicated Channel). E-DCH is the enhanced channel defined in HSUPA 3G standards.

EAP (Extensible Authentication Protocol). EAP is an authentication framework which supports multiple authentication methods. It is defined by the RFC3748 IETF document.

EDGE (Enhanced Data rates for GSM Evolution). EDGE is a high bit rate evolution of 2G/GSM networks.

EMM (EPS Mobility Management). EMM refers to the set of signalling procedures between the terminal and the MME used for user attachment and location management.

eNodeB (evolved NodeB). eNodeB is the radio transmission and receiver node of E-UTRAN Access Network.

EPC (Evolved Packet Core). EPC is the Packet Core Network evolution for 3GPP 2G/3G standards.

EPS (Evolved Packet System). EPS is the 3GPP name for the evolution of UMTS. EPS encompasses both E-UTRAN and EPC.

FDD (Frequency Division Duplex). FDD is a duplex transmission method based on frequency separation.

FDMA (Frequency Division Multiple Access). FDMA is a medium-access method in which different users are allocated different frequencies.

FFT (Fast Fourier Transform). FFT is a mathematical transformation on which OFDM is based.

FLUTE (File Delivery over Unidirectional Transport). FLUTE is a protocol for the unidirectional delivery of files over the Internet, suited to multicast networks. In particular, FLUTE is used in the 3GPP MBMS service.

FOMA (Freedom of Mobile Multimedia Access). FOMA is the commercial name for the 3G/UMTS network initially deployed in Japan by DoCoMo.

GERAN (GPRS EDGE Radio Access Network). GERAN is the 3GPP name for the GSM access network. It refers to as the initial 2G/GSM voice system, as well as GPRS and EGDE evolutions.

GGSN (Gateway GPRS Support Node). The GGSN is a 2G/GSM and 3G/UMTS Core Network entity supporting packet data transmission to and from mobile terminals.

GMSC (Gateway Mobile Switching Center). The MSC is a specific type of MSC, used as a gateway with the PSTN.

GPRS (General Packet Radio Service). GPRS is an evolution of 2G GSM networks for packet data service support.

GSM (Global System for Mobile communications). GSM is the well known 2G cellular telephony system.

GTP (GPRS Tunnelling Protocol). GTP is a protocol used in EPC for establishing data sessions; it also provides data encapsulation between network nodes. GTP has been inherited from 2G/GPRS standard.

HARQ (Hybrid Automatic Repeat Request). HARQ is a method of repeating at the emission and combining at the reception the noncorrectly received packets.

HCR-TDD (High Chip Rate Time Division Duplex). HCR-TDD is the 3GPP UMTS TDD version in which the spreading is realized in 3.84 MHz.

HLR (Home Location Register). The HLR is one of the HSS components.

HSDPA (High Speed Downlink Packet Access). HSDPA is a high-speed enhancement of 3G/UMTS networks for network-to-terminal transmission.

HS-DSCH (High Speed Downlink Shared Channel). HS-DSCH is the high-speed transport channel used in 3G HSDPA.

HSPA (High Speed Packet Access). HSPA designates the combination of HSDPA and HSUPA.

HSS (Home Subscriber Server). The HSS is the master database of 3G networks containing the user subscription-related information.

HSUPA (High Speed Uplink Packet Access). HSUPA is a high-speed enhancement of 3G/UMTS networks for terminal-to-network transmission.

IDMA (Interleaved Division Multiple Access). IDMA is a method of reducing the intercell interference.

IETF (Internet Engineering Task Force). IETF is an international community dedicated to the evolution of Internet and Internet standards.

IEEE (Institute of Electrical and Electronics Engineers). IEEE is a professional association for the advancement of technology. It covers a wide range of technical areas, including wireless telecommunications.

IFFT (Inverse Fast Fourier Transform). IFFT is the reciprocal transform of the FFT.

IK (Integrity Key). IK is the secret key used to ensure signalling transmission integrity between the terminal and the eNodeB. IK is part of the authentication vector (AV).

IOTA (Isotropic Orthogonal Transform Algorithm). IOTA is a pulse Shaping Filter.

ISIM (IMS Subscriber Identity Module). See SIM.

ITU (International Telecommunication Union). ITU is an international organization in charge of global telecom networks and services coordination.

IMS (IP Multimedia Subsystem). IMS is a 3GPP framework, designed for delivering IP multimedia services to end-users.

IMSI (International Mobile Subscriber Identity). The IMSI is a private unique mobile subscriber identifier used in 3GPP networks including GSM, UMTS and EPS technologies.

IMT-2000 (International Mobile Telecommunications 2000). IMT-2000 is the ITU name for post-2G network evolutions. 3GPP/UMTS and 3GPP2/cdma2000 are part of the IMT-2000 standard family.

IR (Incremental Redundancy). IR is a packet repetition method used in HARQ.

ISI (Inter-symbol Interference). ISI is the perturbation brought by late-arrived radio symbols on previous ones in dispersive channels.

LCR-TDD (Low Chip Rate-Time Division Duplex). LCR-TDD is the 3GPP UMTS TDD version in which the spreading is realized in 1.28 MHz.

LTE (Long Term Evolution). LTE is a 3GPP work item aiming at the definition of 3G Access Network evolution, also known as E-UTRAN.

MAC (Medium Access Control). MAC is the E-UTRAN radio interface Layer 2 protocol in charge of priority handling and data scheduling over the radio interface.

MAP (Mobile Application Part). MAP is the general-purpose protocol used between Core Network nodes for communication, subscriber data and mobility management.

MBMS (Multimedia Broadcast and Multicast Service). MBMS is a broadcast/multicast service defined by the 3GPP.

MC-CDMA (Multi-Carrier Code Division Multiple Access). MC-CDMA is an OFDM modulation system operating on a code division multiple access system.

MCC (Mobile Country Code). The MCC is part of the IMSI and uniquely identifies the home country of the subscriber.

MCCH (Multicast Control Channel). The MCCH is a downlink E-UTRAN logical channel supporting control information associated to MBMS.

MCH (Multicast Channel). The MCH is a downlink E-UTRAN transport channel supporting the MCCH and MTCH logical channels.

MCS (Modulation and Code Scheme). MCS represents a coding and modulation scheme applied to a physical channel.

MIB (Master Information Block). The MIB contains key System Information parameters broadcast by E-UTRAN on the BCCH.

MIMO (Multi Input Multi Output). MIMO is a generic term for multi-port systems. It is usually applicable to antenna systems for E-UTRAN.

MTCH (Multicast Traffic Channel). The MTCH is a downlink E-UTRAN logical channel for the transmission of MBMS-related data.

MGCF (Media Gateway Control Function). The MGCF is an IMS node in charge of signalling inter-working with PSTN for voice circuit-switched services.

MGW (Media Gateway). The MGW is responsible for media conversion to and from the PSTN, under the control of the MGCF.

MME (Mobility Management Entity). MME is an entity part of the EPC network, in charge of session and user-mobility management.

MMD (MultiMedia Domain). MMD is the cdma2000 equivalent of the IMS subsystem of UMTS networks.

MMSE (Minimum Mean Square Error). MMSE is an estimation criterion for parameter estimation. Example of use: radio channel identification

MNC (Mobile Network Code). The MNC is part of the IMSI and uniquely identifies the home operator of the subscriber in its country.

MSC (Mobile Switching Center). The MSC belongs to the Circuit Core Network and is responsible for circuit-switched call control.

MSISDN (Mobile Subscriber ISDN Number). The MSISDN is the telephone number for mobile terminals.

MT (Mobile Terminal). MT is the part of the terminal which supports the radio-related protocols and components, as well as the UICC module.

NAI (Network Address Identifier). The NAI is identity, built using the IMSI and used by the subscriber when accessing a 3GPP network through a WLAN.

NAS (Non Access Stratum). The NAS is the set of access-independent network features related to the management subscriber data and communication contexts. In EPC network, the NAS signalling terminates in the MME.

NGN (Next Generation Network). NGN refers to the new packet-based converged network architecture.

NodeB. NodeB is the radio transmission and receiver node of the 3G/UTRAN access network.

OFDM (Orthogonal Frequency Division Multiplexing). OFDM is the radio technology family chosen for the E-UTRAN physical layer. It is based on the mapping of a wide-band signal to be transmitted onto large numbers of narrow-band modulated subcarriers.

OFDMA (Orthogonal Frequency Division Multiple Access). OFDMA is a multi-user multiplexing technique based on OFDM subcarrier arrangement.

P-GW (PDN Gateway). The Packet Data Network Gateway is part of the EPC. It is the functional network entity which terminates the SGi interface towards the PDN.

PARC (Per Antenna Rate Control). PARC is a MIMO technology consisting of encoding the same symbols with different MCS on different antennas.

PAPR (Peak on Average Power Ratio). PAPR is a parameter which dimensions the linearity specification of a power amplifier.

PBCH (Physical Broadcast Channel). PBCH is a downlink physical channel which carries system information.

PCH (Paging Channel). The PCH is a downlink E-UTRAN transport channel associated to the PCCH.

PCCH (Paging Control Channel). The PCCH is a downlink E-UTRAN logical channel supporting the transmission of paging information.

PDCCH (Physical Downlink Control Channel). PDCCH is a downlink physical channel which carries scheduling assignment.

PDSCH (Physical Downlink Shared Channel). PDSCH is a downlink physical channel which carries user data and higher-layer signaling.

PMCH (Physical Multicast Channel). PMCH is a downlink physical channel which carries multicast/broadcast information.

PUCCH (Physical Uplink Control Channel). PUCCH is an uplink physical channel which carries uplink control information.

PUSCH (Physical Uplink Shared Channel). PUSCH is an uplink physical channel which carries user data and higher-layer signaling.

PCEF (Policy and Charging Enforcement Function). PCEF is the generic name for the functional entity which supports service data flow detection, policy enforcement and flow-based charging. For 3GPP IP access types, the PCEF is supported by the GGSN.

PCRF (Policy and Charging Rules Function). The PCRF is actually a concatenation of PDF and CRF network nodes.

PDCP (Packet Data Convergence Protocol). PDCP is the radio interface protocol layer in charge of header compression and security protection.

PDF (Policy Decision Function). The PDF is a PS Core Network element introduced in UMTS for IMS flow policy control.

PDP Context (Packet Data Protocol Context). PDP is the name for the 2G/GPRS or 3G/UMTS packet session context which is set up between the terminal and the Packet Core Network. 'EPS bearer' is the equivalent of the PDP context in EPS networks.

PHY (Physical Layer). PHY is the E-UTRAN physical layer in charge of channel coding and modulation.

PLMN (Public Land Mobile Network). A PLMN is a public mobile network owned by an operator. It is uniquely identified by a concatenation of MNC and MCC codes.

PoC (Push-to-talk over Cellular). PoC is a walkie-talkie-like speech service defined by the OMA.

PS (Packet Switched domain). The PS domain is the subset of the 2G/GSM and 3G/UMTS Core Network domain dedicated to the support of packet-based services.

PSTN (Public Switched Telephone Network). PSTN is the public worldwide circuit-switched-based telephone network.

PSK (Phase Shift Keying). PSK is the generic name of a modulation constellation whose (n) points are separated in phase. If $n = 2$, then BPSK terminology is used. If $n = 4$, then QPSK terminology is used.

PRB (Physical Resource Block).

QAM (Quadrature Amplitude Modulation). QAM is the generic name for a modulation constellation whose (n) points are separated in amplitude and phase.

RACH (Random Access Channel). The RACH is an uplink E-UTRAN transport channel for the transmission of initial access requests from terminals.

RAND (Random). RAND is the random challenge parameter used during the AKA procedure. RAND is part of the authentication vector (AV).

RB (Resource Block).

RLC (Radio Link Control). RLC is the E-UTRAN radio interface Layer 2 protocol in charge of data segmentation and error correction.

RNC (Radio Network Control). The RNC belongs to the UTRAN and is in charge of radio protocols and the control of UTRAN Base Stations.

ROHC (Robust Header Compression). ROHC is a compression header framework suitable for IP-based applications.

RRC (Radio Resource Control). The RRC layer is the E-UTRAN control protocol associated to the radio interface.

RTCP (Real Time Control Protocol). RTCP is an application-layer control protocol developed by the IETF, used together with RTP for the control of real-time application bearers.

RTP (Real Time Protocol). RTP is an application-layer protocol developed by the IETF, used as the transport protocol for real-time-sensitive packet-based services.

RU (Resource Units). RU are uplink radio resources. RU can be localized (LRU) or distributed (DRU).

S-GW (Serving Gateway). The Serving GW is part of the EPC. It is the functional network entity which terminates the interface towards E-UTRAN.

S-TMSI (S-Temporary Mobile Subscriber Identity). The S-TMSI is a temporary user identity used in the terminal–network signaling procedure, so as to preserve IMSI privacy.

SAE (System Architecture Evolution). SAE is a 3GPP work item aimed at the definition of 3G Packet Core network evolution, also known as EPC.

SC-FDMA (Single-Carrier FDMA). OFDM modulation technique which overcomes the PAPR problem. Consists of a DFT followed by a FFT.

SCH (Synchronization Channel).

SCTP (Stream Control Transmission Protocol). SCTP is a reliable connection-oriented transport protocol similar to the well known TCP and used over the S1 E-UTRAN interface.

SDF (Service Data Flow). An SDF is a flow of packet data identified by the IP 5-tuple (source IP address, destination IP address, source port number, destination port number, protocol ID of the protocol above IP).

SGSN (Serving GPRS Support Node). The SGSN is a 2G/GSM and 3G/ UMTS Core Network entity supporting packet data transmission to and from mobile terminals.

SIM (Subscriber Identity Module). Like the USIM (for UMTS) and ISIM (for IMS), the 2G/SIM is an application of the UICC card put in the user terminal and supporting user-related information such as identity, security credentials, phonebook, and application-specific information.

SDP (Session Description Protocol). SDP is an application-layer control protocol developed by the IETF. SDP is the standard for describing and negotiating media components used in a session.

SDMA (Spatial Division Multiple Access). SDMA is a space multiplexing technique making use of antenna processing.

SIB (System Information Block). A SIB is a set of information broadcast by the E-UTRAN on the BCCH.

SIP (Session Initiation Protocol). SIP is an application-layer control protocol developed by the IETF which is used for establishing IP-based multimedia services such as Voice over IP, or the 'Presence' service.

SM (Session Management). SM refers to the set of signalling procedures between the terminal and the MME used for bearer management.

SMS (Short Message Service). SMS is the well known text-message service introduced in 2G/GSM networks.

STTD (Space Time Transmit Diversity). STTD is a transmit diversity technique based on space–time coding. Alamouti's version is a famous one chosen for 3GPP.

SU (Scheduling Unit). A SU consists of one or more SIB and is broadcast by E-UTRAN on the BCCH using its own scheduling period.

TA (Tracking Area). The TA is a group of contiguous cells which helps to track terminal locations in IDLE mode.

TAI (Tracking Area Identity). The TAI uniquely identifies a TA within a network.

TAU (Tracking Area Update). TAU refers to as the procedure performed between the terminal and the network to update the current subscriber's TA.

TCP (Transmission Control Protocol). TCP is a reliable end-to-end connection-oriented data-transport protocol defined by the IETF.

TDMA (Time Division Multiple Access). TDMA is a medium-access method in which different users are allocated different time periods.

TAP (Transferred Account Procedure). TAP is a protocol defined by the GSM Association for the interchange of billing data between different network operators.

TBCP (Talk Burst Control Protocol). TBCP is a protocol layer specific to the PoC service, in charge of speech access arbitration between the participants.

TDD (Time Division Duplex). TDD is a duplex transmission method based on time separation.

TE (Terminal Equipment). TE is the part of the terminal which supports user applications and application-level protocols, as well as user interface.

TISPAN (Telecommunication and Internet converged Services and Protocols for Advanced Networking). TISPAN is the ETSI technical body in charge of next-generation converged network definition.

TTI (Time Transmit Interval). Time spreading of one channel-coded block.

UDP (User Datagram Protocol). UDP is a connectionless end-to-end nonreliable transport protocol defined by IETF.

UE (User Equipment). UE is the 3GPP name for a mobile terminal. A UE is composed of a TE and a ME.

UICC (Universal Integrated Circuit Card). The UICC is a general-purpose circuit card used as a physical support for SIM and USIM applications.

UL (Up-Link). Communication link from mobile to Base Station.

UL-SCH (Uplink Shared Channel). The UL-SCH is an uplink E-UTRAN transport channel for the transmission of user and control data.

UMB (Ultra Mobile Broadband). UMB is the 3GPP2 equivalent of the EPS standard.

UMTS (Universal Mobile Telecommunications System). UMTS is the 3G standard developed by the 3GPP consortium.

USIM (Universal Subscriber Identity Module). See SIM.

UTRAN (Universal Terrestrial Radio Access Network). UTRAN represents the access network (or the radio-specific part) of a 3G UMTS network.

VRB (Virtual Resource Block).

WCDMA (Wide band Code Division Multiple Access). The chosen radio technology for 3GPP UTRAN.

WiMAX (Worldwide Interoperability Microwave Access). The IEEE packet radio standard for short and long-range wireless Internet.

XRES (Expected Result). XRES is the parameter used by the EPC to authenticate the USIM. XRES is part of the authentication vector (AV).

Index

1xEV-DO, 9
3GPP, 14, 24
3GPP2, 14, 24
3GPP specifications, 25
16 QAM, 130
64 QAM, 130

Acknowledged mode, 197
Active mode, 29
ADC, 112
AES, 260
AKA, 238, 254
AMR, 314
ARQ, 132
Attach, 237
AuC, 48
Authentication, 253
AUTN, 255

BCCH, 190
BCH, 125, 148, 191, 230
BLAST, 98
BLER, 143
BM-SC, 300
Broadcast, 298
BTS, 116

CAZAC, 163
CCCH, 191
CCM, 109
CDD, 102
CDMA, 7, 82
cdma2000, 2, 10

cdmaOne, 2, 10
CDR, 53, 58, 293, 308
CEM, 109
Ciphering, 253
CK, 255
COPS, 55
CP, 124
CPRI, 111
CQI, 104, 155
CRC, 145
CRF, 54
CS, 32
CSCF, 50

DAB, 90
DAC, 114
D-BLAST, 99
DCCH, 191
Detach, 240
DFT, 86
DFT-SOFDM, 156
Diameter, 52
Diffserv, 181
DL-SCH, 191
DRX, 265
DS-CDMA, 82
DTCH, 191
DTX, 310
DVB-H, 141

EAP-AKA, 255
EDGE, 7, 32
E-MBMS, 140

Evolved Packet System (EPS) P. Lescuyer and T. Lucidarme
Copyright © 2008 John Wiley & Sons, Ltd.

eNodeB, 38, 172
EPS Bearer, 245
ESP, 252
E-UTRAN, 75

FDD, 119
FDM, 142
FEC, 132
FFT, 77
FLUTE, 302
FSTD, 150

GGSN, 32, 41, 55
GMM, 207
GMSK, 114
GPRS, 6, 32, 243, 248, 268
GSM, 2, 6, 20, 32, 230, 298
GTP, 42, 184

H. 309, 315, 316
HARQ, 8, 143, 169, 197, 311
HCR-TDD, 121
HDLC, 111
HLR, 32, 47
HPSK, 114
HSDPA, 7, 35
HSPA, 8
HSPA+, 8
HSS, 47, 53
HSUPA, 8

IBPTC, 132
I-CSCF, 50
Idle mode, 29, 265
IDMA, 153
IETF, 24
IFDMA, 88
IK, 255
IMS, 6, 18, 33, 50, 209, 260, 312
IMSI, 47, 236, 239, 256
IMT-2000, 12, 20
IMT-Advanced, 13
Integrity, 253
IOTA, 131
IR, 145
IS-95, 2, 9, 10
ISI, 78
ITU, 12, 17, 24

KASUMI, 259
KSI, 259

LA, 268
LCR-TDD, 121
LDCM, 99
LDPC, 132
LFDMA, 88
Logical channel, 190
LPC, 309

MAC, 90, 146, 187, 196
MAP, 4
MBMS, 122, 298
MBSFN, 124
MC-CDMA, 82
MCCH, 191
MCE, 141, 300
MCH, 191
MCS, 128
MGCF, 52
MGW, 52
MIB, 230
MIMO, 76, 91
MISO, 94
MMD, 6
MME, 40, 109, 173
MMSE, 83
MNC, 239
MSC, 32
MSIN, 239
MSISDN, 47
MT, 62
MTCH, 141, 191
MUK, 303
Multicast, 298
MU-MIMO, 106

NAI, 256
NAS, 188
NDS, 181, 250
NGMN, 18
NGN, 17
NodeB, 37

OFDM, 76
OFDMA, 80, 122
OMA, 24, 281
OSTBCM, 98

PA, 110, 113
PAPR, 85
PARC, 104

PBCH, 125, 138, 192
PCCH, 190, 262
PCFICH, 125, 138, 192
PCH, 191
P-CSCF, 50
PDC, 2
PDCCH, 124, 138, 192
PDCP, 187, 200
PDF, 54
PDN GW, 40, 174
PDP context, 53
PDSCH, 124, 137, 192
PDU, 146
PER, 150
PHICH, 125, 138, 192
PHY, 187, 194
Physical channel, 192
PLMN, 155, 239
PMCH, 125, 138, 192
Policy Control, 53
PRACH, 156, 192
Presence, 29, 283, 294
PS, 32
P-SCH, 125
PSTN, 33, 52, 309
PUSCH, 156, 160, 161, 192
Push-to-Talk, 29, 282
PVI, 105

QAM, 77
QPSK, 130

RA, 268
RACH, 191, 234
RAND, 255
RB, 127
Relocation, 276
RF, 111
RLC, 144, 187, 197
RNC, 37
Roaming, 43, 59
ROHC, 200
RRC, 188, 198, 235, 242
RRH, 111
RTCP, 225, 283, 302, 312
RTP, 223, 283, 302, 312

S1, 177
S1-flex, 181
SC-FDMA, 85, 156

SCH, 148
S-CSCF, 50
SDMA, 106
SDP, 220, 287, 306
Serving GW, 40, 174
SFBC, 134
SFN, 141
SGSN, 32, 41
SIB, 230
SigComp, 215
SIM, 62
SIMO, 94
SIP, 210, 287
SM, 207
SNOW, 259
Spectrum, 20, 30
S-SCH, 125
STBC, 134
STCP, 178
STM1, 109
S-TMSI, 237
STTCM, 98
STTD, 98, 100
Supplementary services, 318

TA, 267
TAP, 60
TBCP, 283
TDD, 109, 119
TDM, 142
TDMA, 2
TE, 62
THIG, 50, 217
TISPAN, 17, 318
Transparent mode, 197
Transport channel, 191
TRM, 109
TSTD, 150
TTI, 124

UE, 105
UEP, 310
UICC, 64, 231
UL-SCH, 191
UMB, 12
UMTS, 2, 7, 12, 21, 42, 230, 243, 248, 268
Unacknowledged mode, 197
URA, 268

USIM, 6, 64, 231
UTRAN, 68, 193

V-BLAST, 99
VoIP, 33, 311
VRB, 127

WCDMA, 84
Wifi, 35, 38, 44, 188

WiMAX, 35, 38, 44, 188, 250

X2, 183
XCAP, 313
XDMS, 285
XML, 285
XRES, 255